Proceedings of the 5th International
Mathematica Symposium

CHALLENGING THE BOUNDARIES OF SYMBOLIC COMPUTATION

Proceedings of the 5th International *Mathematica Symposium*

CHALLENGING THE BOUNDARIES OF SYMBOLIC COMPUTATION

Imperial College London 7–11 July 2003

editors

Peter Mitic
Positive Corporation Limited

Philip Ramsden
Imperial College London

Janet Carne
Positive Corporation Limited

Imperial College Press

Published by

Imperial College Press
57 Shelton Street
Covent Garden
London WC2H 9HE

Distributed by

World Scientific Publishing Co. Pte. Ltd.
5 Toh Tuck Link, Singapore 596224
USA office: Suite 202, 1060 Main Street, River Edge, NJ 07661
UK office: 57 Shelton Street, Covent Garden, London WC2H 9HE

British Library Cataloguing-in-Publication Data
A catalogue record for this book is available from the British Library.

CHALLENGING THE BOUNDARIES OF SYMBOLIC COMPUTATION
Proceedings of the 5th International *Mathematica Symposium*

ISBN 1-86094-363-2

This book is printed on acid-free paper.

Printed in Singapore by Uto-Print

Host's Preface

The International *Mathematica* Symposium began, I guess, as a twinkle in the eye of people like Peter Mitic, Gautam Dasgupta and Veikko Keränen, who collectively had the idea at a Wolfram Developer's Conference in Champaign-Urbana, Illinois, back (if I remember rightly) in 1994. How would it be, they wondered, if the worldwide community of *Mathematica* users—in research, in education, in industry and commerce—got together on a regular basis to plot, and scheme, and exchange ideas? How would it be if a Program Committee and a proper peer review process were set up? That way, academic users in particular could gain recognition for their work with *Mathematica*. That way, too, *Mathematica*'s role in research, which orthodox journals and conferences used to treat as something of a shameful secret, (and which some of them, to be honest, still do), could come to the forefront.

Where these flights of imagination led to was Southampton, U.K., in the summer of 1995 and the first IMS of all, which Peter Mitic hosted. Where *that* led was where we are now: back in England again, after travelling via Rovaniemi, Linz and Tokyo. We find ourselves in a computational environment that has changed utterly in those eight years, yet *Mathematica* and IMS are both stronger than ever. In 1995, *Mathematica* development meant writing packages. In 1995, *Mathematica* in Education meant building courses around the Notebook's Front End. Some great truths never change: if you leaf through the papers in this volume, you'll find plenty of packages and plenty of examples of students using Notebooks, but you'll also find *Mathematica* acting in new and exciting ways: powering specialist research, educational and fieldwork tools written by the Symposium's authors, and often available over the Internet, (itself changed in ways we couldn't have imagined as recently as eight years ago).

The word "interdisciplinary" usually strikes dread into me. Too often, it signals either woolly generalities, or specialists talking past one another. But IMS has always been about the best kind of "interdisciplinary". Widely divergent as our fields and backgrounds are, we speak a shared language, and we find ourselves able to strike sparks across disciplinary boundaries. That's the main reason why IMS has become my favourite regular conference of all, and why I'm delighted to see the Symposium in London and even more pleased to welcome IMS to Imperial College.

Phil Ramsden, Imperial College London

Editor's Preface

This year we are fortunate to welcome many new authors to IMS, as well as old friends. Their papers cover, as usual, a wide range of topics, although there seemed to be some concentration on physics and applied maths. Many authors, particularly in education, have webified their applications and made solid advances as a result. There are some unusual papers, too, and they're worth reading just to see what can be done. One of the original aims of IMS was to encourage students and new users to contribute, and I ask you to encourage your students to do just that. IMS is an ideal medium for short precursors to more substantial research or practice.

Before the job of producing final papers started, we spent a lot of time discussing how to present the proceedings. There were two main issues. The first was what format to use, and we settled on the *Mathematica* notebook. That must seem odd, since *Mathematica* is designed for development and to produce technical documents, yet this is the first IMS where we have asked authors to submit notebooks. The results have been positive and uniformity in submissions has eased the task of editing. This issue was originally prompted by the second main issue, which was whether or not to produce the proceedings as a book or on a CD. I was very much in favour of the CD because it has clear advantages, particularly for work in *Mathematica*. Notebooks on CD need not be limited in length, need not suffer from formatting problems, and can be what they should be: live documents that you can use as well as read. It's quite clear that these proceedings are intended almost entirely for the printed page. However, we have found significant problems with the acceptability of proceedings on CD on the part of some academics. They like paper! They are also the ones who control the budgets. Arguing that material on a CD is refereed to the same standard as that in a book, can have an ISBN and a NLC number and can even be printed does not work. I have not yet heard a rational argument against CDs. So, my message is: persuade your senior academics, (yourself if you are one), to agree that a CD is as acceptable as print, if not more so.

Finally, I would like to thank two people. Larry Adelston of Wolfram Research did an excellent job in producing the IMS 2003 stylesheet. He must have been awake all night at times: I often received e-mails when it must have been dark in Champaign IL. My co-editor, Janet Carne, did a meticulous job in editing many papers in these proceedings. You can be sure that if you see extra commas, (she likes those...), or different wording in your text, it has been done for a purpose and is worth emulating.

Peter Mitic, Positive Corporation Ltd.

Contents

Education

Statistics and Probability

Visualisation

Miscellaneous

Linguistic Fractal Analysis of Symbolic Sequences

J.M. Gutiérrez, A.S. Cofiño, University of Cantabria, Spain
gutierjm@unican.es, cofinoa@unican.es

P. Abbott, University of Western Australia
paul@physics.uwa.edu.au

1 Abstract

In this paper we present tools for linguistic fractal analysis of symbolic sequences — sequences of formal symbols from a given alphabet. We introduce a method for graphically representing and analyzing symbolic sequences based on fractal Iterated Function Systems (IFSs). This method gives a visual description (a pattern) of the structure of the subsequences within the original sequence, and permits quantitative characterization of the resulting patterns. We present a number of examples and also demonstrate the use of *J/Link* for importing sequences and speeding up the computations.

2 Introduction

The analysis of symbolic sequences has wide applications to problems in areas such as mathematics (sequences of prime numbers), biology (DNA sequences), environmental sciences (precipitation time series) and finance (trends in stock markets). For some problems, symbolic sequences are the system's natural language. For instance, DNA sequences are naturally defined by a four letter alphabet of nucleotides: adenine (a), cytosine (c), thymine (t), and guanine (g), or by a two letter alphabet of piramidines (a,c) and purines (t,g). In other cases, symbolic sequences are not the natural language for the problem (e.g. precipitation values are real numbers, not symbols), but symbolic analysis can be used to give new insight into the system's structure. For instance, we can consider precipitation as a discrete variable (rain/no rain), giving a symbolic sequence. In this paper we establish an analogy with linguistic analysis, considering symbolic sequences as languages. Each *k*-length subsequence is considered as a "word" (e.g. weekly sequences of precipitation, triplets (codons) of DNA sequences, etc.) which defines a specific symbolic pattern. Thus, each symbolic sequence defines a "language" characterized by the set of words (subsequences) appearing in the sequence. For instance, in the case of precipitation, greater variability in the observed weekly precipitation sequences leads to a more complex language.

We describe the use of IFS models in the analysis of symbolic sequences, obtaining a statistical characterization of the distribution of words within a sequence. A modification of the chaos-game algorithm is used to graphically represent symbolic sequences. The basic idea is to assign an arbitrary linear contraction to each symbol and use the symbolic sequence to drive the chaos-game algorithm, instead of randomly chosen transformations. The resulting patterns characterize the structure embedded in the sequence in a simple and intuitive way. Then we show how the resulting fractal patterns can be characterized using different fractal exponents (dimensions, entropies, etc.).

3 Iterated Function Systems

The study of symbolic sequences deals with the statistical properties of the different k-length sub-sequences $\sigma_1, \ldots, \sigma_k$ (or *words*) within the original sequence. The distribution of word counts can be efficiently represented by means of IFSs [1]. An IFS is a finite collection, $f_1(x), \ldots, f_N(x)$, of affine contractive transformations of the form $f(x) = A \cdot x + b$, where b is a translation vector and A is a 2×2 matrix with eigenvalues λ_1, λ_2, such that $|\lambda_i| < 1$. These transformations shrink distances between points and have a unique fixed point. This guarantees the existence of a unique compact set S, which satisfies the equation of self-similarity: $S = \bigcup_i f_i(S)$. This set is called the attractor of the IFS and is formed by a number of copies of the attractor, but at reduced scales: $f_1(S), \ldots, f_N(S)$. Depending on the mappings considered, the resulting attractor may have a regular form, or a fractal structure.

IFS models provide a convenient framework for symbolic analysis, since there exists a correspondence between the set of words $(s_1, \ldots, s_k)$, $s_i \in \{1, \ldots, N\}$, and a partition of S formed by the so-called "iterator sets" $f_{s_k}(\ldots f_{s_1}(S) \ldots)$ (or, alternatively, with the pixels of the attractor at a given resolution scale).

3.1 Graphical Representation of Symbolic Sequences

The chaos-game algorithm used to render the attractor of IFSs [1] can be adapted to visually represent symbolic sequences. Given a symbolic sequence $s_1, s_2, \ldots, s_n$, we consider the dynamical system $x_{i+1} = f_{s_i}(x_i)$, $i = 1, \ldots, n$, which defines an *orbit* on the attractor of the IFS from an arbitrary initial condition x_0. The orbit only visits those regions of the attractor associated with the words appearing in the symbolic sequence. This method was first introduced by Jeffrey [2] to represent DNA sequences.

This code implements an efficient version of the chaos-game algorithm for an IFS consisting of four transformations. We map a, b, c, d to the corners of a unit square. `FoldList` computes the orbit and compilation is used to speed up the computation.

```
In[1]:= Orbit[s_List, {a_, b_, c_, d_}] := OrbitMap[
          s /. {a → {0, 0}, b → {0, 1}, c → {1, 0}, d → {1, 1}}]
```

```
In[2]:= OrbitMap = Compile[{{m, _Real, 2}},
          FoldList[(#1 + #2) / 2 &, {0, 0}, m]];
```

```
In[3]:= IFSPlot[s_List] :=
         Show[Graphics[{AbsolutePointSize[0.5],
            Point /@ Orbit[s, Union[s]]}], AspectRatio →
          Automatic, PlotRange → {{0, 1}, {0, 1}},
         GridLines → {Range[0, 1, 2^-3], Range[0, 1, 2^-3]}]
```

3.2 *Using J/Link to Import Data via the Internet*

Symbolic sequences can be extremely long, so it is important to efficiently load and process such data. Moreover, if the data is available via the internet there is no need to store it locally. We use J/Link to directly import data.First load the `Graphics` and J/Link packages.The code then opens and reads from a URL stream and also installs the Java enviroment (thanks to Rolf Mertig).

```
In[5]:= Needs["Graphics`"]; Needs["JLink`"];
```

```
In[6]:= ImportURL[url_String] :=
          JavaBlock[Module[{u, s, stream,
             numRead, outFile, buf}, InstallJava[];
            u = JavaNew["java.net.URL", url];
            stream = u[openStream[]];
            If[stream === $Failed, Return[$Failed]];
            buf = JavaNew["[B", 5000]; s = "";
            While[(numRead = stream[read[buf]]) > 0, s =
              s <> FromCharacterCode[(If[# < 0, # + 256, #] &) /@
                 Take[Val[buf], numRead]]];
            stream[close[]]; s]];
```

3.3 *Precipitation*

In the case of precipitation, (p in mm), we use four different thresholds; $p < 0.1$, $0.1 \le p < 2$, $2 \le p < 20$ and $p \ge 20$, with associated symbols 1 (no rain), 2 (light rain), 3 (moderate rain), 4 (heavy rain) respectively. For instance, the sequence of daily rainfall in Santander (North basin) since 1979: 3.7, 0.1, 6.8, 13.0, 16.0, 28.1, 15.4, ..., leads to the symbolic sequence: 3, 2, 3, 3, 3, 4, 3, ….

This code imports the ASCII rainfall symbolic sequence from a URL using `ImportURL` and `ReadList`, then displays the rainfall symbolic sequence using `IFSPlot`. This is the "fingerprint" of precipitation in Santander.

```
In[7]:= rain = ReadList[
          StringToStream[ImportURL["ftp://physics.uwa.edu.
            au/pub/Mathematica/Santander.txt"]], Number];

In[8]:= IFSPlot[rain];
```

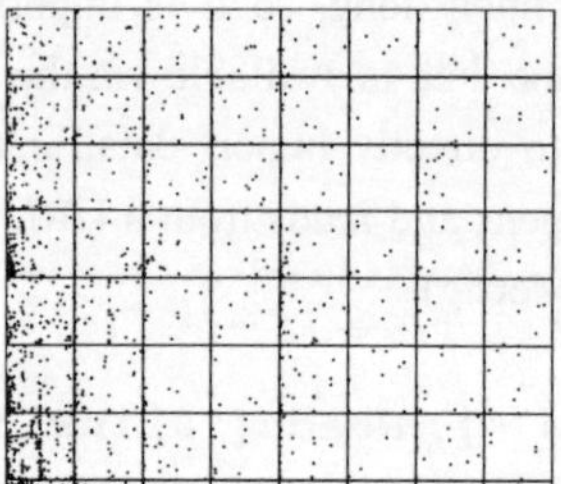

3.4 *DNA sequences*

DNA sequences are available at the GenBank® database of the National Center for Biotechnology Information (NCBI) at www.ncbi.nlm.nih.gov/entrez/viewer.fcgi as plain ASCII files, containing information about the organism, as well as the sequence of nucleotides. Discovering the structure encoded in DNA sequences is of paramount importance. The following code reads in DNA sequences as a string of characters, separately printing out descriptive information.

```
In[9]:= NCBI = "http://www.ncbi.nlm.nih.gov/
            entrez/viewer.fcgi?txt=on&list_uids=";
```

```
In[10]:= DNASequence[id_] := Module[
           {lines = Import[StringToStream[ImportURL[NCBI <>
                ToString[id] <> "&view=fasta"]], "Lines"]},
           Print[First[lines]];
           Characters[StringJoin @@ Rest[lines]]]
```

There are two types of **sequence identification numbers**: *GI numbers,* (a series of digits that are assigned consecutively by NCBI to each sequence it processes), and *version numbers,* (which consist of the accession number followed by a dot and a version number). Either format can be used. Now we use the above command to import the mouse mitochondrion and man mitochondrion complete genomes.

```
In[11]:= mouse = DNASequence["342520"];

           >gi|342520|gb|J01420.1|MUSMTCG
             Mouse mitochondrion, complete genome

In[12]:= man = DNASequence["X93334.1"];

           >gi|1262342|emb|X93334.1|HSMITG Homo
             sapiens mitochondrial DNA, complete genome
```

Both genomes define similar patterns (they are close in the phylogenetic tree: mitochondrial eukaryotes, vertebrata).

```
In[13]:= DisplayTogetherArray[IFSPlot /@ {mouse, man}];
```

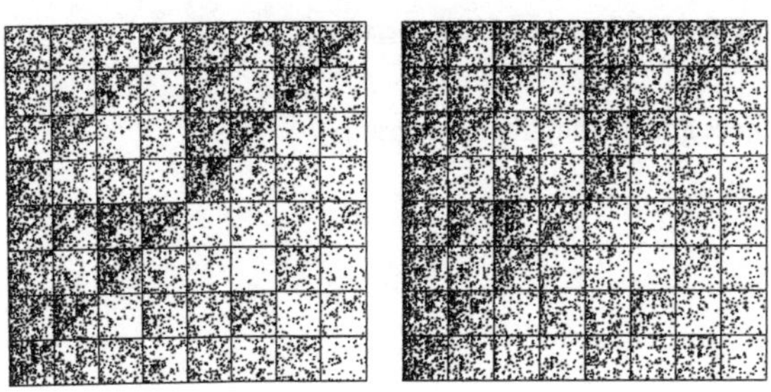

4 Linking to a Java Application

With *J/Link* you can invoke Java objects directly from *Mathematica*.

1. Download the IFS applet: http://physics.uwa.edu.au/pub/Mathematica/IFS.zip.
2. *J/Link* automatically adds any `Java` directories in the `Applications` or `ExtraPackages` directories to the `CLASSPATH`. Create a directory in the `Applications` subdirectory, whose (machine-dependent) name is generated by evaluating the following command:

```
In[14]:= IFSpath =
           If[$VersionNumber ≥ 4.2, {$UserAddOnsDirectory,
             "Applications", "IFS", "Java"}, {$TopDirectory,
             "AddOns", "Applications", "IFS", "Java"}];
         dir = ToFileName[IFSpath]
```

3. Place the IFS applet files into the directory `dir`.

Then we create a new object of the class `IfsGenerator`. This class includes two methods:

1. **`ifsDraw`** loads the selected file, runs the chaos game, and draws the resulting pattern;
2. **`getHistogram`** returns the resulting matrix of frequencies for a given resolution (here $2^9 \times 2^9$ pixels).

```
In[16]:= ifs = JavaNew["IfsGenerator", 2^9];
         ifs@setFile[ToFileName[IFSpath, "ecoli.gbd"]];
         ifs@ifsDraw[];
         histogram = ifs@getHistogram[];
```

Finally, release the object and display the histogram.

```
In[20]:= ReleaseJavaObject[ifs]
         ListDensityPlot[histogram, Mesh → False];
```

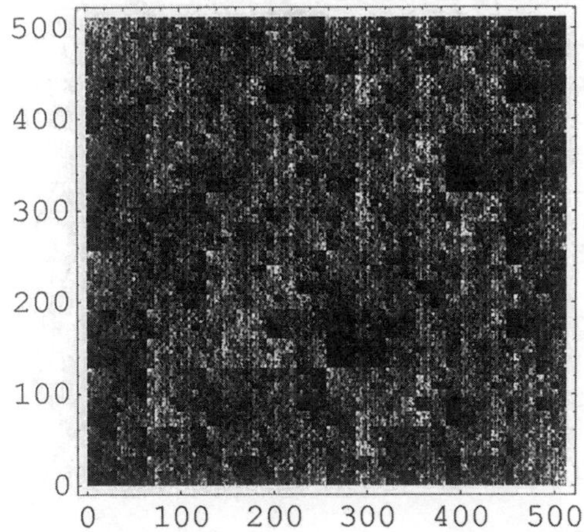

From the histogram we can easily compute fractal exponents related to the fractal pattern, such as Zipf's exponent.

4.1 Quantitative Characterization: Zipf's Law

Zipf analysis [3] is based on the histogram of the total number of occurrences of each word in a large representative text; if all the words are arranged in rank order, from the most to the least frequent, the resulting histogram may present a power law, (it is linear on log-log scale). Such an analysis provides a characteristic scaling exponent associated with the fractal self-similar character of the symbolic sequence. This technique has been applied to natural and artificial languages, allowing the characterization and discrimination of different characteristics of the problem. See [3] for more details.

The histogram values are associated with the frequency of the corresponding word.

```
In[22]:= freqs = Select[
           Sort[N[Flatten[histogram]], Greater], # > 0 &];
```

Display the histogram on a log-log plot.

```
In[23]:= pts = Transpose[
           {Log[10, Range[Length[freqs]]], Log[10, freqs]}];

In[24]:= ListPlot[pts, Axes → True,
           PlotRange → {{1, 6}, Automatic},
           AxesLabel → {"log rank", "log freq"}];
```

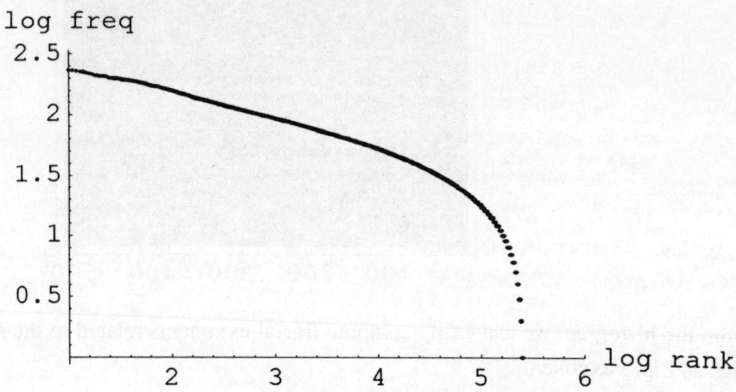

Finally, determine the Zipf exponent, obtaining the slope of the curve in the scaling region.

```
In[25]:= -Coefficient[
            Fit[Take[pts, {10^1, 10^4}], {1, x}, x], x]

Out[25]= 0.238975
```

5 References

[1] Barnsley, M.F. Fractals Everywhere, second edition, 1990, Academic Press.

[2] Jeffrey, H.J. Chaos Game Representation of Gene Structure. Nucleic Acids Research, Vol. 18 (8), 2163-2175, 1990.

[3] http://linkage.rockefeller.edu/wli/zipf/

Automatic Generation of Numerical Code

J. Korelc
University of Ljubljana, Slovenia
jkorelc@fgg.uni-lj.si

1 Abstract

This paper presents a hybrid system for automatic generation of numerical codes using a symbolic approach, with emphasis on finite element formulations where straightforward use of symbolic and algebraic systems leads to exponential growth of derived expressions. The system comprises two major components. The *Mathematica* package *AceGen* is used for symbolic derivation of formulae needed in numerical procedures and automatic code generation. An approach implemented in *AceGen* overcomes the problem of expression growth by combining automatic differentiation, automatic code generation and theorem proving. The second component, named *Computational Templates* is a collection of prearranged modules for automatic creation of the interface between the finite element code and the finite element environment.

2 AceGen Symbolic Code Generator

The symbolic-numeric approach to the finite element (FE) method has been extensively studied recently. Various systems for automatic code generation have been developed. Transformations, differentiations, matrix operations, and the large amount of data involved in the derivation of characteristic finite element quantities (e.g. gradients, tangent matrices, sensitivity vectors) often lead to exponential growth of expressions. An approach to automatic code generation used in the *Mathematica* package *AceGen* is called Simultaneous Stochastic Simplification of numerical codes [1]. This approach combines *Mathematica* with an automatic differentiation technique and automatic theorem proving by examples. To alleviate the problem of expression growth and redundant calculations, simultaneous simplification of symbolic expressions is used. Stochastic evaluation of the formulae is used for determining the equivalence of algebraic expressions, instead of conventional pattern matching technique. The general characteristics of the *AceGen* code generator are:

⇒ simultaneous optimization of expressions immediately after they have been derived;

⇒ automatic differentiation techniques;

⇒ automatic selection of the appropriate intermediate variables;

⇒ the whole program structure can be generated;

⇒ appropriate for large problems where intermediate expressions can be subject to uncontrolled swell;

⇒ improved optimization procedures with stochastic evaluation of expressions;

⇒ generation of characteristic formulae;

⇒ automatic interface to other numerical environments (by using the Splice command of *Mathematica*);

⇒ multi-language code generation (FORTRAN, C/C++, *Mathematica*);

⇒ advanced user interface;

⇒ advanced methods for exploring and debugging generated formulae;

⇒ special procedures are needed for non-local operations.

AceGen was designed to approach especially hard problems, where the general strategy for efficient formulation of numerical procedures has not yet been established. Typically the *AceGen* function takes the expression provided by the user and returns an optimized version of it. This optimized version can result in a newly created auxiliary variable or, parts of an original expression may be replaced by previously created auxiliary variables. In the first instance *AceGen* stores the new expression in an internal database. The database contains a global vector of all expressions, information about dependencies of the symbols, labels and names of the symbols, partial derivatives, etc. The database maintains information during the *Mathematica* session.

3 Typical example of automatic code generation procedure

Let us consider a simple example to illustrate the standard *AceGen* procedure for the generation of a typical numerical sub-program that returns derivative of function Π given by $\Pi(x,n)=\prod_{i=1}^{n}\frac{\mathrm{Sin}[i\,x]}{x}$ with respect to *x*. Due to the numerical problems arising at *x=0* we have to consider exceptions in the evaluation of the function as well as in the evaluation of its derivatives. The syntax of the *AceGen* script language is the same as the syntax of the *Mathematica* script language with some additional functions. The input for *AceGen* (Figure 1) can be divided into six characteristic steps:

1. At the beginning of the session the *SMSInitialize* function initializes the system;
2. The *SMSModule* function defines the input and output parameters of a new subroutine "Test";
3. The *SMSReal* and *SMSInteger* functions assign the input parameters *n$$* and *x$$* of the subroutine to the standard Mathematica symbols n and x. Double $ character indicates that the symbol is an input or output parameter of the generated subroutine;

4. During the description of the problem special operators (⊨, ⊢, ⫤, and ⊣) are used to perform simultaneous optimization of expressions and creation of new auxiliary variables;
5. The results of the derivation are assigned to the output parameters *pi$$* and *dpi$$* of the subroutine by the *SMSExport* function;
6. At the end of the session the *SMSWrite* function writes the contents of the vector of the generated formulae to the file in a prescribed language format.

```
<< AceGen`;
SMSInitialize["Test", "Language" -> "C++"];
SMSModule["Test",
  Integer[n$$], Real[x$$, pi$$, dpi$$]];
x ⊢ SMSReal[x$$]; n ⊢ SMSInteger[n$$]; Π ⫤ 1;
SMSDo[i, 1, n, 1, {Π}];
 func = Sin[i x^2] / x;
 SMSIf[SMSAbs[x] > 10^-10];
  f ⫤ func;
 SMSElse[];
  f ⊣ SMSFreeze[Limit[func, x → 0], "Dependency",
     {{x, Limit[D[func, x] // Evaluate, x → 0]}}];
 SMSEndIf[f];
 Π ⊣ Π f;
SMSEndDo[Π];
δΠ ⊨ SMSD[Π, x];
SMSExport[{Π, δΠ}, {pi$$, dpi$$}];
SMSWrite[];
```

Figure 1: Characteristic AceGen input

```
void Test(double v[1505],int *n,double *x,double *pi,double *dpi){
int i2,i4,b5;
v[1]=*x;
i2=*n;
v[7]=0e0;
v[3]=1e0;
for(i4=1;i4<=i2;i4++){
 if(fabs(v[1])>0.1e-9){
  v[9]=i4*(v[1]*v[1]);
  v[8]=sin(v[9]);
  v[10]=-(v[8]/Power(v[1],2))+2e0*i4*cos(v[9]);
  v[6]=v[8]/v[1];
 } else {
  v[10]=i4;
  v[6]=0e0;
 };
 v[7]=v[10]*v[3]+v[6]*v[7];
 v[3]=v[3]*v[6];
};/* end for */
*pi=v[3];
*dpi=v[7];
};
```

Figure 2: Generated C++ code

New auxiliary variables are labeled consecutively in the same order as they are created, and these labels remain fixed during the *Mathematica* session. This enables free manipulation with the expressions returned by the *AceGen* system. Operators ⊧ and ⊢ are used for introduction of auxiliary variables that will appear only once on the left-hand side of equation. For introduction of auxiliary variables that will appear more that once on the left-hand side the operators ⫤ and ⊣ are used. These operators are replacements for the simple assignment command in Mathematica (lhs=rhs). In principle we can get *AceGen* input simply by replacing = operators in standard Mathematica input by one of the *AceGen* assignment operators. *AceGen* can automatically generate conditionals (SMSIf, SMSElse, SMSEndIf) and loops (SMSDo, SMSEndDo). The program structure specified by the conditionals and loops is created simultaneously during the *AceGen*

session and appears as part of automatically generated code in a specified language.
The *SMSD[exp, variable]* function performs automatic differentiation of one or several expressions with respect to an arbitrary variable or vector of variables by simultaneously enhancing the already derived code. The automatic differentiation procedure can be initiated at any time and at any point of derivation of the formulae and as many times as required. This represents considerable advantage over the standard implementation of automatic differentiation as a code-to-code translator. There are several situations when the formulae and the program structure alone are not sufficient to make proper derivative code. Various exceptions can be defined by the $\texttt{SMSFreeze}\left[exp, \texttt{"Dependency"}, \left\{\left\{var_1, \frac{\partial exp}{\partial var_1}\right\}, \ldots\right\}\right]$ command (e.g. if the evaluation of the derivative code would lead to numerical errors as in example presented of Figure 1).

4 Computational Templates package

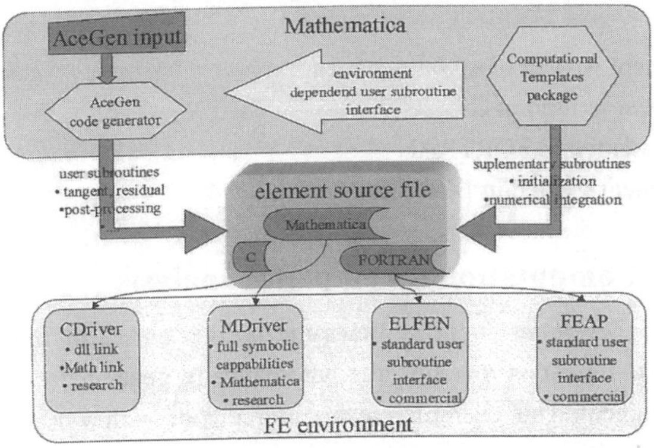

Figure 3: Hybrid system

Using a general FE environment for analysing a variety of problems and for incorporating new elements is now commonplace. However, large FE systems can be awkward for development and testing new numerical procedures. The basic tests which are performed on a single finite element or on a small patch of elements can be done most efficiently by using symbolic-numeric environments like *Mathematica*. Many design flaws, like element instabilities or poor convergence properties can be easily identified if we are able to investigate element quantities on a symbolic level. Unfortunately, symbolic-numeric environments become very inefficient if there are a large number of elements or if we have to perform iterative numerical procedures. In order to assess element performances under real conditions, we do tests on sequential machines with good debugging capabilities. To optimise, we need to generate the same code from multi-language environments. *Mathematica* uses package *Computational Templates* [2], an extension of *AceGen* (Figure 3). Packages *AceGen* and *Computational Templates* together enable multi-language and multi-environment generation and testing of nonlinear FE code from the same symbolic description.

As well as the FE interfaces the *Computational Templates* also contain a model FE environment, *Finite Element Driver*, which is in two equivalent versions. The first version (*MDriver*) is written in Mathematica, so it can use elements like high precision arithmetic and even symbolic evaluation. The second version (*CDriver*) is in C and is connected to *Mathematica* by *MathLink*. Both drivers pre-processing the input data, and post-processing graphics and commands that control the solution procedures. They are not intended to replace efficient general FE environments, but serve to develop new numerical methods. At the exploratory stage, efficient element code is generated from the same symbolic description and incorporated into a professional FE environment. The *Computational Templates* system currently supports the following FE environments:

⇒MDriver is a model FE environment written in a Mathematica
⇒CDriver is a model FE environment written in C
⇒FEAP is a research environment written in FORTRAN
⇒ELFEN is a commercial environment written in FORTRAN

5 Typical example of Computational Templates analysis

The syntax of the *Computational Templates* input is the same as the syntax of the *AceGen* input with some additional functions that link the automatically generated FE codes with the selected FE environment. This example creates source code with a user-defined subroutine that calculates the element tangent matrix and residual (Figure 4).

```
<< "ComputationalTemplates`";
SMSInitialize["Hypersolid", "Language" → "C++"];
SMTInitialize["CDriver", "SMTTopology" → "Q1"]
SMTUserSubroutine["Tangent and residual"];
...
SMSWrite[];
SMSInitialize["GasPressure", "Language" → "C++"];
SMTInitialize["CDriver", "SMTTopology" → "L1"]
...
SMSWrite[];
```

Figure 4: Part of the input that generates FE user subroutines

Two different elements are considered. The first is a 2-D, 4-node element (indicated by the option SMTTopology→"Q1") that can be used for simulating rubber-like solids (e.g. deflating a tyre), and the second is a 2-node element ("SMTTopology"→"L1") that simulates the effect of the gas pressure on a solid surface. The body of the user subroutines is intentionally left empty, and the complete input is in [3]. With the derived code a problem such as inflating a tyre can be analysed using the CDriver FE environment. The goal is to calculate the shape of the tyre in its deflated and inflated states. The tyre is first deflated by increasing the weight on the rubber and then inflated by increasing the inside gas pressure. Figure 5 shows the simulation results and Figure 6 shows the mesh and simulation procedure for the first stage.

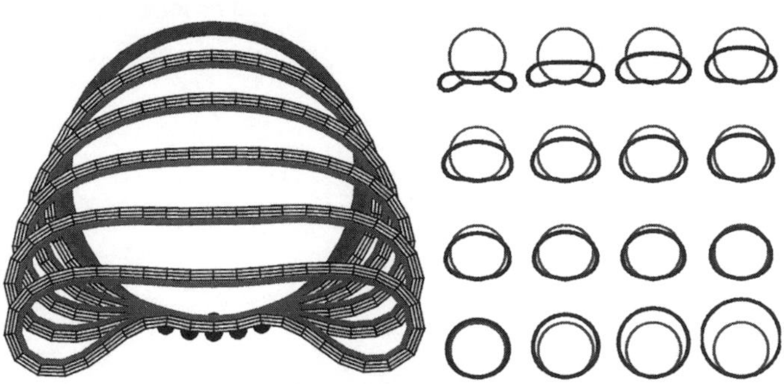

Figure 5 : Simulation of deflating and inflating the tyre

```
<< ComputationalTemplates`;
SMTSetDriver["CDriver"];
SMTDomains = {{"gas", "GasPressure", {0, 1.}},
    {"tyre", "Hypersolid", {1000., .3, 1., 0, 0}}};
points = Table[{Cos[φ], Sin[φ]}, {φ, 0, 2. π, 2. π / 20}];
SMTMesh["tyre", "Q1", {40, 4}, {points, 1.1 points}];
SMTMesh["gas", "L1", {40}, points];
SMTEssentialBoundary = {{ "Y" < -1.02 &, 0, 0}};
SMTStructure[];
Do[SMTNextStep[1, 0];
  SMTDomainData["tyre", "Data", "Weigth", weight];
  While[SMTConvergence[], SMTNewtonIteration[]];
  , {weight, 1, 5, 1}];
```

Figure 6 : Mesh generation and analysis

6 References

[1] Korelc, J. "Automatic generation of finite-element code by simultaneous optimization of expressions", Theoretical Computer Science, 187, 231-248, 1997.

[2] Korelc, J. Hybrid system for multi-language and multi-environment generation of numerical codes, in Proc. ISSAC 2001, 209-216, 2001

[3] www.fgg.uni-lj.si/Symech/

Programming with Sequence Variables: the *Sequentica* Package

Mircea Marin, Radon Institute for Computational Applied Mathematics, Austria
Dorin Tepeneu, University of Tsukuba
{mmarin,dorinte}@score.is.tsukuba.ac.jp

1 Abstract

Sequence variables are an advanced feature of modern programming languages. They enhance the support for writing programs in a declarative and easily understood way. To our knowledge, *Mathematica* provides the best support for programming with sequence variables, but it requires a good understanding of how the interpreter chooses the matcher. This is because matching against patterns with sequence variables is generally not unique. We claim that there is room to improve the programming style with sequence variables. We propose a number of new programming constructs which impose certain strategies on the pattern matching process. Our constructs enable us to control the selection of a matcher by annotating sequence variables with binding priorities and ranges for their lengths, and to compute optimal values characterized by a score function to be optimized. To this end we have developed the package *Sequentica*. With *Sequentica, Mathematica* programmers and users get additional support for defining functions and transformation rules in an easy and convenient way. We outline the algorithmic difficulties to support these extensions and describe how they are implemented in *Sequentica*. The usefulness of these extensions is illustrated with various examples. We regard these extensions as a first step towards identifying a new programming style: programming with sequence variables. Such a programming style is useful to solve problems based on sequence analysis, such as bio-informatics, cryptography or data mining.

2 Motivation

The concept of sequence variables has proved useful in many areas, such as symbolic computation, theorem proving [2], term rewriting [1], unification [5] and computational logic [4]. We feel that the expressive power of this concept is not yet completely understood and that a better formalization is desirable. Sequence variables, as defined in *Mathematica*, are very powerful for proving, solving and computing. The expressive power of our concept is superior to that which *Mathematica* provides. To emphasize this fact, we mention:

1. Matching against patterns with sequence variables is generally not unique. There is no way to specify which matcher to choose. One has to accept the matcher chosen by the underlying interpreter. Often, the programmer wants to control this choice.

Example 1. Assume we want to define a function which returns the longest run of successive integers in a sequence. This would be possible if we could write:

```
GetMaxSublist[___, x___Integer, ___] := {x}
```

and specify a strategy to choose a matcher which yields the longest sequence binding for the sequence variable **x**. *Mathematica* has no programming support for such definitions.

2. Another desirable feature is to select the matcher which yields an optimum value, defined as a matcher which minimizes or maximizes a certain expression.

Example 2. Suppose we have a list of lists of temperature measurements during one year, `L={{`T_{11}`, `T_{12}`, ...},...,{`$T_{12,1}$`, `$T_{12,2}$`, ...}}`, where each sublist contains the temperatures measured in one month. We want to find the sublist of temperatures for the month with largest variation. It would be convenient to be able to write something like:

```
GetMaxVariation[___,{Ts___Real},___]:={Ts}
                 where Max[Ts]-Min[Ts] is maximum
```

to search among all matchers and keep the value of `{Ts}` for which the difference `Max[Ts]-Min[Ts]` is maximum. There is no direct support for such specifications.

3. There is no means to control the lengths of sequences in *Mathematica*. Interval bounds for sequences are supported in the specification of regular expressions and by many tools and languages based on their specification.

Example 3. Assume L is a list of daily temperature measurements over a long period of time, and we want to find the week with highest average temperature by calling HotWeek[L]. In this case, we want to write something like:

```
HotWeek[x___,y___,___]/;
    And[Length[{y}]==7,Mod[Length[{x}],7]==0]:=
Quotient[Length[{x},7],1]
    for which Plus[y]/7 is maximum
```

There is no adequate programming support to specify such a function in *Mathematica*.

To overcome these limitations we propose some new *Mathematica* syntactic constructs for programming with sequence variables and give an account of their *Mathematica* implementation. To this end we have developed the package *Sequentica* which supports the extensions reported in this paper.

3 Pattern Matching With Sequences in *Mathematica*

Pattern matching is the core of the definitional mechanisms of modern functional languages [7] and has wide applications in areas such as cryptography, computational biology, combinatorics and parallel algorithms [6]. Pattern matching with sequences extends conventional pattern matching with a new syntactic category: sequence variables. In *Mathematica*, a sequence variable is a placeholder of a sequence of expressions which are spliced automatically into the body of the function being defined.
We have already mentioned that matching against patterns which contain sequence variables is, in general, not unique. For example, matching `f[1,2,3]` against `f[x___,y__]` yields 3 possible matchers, as shown by the call:

```
In[6]:= ReplaceList[f[1, 2, 3], f[x___, y__] :> {{x}, {y}}]

Out[6]= {{{}, {1, 2, 3}}, {{1}, {2, 3}}, {{1, 2}, {3}}}
```

Because of this situation, when we define functions with sequence variables we must specify the matcher which is being considered. The interpreter of *Mathematica* chooses the matcher which assigns the shortest sequences of arguments to the first sequence variables that it finds when traversing the pattern in a leftmost-innermost manner, e.g.

```
In[9]:= f[x___, y__] := {{x}, {y}}; f[1, 2, 3]

Out[10]= {{}, {1, 2, 3}}
```

4 Extensions

Sequentica [8] is a *Mathematica* package which extends the capabilities of *Mathematica* with constructs for programming with sequence variables. After loading the package with `Get["Sequentica.m"]`, the user can define new functions, f_i, by calling:

```
Sequentica[...; fi[arg1,...,argn]:=<body_i>;...;]
```

or sets, R, of transformation rules by calling:

```
R=Sequentica[{f[arg1,...,argn]:→<body>,...}]
```

The novelty is that the user can control the pattern matching mechanism and the computation with sequence variables by annotating the arguments and body of definitions and transformation rules. We explain how these new programming features work.

Extension 1

This extension addresses three possibilities: Controlling the choice of a particular matcher instead of relying on a built-in pattern matching strategy; Confining the lengths of sequence variable bindings to certain intervals; Imposing equality constraints between the lengths of bindings of sequence variables. These can be achieved by annotating sequence variables with binding priorities which specify the order in which the sequences are assigned bindings upon pattern matching, and bounds for the lengths of their bindings. We propose to extend sequence annotations with the construct `patt←{p,m,M}` to denote a sequence with priority p and length varying from m to M. The bindings assigned to the sequence variables (such as `patt`) are computed in increasing order of priority by looking for the first matcher that is obtained by varying the sequence length from `m` to `M`. For example, the call:

```
Sequentica[GetSubsequence[x___←{2,0,∞},y__←{1,∞,1},y__
,z___←{3,0,∞}]:={y}]
```

defines the function `GetSubsequence[]`, which computes the list of the longest non-empty subsequence in the input which occurs at least twice consecutively. The assigned priorities are: 1 for **y**, 2 for **x** and 3 for **z**, so we first try to bind the sequence variable **y**, next **x** and finally **z**. The bindings for **y** are looked up starting from the greatest possible length (denoted here by ∞) down to length 1.

Example 4. It is convenient to be able to constrain different sequences to have the same length. We achieve this by imposing the condition that sequence variables annotated with the same priority denote sequences of the same length. The function

```
Sequentica[
 GetPalindrome[x___ ← {2, 0, ∞}, y__ ← {1, ∞, 1},
   u_, z__ ← {1, ∞, 1}, t__ ← {3, 0, ∞}] /;
   {y} === Reverse[{z}] := {y, u, z}]
```

computes the longest palindrome of odd length contained in the input sequence.

To avoid excessive annotations, we will assume that (1) it is sufficient to annotate one occurrence of a sequence variable; all other occurrences of that variable are assumed to have the same annotation, (2) anonymous sequence variables have assigned distinct names, and (3) a sequence variable y__ (resp. y___) which is not explicitly annotated has an implicit sequence annotation of the form **←{p,1,∞}** (resp. **←{p,0,∞}**). The default priorities **p** are assigned in a leftmost-innermost manner, and are assumed to be larger than the priorities given explicitly. This means that the non-annotated sequence variables are the last ones to get assigned bindings. With these assumptions in mind, the function **GetMaxSublist[]** from Example 1 can be defined by:

```
Sequentica[
 GetMaxSublist[___, x___Integer, ___] := {x}]
```

Extension 2

Our second extension addresses the possibility of selecting matching substitutions which render an expression optimal. To illustrate, let us reconsider Example 2: we want to select the sublist of a list which has a maximum variation of values. In particular, we suggest the following syntax:

```
Sequentica[GetMaxVariation[___, {x___Real}, ___] :=
  BestFit[{x}, Max[x] - Min[x], Greater]]
```

to find the list {x}, computed by a matching against {___,{x___Real},___}, which produces the greatest value of the expression Max[x]-Min[x].

We suggest the following new definitional mechanisms inside Sequentica[] calls:

```
f[patts]:=BestFit[expr,optim,test]
f[patts]:=BestFits[n,expr,optim,test]
```

The first call defines a partial function f which associates a call f[a1,...,am] with the output θ(expr) produced by the matcher θ between f[a1,...,an] and f[patts], which satisfies test(θ(optim),γ(optim))=True for all other matchers γ between f[a1,...,am] and f[patts].

The second call assumes n>0 and defines a function f which associates a call f[a1,...,am] with the list of instances {θi(expr)|1≤i≤k} produced by the longest list of distinct matchers {θ1,...,θk} between f[patts] and f[a1,...,am] such that k≤n and

1. test[θ_j(optim),θ_{j+1}(optim)]=True if 1≤j<k,
2. test[θ_k(optim),θ(optim)]=True for all matchers $\theta \notin \{\theta_1, \ldots, \theta_k\}$.

In both cases, test[] is assumed to define a total order on the set of values θ(optim) generated by the matchers θ of arbitrary expressions against f[patts].

Example 3 revisited. We can now define HotWeek[] as follows:

```
Sequentica[HotWeek[x___, y___ ← {1, 7, 7}, ___] /;
    Mod[Length[{x}], 7] == 0 :=
  BestFit[Quotient[Length[{x}], 7] + 1,
    Plus[y] / 7, Greater]];
```

5 Algorithmic Aspects and Implementation Issues

Matchers are computed by traversing the sequence pattern outside-in and computing the bindings of the sequence variables in the order given by their priorities, and by varying the bindings of their lengths between their specified limits. For example:

```
In[29]:= Sequentica[f[(x__Integer) ← {1, ∞, 2}, z___,
             x___, y__ ← {2, ∞, 1}] := {{x}, {y}, {z}}];
         f[1, 2, 1, 1, 2, 1, 2, 2]
```

gives `{{1,2,1},{2,2},{}}` as `{x:→Sequence[1,2,1],y:→Sequence[]}` is the first matcher encountered by varying the length of the binding of x from ∞ down to 2 and then the length of y from ∞ down to 1.

If we redefine `f` and call

```
In[34]:= Sequentica[f[(x__Integer) ← {1, 1, ∞}, z___,
             x___, y__ ← {2, ∞, 1}] := {{x}, {y}, {z}}];
         f[1, 2, 1, 1, 2, 1, 2, 2]
```

then we obtain `{{1},{1,2,1,2,2},{2}}` because we have changed the order in which we look for a matcher.

The implementation of such a pattern matching algorithm enumerates the possible lengths of bindings in the order specified by sequence annotations, detects the first combination of lengths for which a matcher exists and evaluates the body of the corresponding function (or transformation rule) for the computed matcher. Note that the algorithm must be recursive on the structure of matchers, because sequence variables can occur at different depths in the pattern specified by the user.
Of crucial importance is the identification of an efficient algorithm which enumerates the possible lengths of sequence variables. For this purpose, *Sequentica* relies heavily on an iterative solver for systems of linear diophantine constraints over finite domains [3]. This means that the solver does not compute all solutions at once, but enumerates them in the order imposed by the sequence annotations. Actually, it would be unreasonable to rely on a solver for linear diophantine constraints which computes all solutions at once because (1) the space of solutions can be very large, and (2) we are not interested in all solutions but only in the first one which realizes a matching (where *first* is defined w.r.t. the order inferred from the priorities attached to sequence variables).
The second extension is of a slightly different nature: here the sequence annotations are irrelevant since the evaluation must check all possible matchers and resume the computation for the matchers which fulfill the requirements of the score function specified by the user. In this case, *Sequentica* generates a *Mathematica* definition which accumulates the optimum value during an exhaustive search of all possible matchers.

6 Conclusion and Further Work

We believe that the support for programming with sequence variables can be further enhanced. We will continue to look for such useful programming idioms and will integrate them into the *Sequentica* package, which can be downloaded from http://www.score.is.tsukuba.ac.jp/~mmarin/Sequentica.
Further implementation details of *Sequentica* are available from http://www.score.is.tsukuba.ac.jp/~mmarin/Sequentica/sequentica.ps.

(Mircea Marin is supported by the Austrian Academy of Sciences)

References

[1] B. Buchberger. *Mathematica* as a Rewrite Language. In T. Ida, A. Ohori, and M. Takeichi, editors, *Proceedings of the 2nd Fuji International Workshop on Functional and Logic Programming*, Shonnan Village Center, 1996.

[2] B. Buchberger, T. Jebelean, F. Kriftner, M. Marin, and D. Vasaru. An Overview of the *Theorema* Project. In *Proc. of ISSAC'97*, pp. 384-391, Maui, Hawaii, 1997. ACM Press.

[3] E. Contejean. Solving Linear Diophantine Constraints Incrementally. In D.S. Warren, editor, *Proc. of the 10th Intl. Conf. on Logic Programming*, pp. 532-549, Budapest, Hungary 1993. MIT Press.

[4] The Logic Group. Knowledge Interchange Format. Technical report, Stanford University, http://logic.stanford.edu/kif/kif.html, 2002.

[5] T. Kutsia. *Solving and Proving in Equational Theories with Sequence Variables and Flexible Arity Symbols*. PhD thesis, Institute RISC-Linz, Johannes Kepler University, Hagenberg, Austria, June 2002.

[6] Pattern Matching Pointers. http://www.dei.unipd.it/~stelo/pattern.html.

[7] S. Thompson. *Haskell: The craft of functional programming*. Addison-Wesley, second edition, 1999.

[8] The *Sequentica* Package. http://www.score.is.tsukuba.ac.jp/~mmarin/Sequentica.

XML and Web Computation with *Mathematica*

T. Wickham-Jones
Wolfram Research Inc.
twj@wolfram.com

1 Abstract

This paper will describe how XML and web technologies can be combined with *Mathematica* in order to provide innovative solutions for computation. It will pay particular attention to the use of MathML, both as a way to render mathematical and technical information in a web browser and as an interchange format between different computer applications. The use of SVG to generate high-quality graphical images that support dynamic animations will also be discussed. A final section will discuss the use of XML as a general data format for exchange between applications that communicate over the web.

2 *Mathematica* and XML

Mathematica 4.2, released in summer 2002, introduced many important XML-related features. The core functionality was the extension of *Mathematica* symbolic computation techniques to XML. This involves the representation of an XML document as a *Mathematica* expression: such expressions are known as symbolic XML. In addition, many commands were introduced for converting between symbolic XML and XML documents. An example of working with XML in *Mathematica* is demonstrated in this section. First, the contents of a simple XML document are displayed.

```
In[1]:= !! books.xml
```

```
<?xml version="1.0"?>
<library><book>
    <author>Jane Austen</author>
    <title>Pride and Prejudice</title>
</book></library>
```

This XML document can be read into *Mathematica* with the **Import** command. The result is symbolic XML.

```
In[2]:= xml = Import["books.xml", "XML"]

Out[2]= XMLObject[Document][
          {XMLObject[Declaration][Version → 1.0]},
          XMLElement[library, {},
           {XMLElement[book, {}, {XMLElement[author,
               {}, {Jane Austen}], XMLElement[title,
               {}, {Pride and Prejudice}]}]}], {}]
```

Symbolic XML expressions are regular *Mathematica* expressions and can be transformed with standard *Mathematica* programming techniques. In addition to supporting a general representation of XML, as shown previously, *Mathematica* also supports some specific XML applications. Two of these, MathML and SVG are particularly useful for web based computation.

2.1 MathML

MathML is designed to allow mathematical, scientific, and other technical information to be served, received, and processed on the World Wide Web. It is an official recommendation of the World Wide Web Consortium (W3C), working group on mathematics. Wolfram Research was a founding member of the mathematics working group and host of the first two official MathML conferences in 2000 and 2002. Support for MathML is continuing to grow; it is now supported by most mainstream browsers (either natively or via a plug-in). More information on MathML can be found at the W3C site [1]. In addition there is a Wolfram Research sponsored web site, http://www.mathmlcentral.com (this uses web*Mathematica*).

The main MathML operations in *Mathematica* are the conversion of *Mathematica* expressions into MathML and the conversion of MathML back into *Mathematica*. Both of these conversion steps can draw on the symbolic XML functionality, as well as the general *Mathematica* functionality for working with typeset mathematics (introduced in *Mathematica* 3.0). To generate MathML in output, one can use a format type of **MathMLForm**. For more detailed control over the MathML which is produced, you can use functions from the XML` context. In the following example, the output has been chosen to use only presentation MathML and to use **mml** as a namespace prefix.

```
In[3]:= mathml = XML`MathML`ExpressionToMathML[
          x^2, "Formats" → {"PresentationMathML"},
          "NamespacePrefixes" →
           {"http://www.w3.org/1998/Math/MathML" → "mml"}]

Out[3]= <mml:math xmlns:mml='
          http://www.w3.org/1998/Math/MathML'
            xmlns='http://www.w3.org/1998/Math/MathML'>
         <mml:msup>
          <mml:mi>x</mml:mi>
          <mml:mn>2</mml:mn>
         </mml:msup>
        </mml:math>
```

To reverse the process, and go from a string of MathML to a *Mathematica* expression, you can use **XML`MathML`MathMLToExpression**.

```
In[4]:= XML`MathML`MathMLToExpression[mathml]
```

Out[4]= x^2

An alternative way to import MathML into *Mathematica* would involve generating Mathematica typesetting boxes and working with these to form a general *Mathematica* expression. This would allow intermediate transformations to ensure that the appropriate *Mathematica* expressions were generated. Techniques such as these and other issues relating to MathML in *Mathematica* are covered in greater detail in [2].

2.2 SVG

SVG is an XML language for describing two-dimensional graphics. It contains some interesting features including a variety of dynamic animation capabilities. SVG can be rendered with a browser plug-in from Adobe, http://www.adobe.com/svg. More information about SVG can be found at the W3C site: http://www.w3.org/Graphics/SVG/.

Mathematica support for SVG involves exporting graphics expressions into SVG. For example, a graphics object in *Mathematica* can be created from the various graphics primitives such as **`Line`**. This graphics object could be rendered into a picture with the *Mathematica* command **`Show`**. Alternatively, it can be converted into SVG with **`ExportString`**. To get the symbolic XML representation of the SVG one can use the function **`XML`SVG`GraphicsToSymbolicSVG`** as shown below.

```
In[5]:= XML`SVG`GraphicsToSymbolicSVG[
          Graphics[Line[ {{0, 0}, {1, 1}}]]]

Out[5]= XMLObject[Document][
          {XMLObject[Declaration][Version → 1.0,
            Encoding → UTF-8], XMLObject[Doctype][
            svg, Public → -//W3C//DTD SVG 1.0//EN,
            System → http://www.w3.org/TR/2001/REC-SVG-
               20010904/DTD/svg10.dtd]}, XMLElement[
           svg, {xmlns → http://www.w3.org/2000/svg,
            width → 288.0pt, height → 177.994pt,
            viewBox → -0.025 -0.0154508 1.05 0.648936,
            preserveAspectRatio → xMidYMid meet},
           {XMLElement[g, {transform →
               scale(1, -1) translate(0,-0.618034),
              style → fill:black; font-family:'Courier
                 New', Courier, Symbol, monospace;
                 font-size:0.0364583pt; stroke:
                 black; stroke-width:0.000911458pt},
             {XMLElement[line, {fill → none, x1 → 0, y1 → 0,
                x2 → 1.0, y2 → 0.618034}, {}]}]}], {}]
```

It would then be possible to modify the symbolic XML representation, for example, inserting the SVG tags that represented an SVG animation.

3 web*Mathematica*

web*Mathematica* is a web version of *Mathematica* that delivers interactive calculations and visualizations over the web. It allows a web site to return results that are marked up with *Mathematica* computations. When a request is made for one of these pages, the *Mathematica* commands are evaluated and any computed results inserted into the page. This is done with the standard Java templating mechanism, JavaServer Pages. This is described further at http://java.sun.com/products/jsp/.

web*Mathematica* technology uses the request/response standard followed by web servers. Input can come from HTML forms, applets, javascript, and web-enabled applications. It is also possible to send data files to a web*Mathematica* server for processing. Output can use many different formats, such as HTML, images, *Mathematica* notebooks, MathML, SVG, XML, PostScript, and PDF.

The latest release of web*Mathematica*, Version 2.0, was released in early 2003; it has many important features including the XML features of *Mathematica* 4.2. More information on web*Mathematica* can be found at [3].

3.1 MathML

MathML can be used in web*Mathematica* sites in two ways. First, it can be generated dynamically by a web*Mathematica* web site and displayed by an appropriate browser. MathML is now quite well supported by browsers, either natively or via a plug-in, the latest advances are discussed in [1]. Secondly, it can be entered into a browser with a suitable input tool and sent to a web*Mathematica* web site.

One way to enter MathML into a browser is WebEQ, http://www.dessci.com/en/products/webeq/, which provides a suite of tools including a Java applet that functions as an easy-to-use graphical equation editor in a web page. This can allow users to build up a mathematical formula, using the editing tools of WebEQ. The formula can then be sent to web*Mathematica* and used for computation. MathML can be useful in education and other areas where presentation is particularly important. More information on the use of MathML in web*Mathematica* is the web*Mathematica* documentation, [3]. An example of using WebEq with web*Mathematica* is shown below.

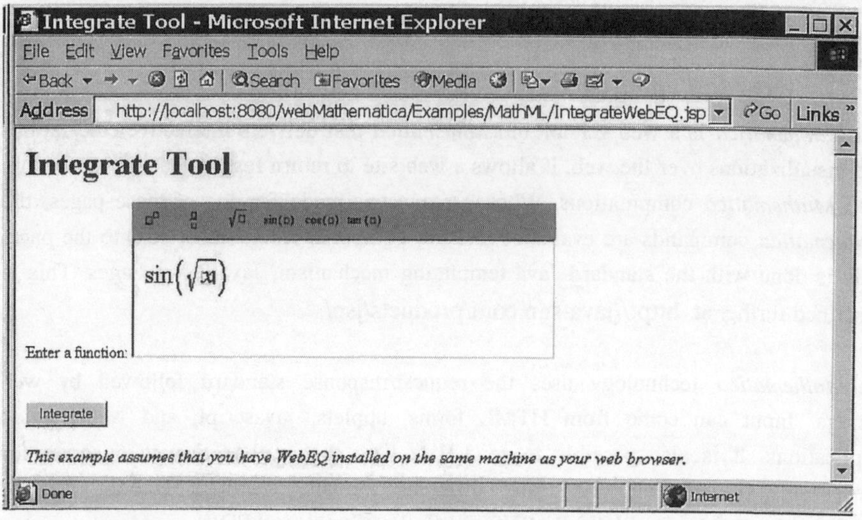

3.2 SVG

SVG provides a third ingredient to work with HTML and MathML to form web documents that contain a mixture of text, mathematics, and graphics. The *Mathematica* tools for working with SVG were reviewed in a previous section and these are all available for use in web*Mathematica*. One particularly interesting use of SVG is to generate an animation; *Mathematica* instructions for doing this are found at http://library.wolfram.com/infocenter/MathSource/4812.

More information on the use of SVG in web*Mathematica* is available in the web*Mathematica* documentation, [3].

3.3 Raw XML

A last example of XML for web computation in web*Mathematica* is its use as a data format for application-to-application communication. This involves a server returning a result to an application instead of to a browser, essentially using the web as a communication mechanism for distributed computation. XML is particular suitable as an interchange format for this type of process, since it is easy to parse and validate, and, in addition, many environments provide support for XML.

An example of a web*Mathematica* script that returns XML in response to a request is shown below. The details of web*Mathematica* programming are discussed elsewhere [3]. Briefly, the contents of the **<msp:allocateKernel>** tags are used to allocate a *Mathematica* kernel and the contents of the **<msp:evalaute>** tags are evaluated by *Mathematica* and returned in the result.

```
<%@ taglib uri="/webMathematica-taglib"
prefix="msp" %>
<msp:allocateKernel>
<msp:evaluate>
 xml = Import[ ToFileName[ MSPPageDirectory[],
                                "phone.xml"], "XML"] ;
 If[ MSPValueQ[ $$name],
  patt = "*" <> $$name <> "*";
  xml = DeleteCases[xml,
  XMLElement[ "Person", {___, "Name"->n_/;
    !StringMatchQ[n, patt],___},_ ], Infinity]] ;
 MSPReturn[ ExportString[ xml, "XML"], "text/xml"]
</msp:evaluate>
</msp:allocateKernel>
```

In this particular example an XML file is loaded into *Mathematica* and any XML tags that do not match a given string pattern removed. The resulting XML is then returned as the result of the request, with a return type of text/xml.

Any XML-aware application can call to a web*Mathematica* site that deploys this code. One way to call this site would be from *Mathematica* itself and suitable *Mathematica* code is shown below. The parameter to the web request is passed by adding **?argument=value** to the request URL.

```
In[6]:= Contact[query_String] :=
         Cases[XML`Parser`XMLGet[
           "http://localhost:8080/webMathematica/
              Examples/XML/Phone.jsp?name=" <> query],
          XMLElement["Person", x_List, {}] :→ x, Infinity]
```

This new *Mathematica* function can now be deployed. It calls the web*Mathematica* site and returns the result.

```
In[7]:= Contact["Tom"]

Out[7]= {{Name → Tom Jones, Email → tomj, Phone → 235-1231}}
```

This is a simple example of a web service, it demonstrates interoperability since it uses standards such as XML and HTTP. A more sophisticated web service could be built from specific web service standards such as SOAP, http://www.w3.org/2002/ws/.

4 Summary

This paper has discussed the use of XML for web computation with *Mathematica* technology. It discussed web sites that made use of MathML and SVG. It then covered the use of XML as a generic interchange format to build web services that use *Mathematica* as a computation tool. MathML and SVG will be relevant where the quality and scope of presentation are particularly important, such as education, which often favours the use of familiar notation. Computational web services are important for distributed network environments and these services can benefit from the use of XML as an interchange format. In addition, this allows them to benefit from the large number of XML applications in many areas such as finance and engineering, that are becoming available. See http://www.oasis-open.org/cover/xml.html.

References

[1] http://www.w3.org/Math/

[2] Sandhu, Pavi. The MathML Handbook,, Charles River Media, 2002.

[3] http://documents.wolfram.com/webmathematica/

Analytical solutions of fundamental problems of plane elasticity theory

N.Bonfoh and A.Carmasol
Ecole Nationale Supérieure d'Ingénieurs de Metz-Ile du Saulcy, France

{bonfoh,carmasol}@enim.fr

1 Abstract

The method of Kolossov and Muskhelishvili, based on complex representation theory, provides a general solution of fundamental problems of plane elasticity theory. Up to now, its effective application was restricted by computation difficulties. In this study, we use *Mathematica* to find analytical solutions of some fundamental problems of the infinite plate with elliptic holes, and subjected to different boundary conditions. The results obtained enable us to assess the validity of asymptotic solutions, widely used in fracture mechanics.

2 Introduction

To overcome mathematical difficulties arising when one attempts to solve fundamental problems of plane elasticity theory analytically, some numerical techniques such as finite element methods are often used. Thus, to solve equations of plane elasticity theory, the complex representation introduced by G.V. Kolossov [6] and developed by N. I. Muskhelishvili [7] has been applied by several authors to study problems involving an infinite plate containing holes. See Sneddon [9], Sokolnikoff [10], Green and Zerna [3] and Aifantis et al. [1]. Iokamidis et al. [4] proposed numerical and semi-numerical results. These initial investigations were only partially successful because of the complexity of the analytical calculations.

3 Plane elasticity problems

In the theory of plane elasticity, in the absence of body forces, the stress components can be expressed using the Airy function $\mathcal{A}(x, y)$:

$$\sigma_{xx} = \mathcal{A}_{,yy}\,, \quad \sigma_{yy} = \mathcal{A}_{,xx}\,, \quad \tau_{xy} = -\mathcal{A}_{,xy} \tag{1}$$

The Airy function, which is biharmonic, must satisfy the boundary and compatibility conditions of the problem. According to the theorem of E. Almansi, any poly-harmonic function of order N is representable by N harmonic functions. The solution of plane elasticity problems therefore arises from two functions (φ and χ) of the complex variable $z= x+i\,y$. These harmonic functions must also satisfy the boundary condition expressed on the elliptic hole contour (3).

$$\mathcal{A}(z) = \mathrm{Re}[\bar{z}\,\varphi(z) + \chi(z)] \tag{2}$$

$$f(x, y) = i\,(X + i\,Y) + \mathrm{const} = \varphi(z) + \overline{z\varphi'(z)} + \overline{\psi(z)} \tag{3}$$

(X,Y) represents the component of the resultant vector of the forces acting on the contours in the case of multiply connected regions and $\psi(z)=\chi'(z)$. The Kolossov-Muskhelishvili potentials (φ and χ) depend on the geometry of the selected region: finite or infinite, simply or multiply connected. The subsequent stress and displacement fields of the problem are given by (4,5), where $\phi(z) = \varphi'(z)$ and $(\mu,\ \lambda)$ are the Lamé constants of the isotropic elastic body .

$$\begin{aligned} \sigma_{xx} + \sigma_{yy} &= 2\,\mathrm{Re}[\phi(z)] \\ \sigma_{yy} - \sigma_{yy} + 2\,i\tau_{xy} &= 2\,(\bar{z}\,\phi'(z) + \psi'(z)) \\ 2\,\mu\,(u_x + iu_y) &= \kappa\varphi(z) - z\,\overline{\phi(z)} - \psi(z) \end{aligned} \tag{4}$$

$$\begin{aligned} \kappa &= \frac{\lambda + 3\,\mu}{\lambda + \mu} \quad \text{(plane strain state)} \\ \kappa &= \frac{5\,\lambda + 6\,\mu}{3\,\lambda + 2\,\mu} \quad \text{(plane stress state)} \end{aligned} \tag{5}$$

3.1 Multiply connected region: Conformal mapping

For the problem of an infinite plate weakened by elliptic holes, it is possible to reduce this case to the less difficult case of a plate with a circular hole and then apply a judicious mapping to derive the solution of the initial problem. Conformal mapping (6) can be used, where the semi-axes of the elliptic hole are given by (7). ($m = 0$) corresponds to a circular hole, while the asymptotic case ($m = 1$) represents an elliptic crack in the Ox axis. The boundary condition (3) can then be rewritten as (8). In (8), $\tau = e^{i\theta}$ is the current point on the unit circle γ, representing the hole boundary in the infinite plate. In a multiply connected region, (9) holds true (see Muskhelishvili, 1953).

$$z = \omega(\zeta) = R\left(\zeta + \frac{m}{\zeta}\right); \quad (R > 0, \ 0 \le m < 1); \quad \zeta = r\,e^{i\theta} \tag{6}$$

$$a = R(1 + m); \quad b = R(1 - m) \tag{7}$$

$$\begin{aligned} &\varphi(\tau) + \frac{\tau^2 + m}{\tau(1 - m\tau^2)}\overline{\varphi'(\tau)} + \overline{\psi(\tau)} = f(\tau) \\ &\overline{\varphi(\tau)} + \frac{\tau(1 + m\tau^2)}{\tau^2 - m}\varphi'(\tau) + \psi(\tau) = \overline{f(\tau)} \end{aligned} \tag{8}$$

$$\begin{aligned} \varphi(z) &= -\frac{X + iY}{2\pi(1 + \kappa)}\,\mathrm{Log}(z) + \Gamma_1 z + \varphi_0(z) \\ \psi(z) &= \frac{\kappa(X - iY)}{2\pi(1 + \kappa)}\,\mathrm{Log}(z) + \Gamma_2 z + \psi_0(z) \end{aligned} \tag{9}$$

Γ_1 and Γ_2 are related to the remote stress field (10). By introducing the conformal mapping relation defined by (6), we derive (11), which contains functions $\varphi_0(\zeta)$ and $\psi_0(\zeta)$ that satisfy boundary conditions (3) if the function f is replaced by f_0, defined by (12). The function f_0 is known since boundary conditions at infinity and on the hole contour are known.

$$\Gamma_1 = \frac{\sigma_{yy}^{\infty} + \sigma_{xx}^{\infty}}{4}; \quad \Gamma_2 = \frac{\sigma_{yy}^{\infty} - \sigma_{xx}^{\infty}}{2} + i\tau_{xy}^{\infty} \tag{10}$$

$$\begin{aligned} \varphi(\zeta) &= -\frac{X + iY}{2\pi(1 + \kappa)}\,\mathrm{Log}(\zeta) + \Gamma_1 R\zeta + \varphi_0(\zeta) \\ \psi(z) &= \frac{\kappa(X - iY)}{2\pi(1 + \kappa)}\,\mathrm{Log}(\zeta) + \Gamma_2 R\zeta + \psi_0(z) \end{aligned} \tag{11}$$

$$f_0(\tau) = f(\tau) + \frac{X + i\,Y}{2\pi}\,\text{Log}(\tau) - \Gamma_2\,R\left(\tau + \frac{\tau^2 + m}{\tau\,(1 - m\tau^2)}\right) + \frac{X - i\,Y}{2\pi\,(1 + \kappa)}\,\frac{\tau^2 + m}{1 - m\tau^2} - \frac{\overline{\Gamma_2}\,R}{\tau} \qquad (12)$$

4 Applications

This general procedure is applied to the problem of an elastic plate subjected to different boundary conditions and containing holes. Whatever the hole shape may be, it is possible to introduce a suitable conformal mapping to a circular hole.

4.1 Infinite plate weakened by an elliptic hole and subjected to uniaxial tension at infinity

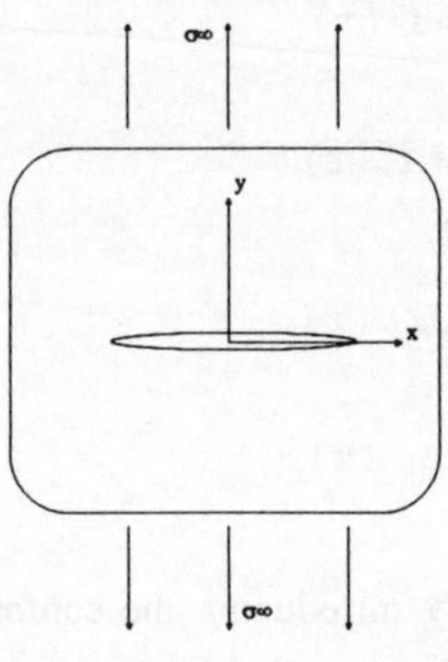

Figure 1

The elliptic hole will be assumed to be free of stress: $X = Y = 0$ and $f = 0$. To derive the complex functions φ_0 and ψ_0 one has to apply the Cauchy integral to expressions (13). The stability of approximation methods for the Muskhelishvili integral equations, subjected to different boundary conditions, has been studied by Didenko et al [2]. It is then possible to derive the complete expressions of stress and displacement fields (14). These expressions have been obtained by using some *Mathematica* functions necessary to solve equations such as (4, 5).

$$\varphi(\zeta) = -\frac{\sigma^{\infty}}{4\,\zeta} R\left(\zeta - \frac{m+2}{\zeta}\right);$$
$$\psi(\zeta) = \frac{\sigma^{\infty} R}{2}\left(\zeta - \frac{1 + (1 + m + m^2)\,\zeta^2}{\zeta\,(\zeta^2 - m)}\right) \tag{13}$$

$$\sigma_{rr} = -((r^2-1)\,\sigma\infty\,(-r^6 + m^2 r^4 - m\,r^4 + m^2 r^2 + 3\,m\,r^2 - m\,(r^2-1)\cos(4\,\theta)\,r^2 - m^4 - 2\,m^3 + ((r^2-3)\,r^4 + m^2\,(3\,r^2-1))\cos(2\,\theta)))\Big/\left(2\,(r^4 - 2\,m\cos(2\,\theta)\,r^2 + m^2)^2\right);$$
$$\sigma_{\theta\theta} = (\sigma\infty\,((-4\,m\,r^6 - (r^4+3)\,r^4 + 4\,m^3\,r^2 - m^2\,(3\,r^4 + 8\,r^2 + 1))\cos(2\,\theta) + (r^2+1)\,(r^6 + m^2\,r^4 + m\,r^4 - m^2\,r^2 + 3\,m\,r^2 + m\,(r^2+1)\cos(4\,\theta)\,r^2 - m^4 + 2\,m^3)))\Big/\left(2\,(r^4 - 2\,m\cos(2\,\theta)\,r^2 + m^2)^2\right); \tag{14}$$
$$\tau_{r\theta} = -((r^2-1)\,\sigma\infty\,(r^6 - 2\,m\,r^4 + 3\,r^4 + 2\,m^3\,r^2 - m^2\,r^2 - 2\,m\,(r^2+1)\cos(2\,\theta)\,r^2 + m^2)\sin(2\,\theta))\Big/\left(2\,(r^4 - 2\,m\cos(2\,\theta)\,r^2 + m^2)^2\right);$$

For example, to evaluate the Real and imaginary parts of some rational expressions arising in calculations, the following *Mathematica* functions have been introduced:

```
expr[x_] := x /. z -> r Cos[θ] + I r Sin[ θ];
complfunct[z_] :=
  Expand[z] /. Complex[x_, y_] -> (x + y j);
complist[z_] :=
  CoefficientList[complfunct[z], j] /;
   Length[CoefficientList[complfunct[z], j]] == 2;
complist[z_] := {complfunct[z], 0} /;
   FreeQ[complfunct[z], j];
realpart[z_] := TrigReduce[First[complist[z]]];
imagpart[z_] := TrigReduce[Last[complist[z]]];
conjug[z_] := {1, -I}.complist[z];
conjug[z_^n_] := (leconjug[z])^n;
```

```
decomp[z_] := Module[{bas, haut},
bas =
    Simplify[complist[Denominator[Together[z]]].
     complist[Denominator[Together[z]]]];
haut = conjug[Denominator[Together[z]]]
     Numerator[ Together[z]];
    Return[{realpart[haut] / bas,
     imagpart[haut] / bas}] ];
imagfrac[z_] := Last[decomp[z]];
realfrac[z_] := First[decomp[z]];
conjugfrac[z_] := {1, -I}.decomp[z];
```

The expressions for stress and displacement components (14) are fully analytical. No numerical assumption has been made for calculations. As stated previously, by assigning the value 1 to the shape parameter *m*, we can obtain a solution for an elliptic crack in an infinite elastic plate under uniaxial tension. Some investigations have been devoted to the problem of elastic plates with an elliptic crack. These initial studies, such as the method of Westergaard, lead to asymptotic expressions of stress and displacement fields, only valid near the vicinity of the elliptic crack. Figure 2 shows the evolution of the stress component in the traction direction as a function of the radial distance to the crack vicinity (*a* is the semi-length of the linear crack).

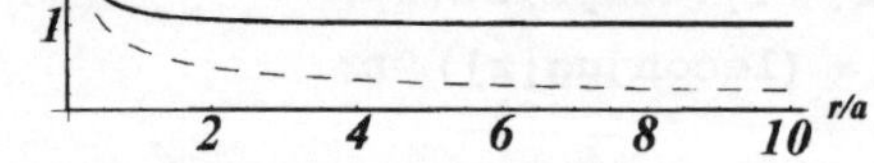

Figure 2

It is noticed that the expressions provided are similar to known asymptotic expressions near the crack. However, for $\frac{r}{a} > 1$, asymptotic expressions seem to be invalid. The same observations have been noticed for all stress and displacement components. Figure 3 is a contour plot of the Von-mises equivalent stress σ_{eq} near the crack tip.

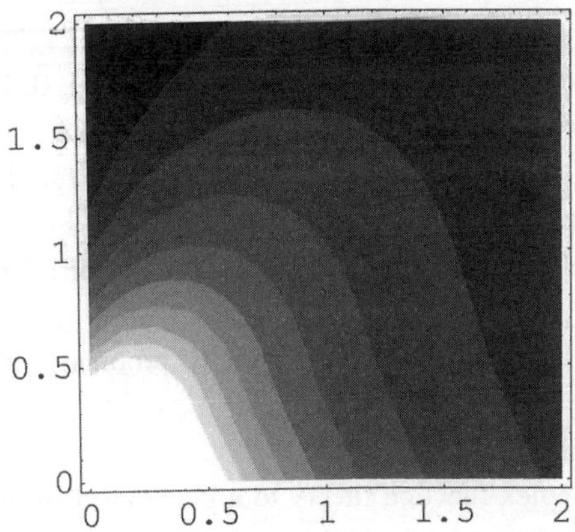

Figure 3

This method has also been applied for mode II and III problems as well as specific problems of an infinite plate containing two cracks. For this last problem, we derived expressions of stress concentration factors at the inner and outer tips of cracks. The results obtained were similar to those of Aifantis et al.[1]. The method suggested is applicable to all types of fundamental problems of plane elasticity.

5 Conclusion

Computer Algebra systems such as *Mathematica* are a powerful tool for use in complex function theory, initiated by Kolossov and Muskhelishvili. Up to now, most investigations devoted to solutions of plane elasticity problems based on this method have been restricted to expressions of two complex functions. Full analytical expressions of stress and displacement fields are rare, unless they are asymptotic expansions. With *Mathematica*, analytical solutions of such problems of physics and mechanics are no longer a task reserved for courageous mathematicians. For fracture calculations, these analytical expressions increase the precision of engineering analysis.

6 References

[1] Aifantis, E.C. and Gerberich, W.W. A new form of exact solutions for mode I, II, III crack problems and implications in Engineering, Fracture Mechanics, vol.10, 1978.

[2] Didenko, V.D. and Silbermann, B. On stability of approximation methods for the Muskhelishvili equation, J. Comput. and Appl. Math. 146, pp.419-441, 2002.

[3] Green, A.E. and Zerna, W. Theoretical elasticity, Oxford University Press, Fairlawn, NJ, 1954.

[4] Ioakimidis, N.I. and Anastasseos, G.T. Solution of plane elasticity problem with *Mathematica*, Comput. Struct. vol. 55, No. 2, pp.229-236, 1995.

[5] Irwin, G.R. Fracturing of metals, ASM, Cleveland, Ohio, 1949.

[6] Kolossov, G.V. On an application of complex function theory to a plane problem of the mathematical theory of elasticity, Communication at the University of Dorpat, 1909.

[7] Muskhelishvili N.I. Some basic problems of mathematical elasticity, translated by Radok, J.R.M., Leyden: Noordhoff International Publishing, 1975.

[8] Schinzinger, R. and Laura, P.A.A. Conformal mapping: methods and applications, Elsevier Science Publishers B.V., Netherlands, 1991.

[9] Sneddon, I.N. The distribution of stress in the neighbourhood of a crack in an elastic solid, in Proc. R. Soc. London, A187, 1946.

Understanding the Kinematics of Eye Movements with *Mathematica*

Richard A. Clement
Visual Sciences Unit, Institute of Child Health, University College London

r.clement@ich.ucl.ac.uk

1 Abstract

The position of the eye is described by Listing's law, which states that any movement of the eye away from the straight ahead position is accomplished by a rotation about an axis which is perpendicular to the straight ahead direction. *Mathematica* is used to provide concise derivations of the implications of this law for vision.

2 Introduction to rotations of the eye

In the context of movements of the eyes the most important visual direction is given by the line of fixation, which passes through the centre of the fovea and the centre of rotation of the eye. The position of an eye can be specified in terms of the direction of the line of fixation and a rotation around that direction. The rotation can be investigated by using after-images, which have a fixed location on the retina and so move with the eye. A straightforward procedure involves forming an after-image of a cross with the eyes in the straight ahead position looking at a far point on the horizon with the head erect, and then measuring the apparent orientation of the cross whilst looking at a wall perpendicular to the primary position, as shown in figure 1. The cross is not distorted when the point of fixation lies on the horizontal or vertical meridian. In general, an after-image of an oblique line does not appear to alter its orientation at points along the associated oblique meridian, so the axis of rotation must be perpendicular to the oblique meridian. This finding is summarised by Listing's law, which states that if the eye moves about a centre O so that the line of fixation moves away from the straight ahead position OA to another position OB, then the displacement of the eyeball is equivalent to rotating it around an axis perpendicular to the plane AOB [1].

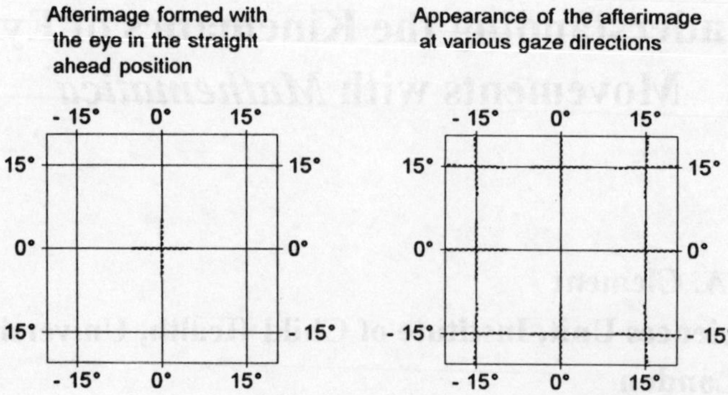

Figure 1. Procedure for investigating the orientation of an eye using after-images. The after-image of a cross is formed with the eye in the straight ahead position. When the eye is subsequently turned to look at various gaze directions the cross appears distorted in diagonal positions.

An appropriate system of co-ordinates for describing the rotation of the eye is provided by the three angles referred to as Euler angles. The first two angles specify the orientation of the line of fixation and the third specifies the rotation of the eye about the line of fixation. The Euler angles are defined by a sequence of rotations which transforms one set of Cartesian base vectors into another. One of the sets, $(\mathbf{e}_1, \mathbf{e}_2, \mathbf{e}_3)$, is fixed in space and the other, $(\mathbf{e}_1', \mathbf{e}_2', \mathbf{e}_3')$, is fixed with respect to the body. The set $(\mathbf{e}_1, \mathbf{e}_2, \mathbf{e}_3)$ is fixed with respect to the head and the set $(\mathbf{e}_1', \mathbf{e}_2', \mathbf{e}_3')$ is fixed with respect to the eye. The head-based system of axes is oriented so that the $\mathbf{e}_3$ direction lies along the primary position of the line of fixation. For convenience, the $\mathbf{e}_1$ base vector is set to lie in a horizontal plane and the $\mathbf{e}_2$ base vector is set to lie in a vertical plane. When the eye is in the primary position the two sets of base vectors are coincident.

The first two Euler angles, (α, β), specify the line of fixation direction $\mathbf{e}_3'$ with respect to the fixed set of base vectors, and the third angle, (γ), specifies the rotation of the eye about the $\mathbf{e}_3'$ direction. The angle α is given by the clockwise rotation about $\mathbf{e}_3$ needed to rotate the plane spanned by the $\mathbf{e}_2$ and $\mathbf{e}_3$ base vectors into the plane spanned by the $\mathbf{e}_3$ and $\mathbf{e}_3'$ base vectors. The angle β is determined by the clockwise rotation about the new $\mathbf{e}_1$ direction which aligns $\mathbf{e}_3$ with $\mathbf{e}_3'$. Finally, the angle γ specifies the clockwise rotation about the $\mathbf{e}_3'$ direction. With this specification of the position of the eye the constraint implied by Listing's law is that $\gamma = -\alpha$.

The rotational state of the eye can be estimated photograpically by comparing the positions of at least two scleral landmarks, before and after a rotation away from the primary position [2]. The pixel co-ordinates of the points in the picture can be converted to 3-D co-ordinates by assuming that they lie on the surface of a standard eye with a radius of 12.43 mm, and the orientation angles which best describe the rotation can be found numerically.

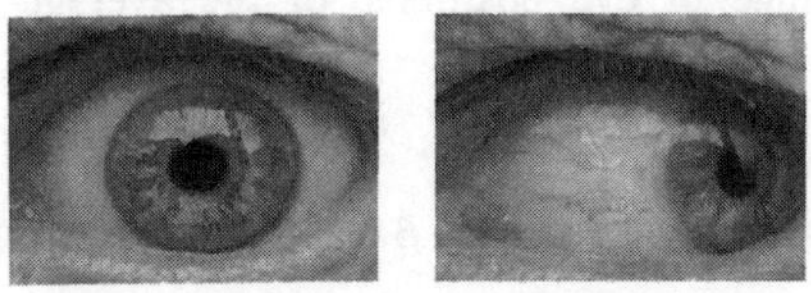

Figure 2. Photographic measurement of eye position

```
In[1]:= << Statistics`NonlinearFit`;
        << Geometry`Rotations`;

        x1 = {-8.37, -2.41, 8.87};
        x2 = {-6.36, 0.19, 10.68};
        y1 = {-1.98, -2.79, 11.95};
        y2 = {0.72, -0.14, 12.41};

        L = RotationMatrix3D[α, β, γ];
        solution = FindMinimum[
           (y1 - L .x1).(y1 - L .x1) + (y2 - L .x2).(y2 - L .x2),
           {α, 0}, {β, 0}, {γ, 0}];
        Evaluate[{α, β, γ} /. solution[[2]]]

Out[9]= {-1.62211, 0.593828, 1.60525}
```

3 Why Listing's law?

Listing's law minimises the rotation of the eye about the line of fixation, as can be proved by application of the calculus of variations to the Euler angle description of the position of the eye [3]. If the position of the eye is described by Euler angles α, β and γ, then any small change in the position of the eye, described by small changes $\delta\alpha$, $\delta\beta$ and $\delta\gamma$ in the Euler angles, involves a rotation which has a component about the line of fixation of $\delta\alpha \cos \beta + \delta\gamma$. Let ds be the displacement of the line of fixation on a unit sphere located at the centre of rotation, and let this displacement make an angle ϵ with the plane spanned by the primary direction of the line of fixation and the direction of the line of fixation specified by the Euler angles α, β and γ, so that $\delta\beta = \cos\epsilon\, \delta s$ and $\sin\beta\, \delta\alpha = \sin\epsilon\, \delta s$. By applying these equalities, the component of rotation about the line of fixation can be written as:

$$\delta\alpha \cos\beta + \delta\gamma = \left(\left(\frac{\delta\gamma}{\delta\alpha} + \cos\beta \right) \frac{\sin\epsilon}{\sin\beta} + \frac{\delta\gamma}{\delta\beta} \cos\epsilon \right) ds \qquad (1)$$

The integral, over the entire range of eye positions, of the average square of this component is:

$$\iint \left(\frac{1}{\sin\beta} \left(\frac{\delta\gamma}{\delta\alpha} + \cos\beta \right)^2 + \sin\beta \left(\frac{\delta\gamma}{\delta\beta} \right)^2 \right) d\beta\, d\alpha \qquad (2)$$

It follows that the condition for a stationary value of the integral is:

$$\frac{1}{\sin\beta} \frac{\partial^2\gamma}{\partial\alpha^2} + \cos\beta \frac{\partial\gamma}{\partial\beta} + \sin\beta \frac{\partial^2\gamma}{\partial\beta^2} = 0 \qquad (3)$$

which is satisfied if γ is a linear function of α only. To ensure that the rotation about the line of fixation is zero as $\beta \to 0$, it is necessary that $\gamma = -\alpha$, which is Listing's law.

```
In[18]:= << Calculus`VariationalMethods`;

         EulerEquations[ 1/Sin[β] * (∂α f[α, β] + Cos[β])^2 +
            Sin[β] * (∂β f[α, β])^2, f[α, β], {α, β}]

Out[19]= -2 (Cos[β] f^(0,1)[α, β] +
            Sin[β] f^(0,2)[α, β] + Csc[β] f^(2,0)[α, β]) == 0
```

4 Binocular eye movements

The difference between the positions of the images in the two eyes, referred to as disparity, gives rise to an impression of depth. This impression is most accurate for points with least disparity. Given the way in which the eyes move, it is straightforward to calculate the locus of points, referred to as the horopter, which have zero disparity. The position of the retinal images of any point can be calculated by describing the transformation from the head-based system of co-ordinates to the eye-based system. Consider three systems of right-handed Cartesian base vectors, located at the centres of rotation of the left and right eyes, and at the midpoint of the line joining the centres of rotation of the two eyes. Let the base vectors $\mathbf{e}_1$, $\mathbf{e}_2$ and $\mathbf{e}_3$ at the midpoint be fixed with respect to the head, as shown in figure 3. Let the base vectors $\mathbf{e}_1'$, $\mathbf{e}_2'$ and $\mathbf{e}_3'$ and $\mathbf{e}_1''$, $\mathbf{e}_2''$, and $\mathbf{e}_3''$, associated with the right and left eyes respectively, be fixed with respect to the eyes, with the third base vector pointing along the optic axis of the eye in each case. Let a point p in the head-based system of cartesian base vectors have corresponding images p' and p'' in the right and left eye systems of base vectors respectively.

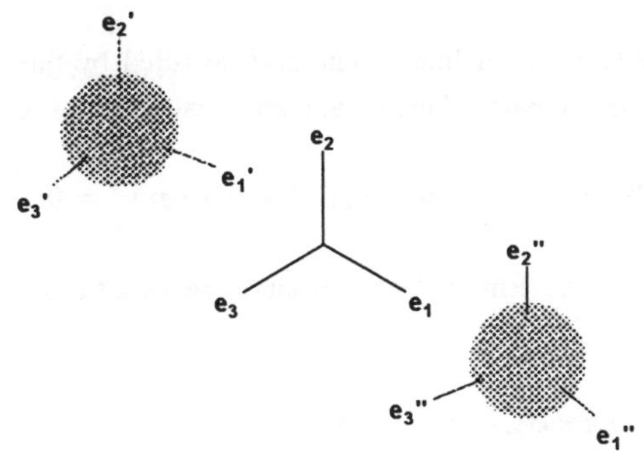

Figure 3. Co-ordinate systems used to relate a physical point to the two retinal images.

To obtain the co-ordinates of a retinal image, the point p is translated along the line joining the centres of rotation of the two eyes, and then rotated through the angle the optic axis of the eye makes with the midline. Let the interocular distance be equal to s, so that a translation from the head-fixed system of co-ordinates to the right eye system of co-ordinates is specified by the vector:

$$\mathbf{s} = \begin{pmatrix} \frac{s}{2} \\ 0 \\ 0 \end{pmatrix} \tag{4}$$

If it is assumed that the eyes move according to Listing's law then the rotation of the right eye can be described by the rotation matrix $L[\alpha_r, \beta_r, -\alpha_r]$, where α_r and β_r are the Euler angles which specify the position of the right eye. The left eye is described similarly. The overall transformations from head-fixed co-ordinates to eye-fixed co-ordinates are then given by a pair of matrix equations:

$$\boldsymbol{p}' = L[\alpha_r, \beta_r, -\alpha_r][\boldsymbol{p}+\boldsymbol{s}]$$
$$\boldsymbol{p}'' = L[\alpha_l, \beta_l, -\alpha_l][\boldsymbol{p}-\boldsymbol{s}] \tag{5}$$

To calculate the physical stimulus which will produce a set of lines which are invariant under the stereoscopic transformation, consider two pairs of planes which are specified by the same equation in each eye:

$$\boldsymbol{a}\cdot\boldsymbol{p}' = 0,\ \boldsymbol{a}\cdot\boldsymbol{p}'' = 0 \text{ and}$$
$$\boldsymbol{b}\cdot\boldsymbol{p}' = 0,\ \boldsymbol{b}\cdot\boldsymbol{p}'' = 0 \tag{6}$$

These two planes specify two retinal lines. The surface ruled by the intersections of linear combinations of these two pairs of identical planes satisfies the two equations:

$$\boldsymbol{a}\cdot\boldsymbol{p}' + s\boldsymbol{b}\cdot\boldsymbol{p}' = 0 \text{ and } \boldsymbol{a}\cdot\boldsymbol{p}'' + s\boldsymbol{b}\cdot\boldsymbol{p}'' = 0 \tag{7}$$

where s is a scalar parameter. Eliminating s from these equations, one obtains the equation:

$$(\boldsymbol{a}\times\boldsymbol{b})\cdot(\boldsymbol{p}''\times\boldsymbol{p}') = 0 \tag{8}$$

The surface in space which is defined by a set of identical lines is referred to as a line horopter. The equations for describing the surfaces are simplest when the point of fixation lies in the horizontal plane passing through the two centres of rotation of the eyes [2]. If $\boldsymbol{a}\times\boldsymbol{b} = (1,0,0)$ then the equation for the horizontal line horopter is obtained:

$$\mathrm{p}_2\left(\frac{1}{2}\,\mathrm{s}\,(\mathrm{Sin}[\beta_l]+\mathrm{Sin}[\beta_r]) + (-\mathrm{Sin}[\beta_l]+\mathrm{Sin}[\beta_r])\,\mathrm{p}_1 + (-\mathrm{Cos}[\beta_l]+\mathrm{Cos}[\beta_r])\,\mathrm{p}_3\right) = 0 \tag{9}$$

Similarly, if $\boldsymbol{a}\times\boldsymbol{b} = (0,1,0)$ then the equation for the vertical line horopter is obtained:

$$\frac{1}{4}\,\mathrm{s}\,(\mathrm{s}\ \mathrm{Sin}[\beta_l-\beta_r] - 4\,\mathrm{Cos}[\beta_l-\beta_r]\,\mathrm{p}_3) - \mathrm{Sin}[\beta_l-\beta_r]\,(\mathrm{p}_1^2+\mathrm{p}_3^2) = 0 \tag{10}$$

All the points common to both the horizontal and vertical line horopters have the same direction in each eye. The curve formed by all such points is the point horopter. This consists of a single vertical line and a circle, called the Vieth-Müller circle, which passes through the point of fixation and the centres of rotation of the eyes. When the point of fixation is out of the horizontal plane, the horizontal line horopter changes from a pair of intersecting planes to a hyperboloid of two sheets and the vertical line horopter changes from a cylinder to a hyperboloid of one sheet. These changes can be avoided if the eyes do not move in accordance with Listing's law, but move so that their horizontal meridians are coincident. A plausible binocular extension of Listing's law has been proposed by Tweed [4].

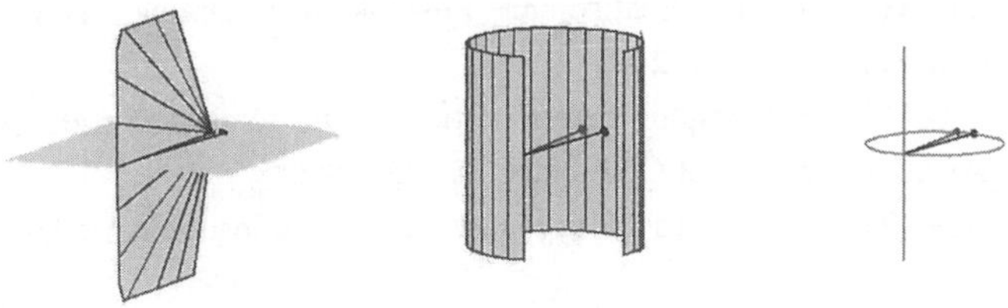

Figure 4. Cut-away drawings of line horopters with a point of fixation located on the line which is parallel to the primary directions of the lines of fixation and midway between them. In this case the point horopter consists of a vertical line and the Vieth-Müller circle, which intersect at the point of fixation.

In[7]:= `<< Geometry`Rotations`;`

$$PL = \text{RotationMatrix3D}\left[\frac{\pi}{2},\ \beta_l,\ -\frac{\pi}{2}\right].\left\{p_1 - \frac{s}{2},\ p_2,\ p_3\right\};$$

$$PR = \text{RotationMatrix3D}\left[\frac{\pi}{2},\ \beta_r,\ -\frac{\pi}{2}\right].\left\{p_1 + \frac{s}{2},\ p_2,\ p_3\right\};$$

```
i = {1, 0, 0};
j = {0, 1, 0};
k = {0, 0, 1};
FullSimplify[(PL.j) * (PR.k) - (PL.k) * (PR.j)]
FullSimplify[(PL.i) * (PR.k) - (PL.k) * (PR.i)]
```

Out[13]= $$p_2 \left(\frac{1}{2} s\,(\text{Sin}[\beta_l] + \text{Sin}[\beta_r]) + (-\text{Sin}[\beta_l] + \text{Sin}[\beta_r])\,p_1 + (-\text{Cos}[\beta_l] + \text{Cos}[\beta_r])\,p_3\right)$$

Out[14]= $$\frac{1}{4} s\,(s\,\text{Sin}[\beta_l - \beta_r] - 4\,\text{Cos}[\beta_l - \beta_r]\,p_3) - \text{Sin}[\beta_l - \beta_r]\,(p_1^2 + p_3^2)$$

5 Conclusions

Mathematica provides a coherent environment in which one can readily combine analytic calculations of the implications of Listing's law, numerical analysis of experimental tests of the law and graphical display of the results.

6 References

[1] Helmholtz H. von (1910) Treatise on Physiological Optics. Translated by J.P.C.Southall, New York, Dover, 1924.

[2] Nakayama K. (1974) Photographic determination of the rotational state of the eye using matrices. American Journal of Optometry, 51, 736-742.

[3] Lamb H. (1919) The kinematics of eye movements. Philosophical Magazine 38, 685-695.

[4] Tweed D. (1997) Visual-motor optimization in binocular control. Vision Research 37, 1939-1951.

Applying Jacobian Elliptic Functions to Solve Linear and Nonlinear Differential Equations With Mathematica

Alex Elías-Zúñiga
ITESM, Monterrey, México
aelias@itesm.mx

1 Abstract

We use *Mathematica* to illustrate the applications of Jacobi elliptic functions to obtain the approximate solution of a two degree of freedom, undamped, homogeneous dynamical system having small cubic nonlinearities. We also illustrate in this article how *Mathematica* may help to obtain the closed-form solution of Lamé's equation.

2 Introduction

In many physical systems, a pair of coupled, homogeneous nonlinear ordinary differential equations model the system's dynamical response. In general, however, there is no known exact solution of this sort of system of equations of motion, and hence numerical integration or perturbation techniques are applied to obtain their approximate solution. Here we apply the Elliptic Balance Method (EBM) introduced by Elías-Zúñiga in [1] to obtain the approximate solution of

$$\begin{pmatrix} u_1'' \\ u_2'' \end{pmatrix} + \begin{pmatrix} \omega_n^2 & 0 \\ 0 & \omega_n^2 \end{pmatrix} \begin{pmatrix} u_1 \\ u_2 \end{pmatrix} - \varepsilon \begin{pmatrix} \varphi_1 u_1^3 + \varphi_2 u_1^2 u_2 + \varphi_3 u_1 u_2^2 + \varphi_4 u_2^3 \\ \varphi_5 u_1^3 + \varphi_6 u_1^2 u_2 + \varphi_7 u_2^2 + \varphi_8 u_2^3 \end{pmatrix} = \begin{pmatrix} 0 \\ 0 \end{pmatrix} \tag{1}$$

where the prime denotes the derivative with respect to the time t, u_1 and u_2 are finite modal amplitudes, ω_{n1} and ω_{n2} are the system eigenvalues, ε is a small, positive dimensionless parameter, and $\varphi_1 \ldots \varphi_8$ are parameters related to the physics of the system. One advantage of the EBM is based on its ability to predict the solution phase for both weak and moderate nonlinear homogeneous systems. As pointed out in [2], the Elliptic Balance Method (EBM) may be considered as a generalization of the method of harmonic balance because the trigonometric functions utilized in the latter approximate method are particular cases of the Jacobi elliptic functions, and hence the conditions for which the method of harmonic balance may be applied must also be applicable to the EBM.

In the first part of this article, we focus, by the use of the concept of averaging, on the application of the EBM to obtain the approximate solution of Equation (1), and in the second part we use *Mathematica*'s power and range to obtain the exact solution of Lamé's equation. We next illustrate the application of the EBM.

3 The Elliptic Balance Method

The approximate solution of Equation (1) by the EBM will be developed in terms of the Jacobian elliptic functions *cn*, *sn* and *dn*. The approximate solution will be obtained by following the same idea used in the method of harmonic balance [2], [3], but instead of balancing trigonometric functions, we now balance Jacobian elliptic functions. This will require use of several identities introduced as their need arises. To begin with the solution of the system (2.1) by application of Jacobian elliptic functions, we shall assume its approximate solution to be of the form

$$u_1 = a_1 \operatorname{cn}(\omega_1 t + \phi_1, k_1^2) \tag{2}$$

$$u_2 = a_2 \operatorname{cn}(\omega_2 t + \phi_2, k_2^2) \tag{3}$$

where ω_j, k_j, ϕ_j, and a_j are constants to be found. For simplicity in the notation, let us write the Jacobian elliptic functions in Equations (2) and (3) as $cn_j \equiv cn(\omega_j t + \phi_j, k_j^2)$, for j = 1, 2. Similar notation is defined for the functions sn_j and dn_j. Substituting (2) and (3) into (1) and using *Mathematica*, we get

```
In[1]:= u₁ = a₁ JacobiCN[ω₁ t + ϕ₁, k₁²];
        u₂ = a₂ JacobiCN[ω₂ t + ϕ₂, k₂²];
        eq1 = D[u₁, {t, 2}] + u₁ ωn₁² +
          ε * (u₁³ φ₁ + u₁² u₂ φ₂ + u₁ u₂² φ₃ + u₂³ φ₄);
        eq2 = D[u₂, {t, 2}] + u₂ ωn₂² +
          ε * (u₁³ φ₅ + u₁² u₂ φ₆ + u₁ u₂² φ₇ + u₂³ φ₈);
        JacobiCN[ω₁ t + ϕ₁, k₁²] := cn1;
        JacobiSN[ω₁ t + ϕ₁, k₁²] := sn1;
        JacobiDN[ω₁ t + ϕ₁, k₁²] := Sqrt[1 - k₁² * sn1^2];
        JacobiCN[ω₂ t + ϕ₂, k₂²] := cn2;
        JacobiSN[ω₂ t + ϕ₂, k₂²] := sn2;
        JacobiDN[ω₂ t + ϕ₂, k₂²] := Sqrt[1 - k₂² * sn2^2];
        sn1 = Sqrt[1 - cn1^2];
        sn2 = Sqrt[1 - cn2^2];
        Collect[eq1, cn1]
        Collect[eq2, cn2]
```

Out[13]= $\mathrm{cn1}^2\,\mathrm{cn2}\,\varepsilon\,a_1^2\,a_2\,\varphi_2 + \mathrm{cn2}^3\,\varepsilon\,a_2^3\,\varphi_4 + \mathrm{cn1}^3\,(\varepsilon\,a_1^3\,\varphi_1 - 2\,a_1\,k_1^2\,\omega_1^2) + \mathrm{cn1}\,(\mathrm{cn2}^2\,\varepsilon\,a_1\,a_2^2\,\varphi_3 - a_1\,\omega_1^2 + 2\,a_1\,k_1^2\,\omega_1^2 + a_1\,\omega n_1^2)$

Out[14]= $\mathrm{cn1}^3\,\varepsilon\,a_1^3\,\varphi_5 + \mathrm{cn1}\,\mathrm{cn2}^2\,\varepsilon\,a_1\,a_2^2\,\varphi_7 + \mathrm{cn2}^3\,(\varepsilon\,a_2^3\,\varphi_8 - 2\,a_2\,k_2^2\,\omega_2^2) + \mathrm{cn2}\,(\mathrm{cn1}^2\,\varepsilon\,a_1^2\,a_2\,\varphi_6 - a_2\,\omega_2^2 + 2\,a_2\,k_2^2\,\omega_2^2 + a_2\,\omega n_2^2)$

(4)

We next apply the averaging procedure introduced by Elías-Zúñiga and Beatty in [4]. In this process, the arguments in Equations (2) and (3) are treated as independent variables. Because the Jacobian elliptic functions *cn* and *sn* are of period $4\,K\,(k^2)$, we wish to average Equations (4) over this period. Notice, however, that these equations depend on Jacobian elliptic functions whose moduli are different. Therefore, we compute the average of equation $(4)_1$ with respect to $4\,K_2$ and the average of Equation $(4)_2$ with respect to $4\,K_1$ where $K_j = K\,(k_j^2)$ is the complete elliptic integral of the first kind for the modulus k_j^2. Notice that the average value of both *cn* and cn^3 is zero [5]. Thus, with Equations (2) and (3) in mind, Equations (4) become

$$u_1\,(\Delta_1 - \varepsilon\varphi_3\,a_2^2\,I_2) - u_1^3\left(\frac{2\,\omega_1^2\,k_1^2}{a_1^2} + \varepsilon\varphi_1\right) = 0,$$

$$u_2\,(\Delta_2 - \varepsilon\varphi_6\, a_1^2\, I_1) - u_2^3\left(\frac{2\,\omega_2^2\,k_2^2}{a_2^2} + \varepsilon\varphi_8\right) = 0, \tag{6}$$

where

$$\Delta_j \equiv \omega_{nj}^2 - \omega_j^2\,(1 - 2\,k_j^2),\;\; j = 1, 2, \tag{7}$$

$$I_j \equiv \frac{1}{4\,K_j}\int_{\phi_j}^{4\,K_j+\phi_j} cn_j^2\, d(\omega_j\, t + \phi_j) = 1 - \frac{1}{k_j^2}\left(1 - \frac{E_j}{K_j}\right), \tag{8}$$

in which E_j is the complete elliptic integral of the second kind for the modulus k_j. Thus Equations (5) and (6) hold for all u_1, and u_2 if and only if

$$\Delta_1 = \varepsilon\varphi_3\, a_2^2\, I_2, \quad \Delta_2 = \varepsilon\varphi_6\, a_1^2\, I_1, \tag{9}$$

$$k_1^2 = -\frac{\varepsilon\varphi_1\, a_1^2}{2\,\omega_1^2}, \quad k_2^2 = -\frac{\varepsilon\varphi_8\, a_2^2}{2\,\omega_2^2}. \tag{10}$$

Use of equation (7) in Equations (9) delivers the following exact relations for ω_j^2:

$$\omega_1^2 = \omega_{n1}^2 - \varepsilon a_1^2\,\varphi_1 - \varepsilon a_2^2\,\varphi_3\, I_2, \quad \omega_2^2 = \omega_{n2}^2 - \varepsilon a_2^2\,\varphi_8 - \varepsilon a_1^2\,\varphi_6\, I_1. \tag{11}$$

Equations (10) and (11) provide a set of four equations for the parameters ω_j and k_j. The constants a_1, a_1, ϕ_1, and ϕ_2 may be evaluated from the assigned initial conditions that, for convenience and without loss of generality, are assumed to be given by: $u_1(0) = u_{10}$, $u_2(0) = u_{20}$, $u_1'(0) = u_2'(0) = 0$. Use of these in Equations (2) and (3) shows that $a_1 = u_{10}$, $a_2 = u_{20}$, and $\phi_1 = \phi_2 = 0$. An analytical assessment of the accuracy of EBM compared to the numerical integration solution of equation (1) is discussed in references [1], [2], and [4].

4 The Exact Solution of Lamé's Equation

In this section we derive, with the help of *Mathematica,* the exact solution of Lamé's equation which is given by

$$\frac{d^2 y}{dt} + y(d_1 + d_2\, cn^2[t, k^2]) = 0, \tag{12}$$

where t, y, d_1, d_2, and k are physical parameters. We can see that this is a second order linear, ordinary differential equation whose exact solution has two linearly independent solutions. We assume here that one of these linearly independent solutions is of the form [6]

$$y_1 = cn[t, k^2] \tag{13}$$

Substitution of Equation (13) into Equation (12) gives

```
In[1]:= y1 = JacobiCN[t, k^2];
        equation1 =
         D[y1, {t, 2}] + (d2 JacobiCN[t, k^2]^2 + d1) y1;
        JacobiDN[ t , k^2] :=
         Sqrt[1 - k^2  + k^2 JacobiCN[ t , k^2]^2];
        JacobiSN[ t , k^2] := Sqrt[1 - JacobiCN[t, k^2]^2];
        equation2 = Collect[equation1, JacobiCN[t, k^2]] == 0

Out[5]= JacobiCN[t, k^2] (-1 + 2 k^2 + d1) +
          JacobiCN[t, k^2]^3 (-2 k^2 + d2) == 0
```

(14)

Equation (14) holds for all time t if and only if each coefficient vanishes, *i. e.* provided that

$$d_1 = 1 - 2k^2;\ d_2 = 2k^2 \tag{15}$$

The second linearly independent solution may be found by using the solution given by Equation (13) and applying the method described in [7]. In accordance with this, we assume that the second linearly independent solution is of the form:

$$y_2 = y_1(t) \int \frac{1}{y_1(t)^2}\, dt, \tag{16}$$

Substitution of equation (13) into equation (16) yields:

$$y_2 = \text{JacobiCN}[t, k^2] \int \frac{1}{\text{JacobiCN}[t, k^2]^2} \, dt$$

Out[3]=

$$\frac{1}{-1+k^2}\left(\text{JacobiCN}[t, k^2]\left(-\frac{\text{JacobiDN}[t, k^2]\,\text{JacobiSN}[t, k^2]}{\text{JacobiCN}[t, k^2]} + k^2\left(t - \frac{t}{k^2} + \left(\text{EllipticE}[\text{JacobiAmplitude}[t, k^2], k^2]\left(-1 + \frac{1}{k^2} + \text{JacobiCN}[t, k^2]^2\right)\right)\Big/\left(\text{JacobiDN}[t, k^2]\sqrt{1 - k^2\,\text{JacobiSN}[t, k^2]^2}\right)\right)\right)\right) \tag{17}$$

Thus, the general exact solution of Lamé's equation becomes:

In[4]:=

```
FullSimplify[y = C1 y1 + y2 C2]
```

Out[4]=

$$\text{JacobiCN}[t, k^2]\left(C_1 + \frac{1}{-1+k^2}\left(\left((-1+k^2)\,t - \frac{\text{JacobiDN}[t, k^2]\,\text{JacobiSN}[t, k^2]}{\text{JacobiCN}[t, k^2]} + \left(\text{EllipticE}[\text{JacobiAmplitude}[t, k^2], k^2]\left(1 - k^2 + k^2\,\text{JacobiCN}[t, k^2]^2\right)\right)\Big/\left(\text{JacobiDN}[t, k^2]\sqrt{1 - k^2\,\text{JacobiSN}[t, k^2]^2}\right)\right) C_2\right)\right) \tag{18}$$

where C_1 and C_2 are integration constants that can be determined from the assigned initial conditions. We now compare the closed-form solution given by Equation (18) with the numerical integration of Lamé's Equation given by (12) for the parameter value of $k = 1/\sqrt{2}$ and initial conditions $y(0) = 1$, and $y'(0) = 1$. In order to do this, we use *Mathematica* coding to write the following program:

```
In[1]:= k = 1/√2.; C1 = 1; C2 = 1.; d1 = 1 - 2 k^2; d2 = 2 k^2;
        y = JacobiCN[t, k^2]
          (C1 + 1/(-1 + k^2) (((-1 + k^2) t - (JacobiDN[t, k^2] JacobiSN[t, k^2])/JacobiCN[t, k^2] +
                (EllipticE[JacobiAmplitude[t, k^2], k^2]
                   (1 - k^2 + k^2 JacobiCN[t, k^2]^2)) /
                (JacobiDN[t, k^2] Sqrt[1 - k^2 JacobiSN[t, k^2]^2])) C2));

        solution1 = NDSolve[{(d2 JacobiCN[x, k^2]^2 + d1) y1[x] + y1''[x] == 0,
            y1[0] == C1, y1'[0] == C2}, y1, {x, 0, 150}, MaxSteps → ∞];

In[4]:= Show[Plot[Evaluate[y1[x] /. solution1], {x, 0, 150},
            PlotStyle → {GrayLevel[.5]}, AxesStyle → {AbsoluteThickness[1]},
            DisplayFunction → Identity], Plot[ y, {t, 0, 150},
            PlotStyle → {AbsoluteThickness[1], AbsoluteDashing[{5, 5}]},
            AxesStyle → {AbsoluteThickness[1]}, DisplayFunction → Identity],
          DisplayFunction → $DisplayFunction];
```

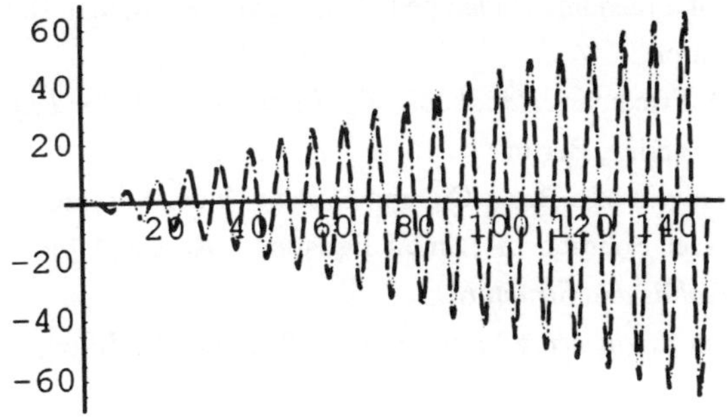

Figure 1. Schematic of the exact solution of Lamé's equation. The solid line represents the numerical integration solution and the dashed line the derived closed-form solution.

We can see from Figure 1, that both solutions are identical for the time interval shown . The agreement between both solutions is expected since Equation (18) represents the exact solution of Lamé's Equation.

5 Conclusions

The examples presented in this paper give supporting evidence that applications of Jacobi elliptic functions to find their solutions were simplify by *Mathematica* since this engineering software includes the common elliptic functions. It is also evident that for similar types of differential equations, we may derive with the help of *Mathematica* solutions that could also be based on elliptic functions. At present, the author is currently engaged in work on the solution of some of these with good preliminary results.

6 References

[1] A. Elías Zúñiga, *Absorber Control of the Finite Amplitude Nonlinear Vibrations of a Simple Shear Suspension System*. University of Nebraska-Lincoln, Ph. D. Dissertation, 1994.

[2] S. Bravo Yuste, "Comments on the Method of Harmonic Balance in Which Jacobi Elliptic Functions are Used," *Journal of Sound and Vibrations* **139**, 1990 pp. 381-390.

[3] A. Elías Zúñiga, "On the Elliptic Balance Method," *Mathematics and Mechanics of Solids* **8**, 2003. To appear.

[4] A. Elías Zúñiga and M. F. Beatty, "On the Application of the Elliptic Balance Method to Two Degree of Freedom, Undamped, Homogeneous System Having Cubic Nonlinearities." In preparation.

[5] P. F. Byrd, and M. D. Friedman, *Handbook of Elliptic Integrals for Engineers and Physicists*, 1953, Springer-Verlag, Berlin.

[6] A. Elías Zúñiga, "Analysis of a Beam-Column System Under Varying Axial Forces of Elliptic Type. The Exact Solution of Lamé's Equation." Accepted for Publication. *International Journal of Solids and Structures*.

[7] P. V. O'Neil, *Advanced Engineering Mathematics*, 1991, Third Edition, Wadsworth, Inc.

A Fractional Calculus Model of Semilunar Heart Valve Vibrations

Moustafa ElShahed
Misr University for Science&Technology,Giza,Egypt

elshahedm@yahoo.com

1 Abstract

The objective of this paper is to solve the equation of motion of semilunar heart valve vibrations. The vibration of the closed semilunar valves are modeled by a Caputo fractional derivative of order α.Using a Laplace transformation, a closed-form solution is obtained for the equation of motion in terms of a Mittag-Leffler function.An alternative method for a semi-differential equation $\alpha = 1/2$ when is also examined. The simplicity of theMathematica solution makes it ideal for testing the accuracy of numerical methods.This solution can be of some interest for fitting experimental data.

2 Introduction

Heart valves are mechanical devices that permit the flow of blood in one direction only. Four sets of valves are of importance to the normal functioning of the heart. Two of these, the atrioventricular valves, guard the opening between the atria and the ventricles.The other two,the semilunar valves, are located where the pulmonary artery and the aorta arise from right and left ventricles, respectively[10]. The semilunar valves consist of half-moon shaped flaps growing out from the lining of the pulmonary artery and aorta. When these valves are closed, blood fills the spaces between the flaps and the vessel wall.Each flap then looks like a tiny filled bucket.Inflowing blood smooths the flaps against the blood vessel walls, collapsing the buckets and thereby opening the valves.Closure of the semilunar valves simultaneously prevents back flow and ensures forward flow of blood in places where there would otherwise be considerable back flow[10]. Iggers (1915) suggested that there is silent closure of the semilunar valves and that afterwards the closed valve vibrates and the column of blood causes the second heart sound. Blick[2] modeled the aortic valve as a circular membrane of radius α.In his model the aortic valve is assumed to be an elastic, homogeneous membrane secured around a circular edge and undergoing a parabolic displacement. The valve closes because the intraventricular pressure falls below that of the aortic pressure. The force that drives the closed valve to vibrate is the pressure difference Δp that occurs across the valve. The equation of motion for forced and damped vibrations of a membrane at time t, with one degree of freedom is given in [2]:

$$D_t^2\, x(t) + \frac{c}{m} D_t\, x(t) + \frac{k}{m} x(t) = f(t), \qquad \text{................(1)}$$

where m, c and k represent the effective mass of vibration, the damping force coefficient and stiffness factor, respectively. $f(t) = \frac{\Delta p\, \pi\, a^2}{m}$ is an external force.

3 Fractional calculus model

The fractional derivative approach provides a powerful tool for modeling damped systems. The model based on fractional derivatives has been shown to be one of the most effective approaches [9]. If the fractional derivative model is used to represent the damping characteristic, the equation of motion (1) assumes the form :

$$D_t^2\, x(t) + \frac{c}{m} D_t^\alpha\, x(t) + \frac{k}{m} x(t) = f(t), \qquad \text{................(2)}$$

The operator D_t of the conventional model is replaced by D_t^α. There are several definitions of derivative of fractional order. The fractional operator D_t^α based on Caputo's definition can be written as [1]:

$$D_t^\alpha\, x(t) = \frac{1}{\Gamma(n-\alpha)} \int_0^\infty \frac{x^n(u)}{(t-u)^{\alpha-n+1}}\, du, \quad n-1 < \alpha < n, \text{.......................(3).}$$

where $n \in N$ and Γ is the gamma function. $x^n(t)$ represents the mth ordinary derivative of $x(t)$. A 2-parameter function of the Mittag -Leffler type is defined by the series expansion [4]:

$$E_{\alpha,\beta}(z) = \sum_{j=0}^{\infty} \frac{z^j}{\Gamma(\alpha j+\beta)}, \qquad \text{..............(4)}$$

4 Analysis

One of the most efficient and elegant methods to solve (2) is by using the Laplace transform. The Laplace transform of the Riemann-Liouville fractional derivatives is well known. However, its practical applicability is limited by the absence of the physical interpretation of the limit values of fractional derivatives at $t=0$. The formula for the Laplace Transform of the Caputo fractional derivative is

$$\int_0^\infty e^{-st} D_t^\alpha\, x(t)\, dt = s^\alpha x(s) - \sum_{j=0}^{\infty} s^{\alpha-j-1} D_t^j\, x(0), \quad (n-1 < \alpha < n), \text{ .. (5)}$$

where $x(s)$ is the Laplace transform of $x(t)$. Since this formula involves $x(0)$ and $x'(0)$, where a certain physical interpretation exists, we can expect that it can be useful for solving applied problems leading to linear fractional differential equations with constant coefficients with accompanying initial conditions in traditional form [4, p.106].

Applying the Laplace transform to equation(2) we obtain:

$$x(s) = \frac{f(s)}{s^2 + \frac{c}{m} s^{\alpha} + \frac{k}{m}}, \qquad (6)$$

where $f(s)$ is the Laplace transform of $f(t)$. Equation (6) can be written in the form:

$$x(s) = \sum_{j=0}^{\infty} \frac{\left(\frac{-k}{m}\right)^j s^{-\alpha(j+1)} f(s)}{\left(s^{2-J} + \frac{c}{m}\right)^{j+1}}, \qquad (7)$$

The inversion of (7) takes the form: $x(t) = \int_0^t G(t-u)\, f(u)\, du$,

where, $$G(t) = \sum_{j=0}^{\infty} \frac{1}{j!} \left(\frac{-k}{m}\right)^j t^{2j+1} E^{j}_{2-\alpha,2+\alpha j}\left(\frac{-c}{m} t^{2-\alpha}\right), \qquad (8)$$

Equation (8) can be written in the form:

$$G(t) = \sum_{j=0}^{\infty} \left(\frac{-k}{m}\right)^j \sum_{i=0}^{\infty} \frac{(i+j)!}{j!\, i!} \left(\frac{-c}{m}\right)^i \frac{t^{2j+2i-\alpha i+1}}{\Gamma(2j+2i-\alpha i+2)}, \qquad (9)$$

Measurements during catheterization of the instantaneous pressure gradient across the semilunar valve during diastole indicate that, to a first order approximation, the pressure gradient increases linearly with time until a time t_1, following which the pressure gradient remains essentially constant [2]. When $t < t_1$, the deflection of the center line can be expressed as:

$$x(t) = \frac{\pi a^2 A}{m} \sum_{j=0}^{\infty} \left(\frac{-k}{m}\right)^j \sum_{i=0}^{\infty} \frac{(i+j)!}{j!\, i!} \left(\frac{-c}{m}\right)^i \frac{t^{2j+2i-\alpha i+3}}{\Gamma(2j+2i-\alpha i+4)}, \qquad (10)$$

where $A = \frac{d\Delta p}{dt}$ and the deflection of the centerline for $t > t_1$ is given by

$$x(t) = \frac{\pi a^2 B}{m} \sum_{j=0}^{\infty} \left(\frac{-k}{m}\right)^j \sum_{i=0}^{\infty} \frac{(i+j)!}{j!\, i!} \left(\frac{-c}{m}\right)^i \frac{t^{2j+2i-\alpha i+2}}{\Gamma(2j+2i-\alpha i+3)}, \qquad (11)$$

where B is the constant value of Δp at $t > t_1$.

The velocity of the centerline deflection of the membrane can be obtained by differentiating (10) and (11). Instead of approximating the pressure difference across the valve by a ramp function, Mazudard [5] used an exponential function $\Delta p = C(1 - e^{-bt})$ as a better approximation to the physiological pressure gradient. The deflection of the centerline in this case can be expressed as

$$x(t) = \frac{\pi a^2 C}{m} \sum_{j=0}^{\infty} \left(\frac{-k}{m}\right)^j \sum_{i=0}^{\infty} \frac{(i+j)!}{j!\, i!} \left(\frac{-c}{m}\right)^i \frac{F(t)}{\Gamma(2j+2i-\alpha i+2)}, \qquad (12)$$

where:

$$F(t) = t^{-i\alpha}(-b)^{-2(i+j)}\left[\frac{t^2(-bt)^{2(i+j)}}{2j+2i-\alpha i+2} + \frac{e^{-tb}(-tb)^{i\alpha}\left(\Gamma(2j+2i-\alpha i+2) + \Gamma(2j+2i-\alpha i+2, -tb)\right)}{b^2}\right].$$

5 Alternative Method for α=1/2

When $\alpha = \frac{1}{2}$, Eq (6) takes the form

$$x(s) = \frac{f(s)}{s^2 + \frac{c}{m}\sqrt{s} + \frac{k}{m}} = h(s)\, f(s), \qquad (13)$$

where $h(s) = \frac{1}{s^2 + \frac{c}{m}\sqrt{s} + \frac{k}{m}}$ is the transfer function of the oscillator with fractional damping. In order to obtain the inversion of (13), we first write the denominator as $D(s) = s^2 + \frac{c}{m}\sqrt{s} + \frac{k}{m}$, and set $\lambda^2 = s$ to obtain $P(\lambda) = \lambda^4 + \frac{c}{m}\lambda + \frac{k}{m}$. Denoting the roots of $P(\lambda)$ by $\lambda_1, \lambda_2, \lambda_3, \lambda_4$, the roots of $D(s)$ can be found. Suarez and Sakakbira[

7, 8, 9] used a difficult method to obtain the roots of the characteristic polynomial. However, in the present paper we have been able to use an easy method. We use *Mathematica* to obtain the roots of $P(\lambda)$ as:

```
In[1]:=  Rt = Solve[λ^4 + (c/m) λ + (k/m) == 0, λ];

In[2]:=  λ1 = Rt[[1, 1, 2]];
         λ2 = Rt[[2, 1, 2]];
         λ3 = Rt[[3, 1, 2]];
         λ4 = Rt[[4, 1, 2]];
```

It is easy to see that the roots satisfy the following conditions $\sum_{i=1}^{4} \lambda_i = 0$ and $\lambda_1 \lambda_2 \lambda_3 \lambda_4 = \frac{k}{m}$.

```
In[6]:=  λ1 + λ2 + λ3 + λ4

Out[6]=  0

In[7]:=  FullSimplify[λ1 λ2 λ3 λ4]

Out[7]=  k/m
```

Then $h(s)$ takes the form:

$$h(s) = \sum_{i=1}^{4} \frac{1}{(\sqrt{s}-\lambda_i)\prod_{j=1}^{4}(\lambda_i-\lambda_j)}, \quad i \neq j, \qquad \ldots\ldots\ldots\ldots\ldots\ldots\ldots (14)$$

and the inversion of $h(s)$ takes the form:

$$h(t) = \sum_{i=1}^{4} \frac{1}{\prod_{j=1}^{4}(\lambda_i-\lambda_j)}\left[\frac{1}{\sqrt{\pi t}} + \lambda_i\, e^{t\lambda_i^2}\left(1+\operatorname{erf}\left(\lambda_i\sqrt{t}\right)\right)\right], \quad i \neq j, \quad \ldots\ldots\ldots\ldots (15)$$

Using the convolution theorem, the inversion of Eq (15) takes the form:

$$x(t) = \int_0^t h(t-u)\, f(u)\, du, . \qquad \ldots\ldots\ldots\ldots (16)$$

which is the centerline deflection of the membrane. Equations (15) and (16) contain terms that go to infinity at t=0. These terms can be eliminated as:

```
In[9]:= FullSimplify[1/((λ1 - λ2) (λ1 - λ3) (λ1 - λ4)) +
    1/((λ2 - λ3) (λ2 - λ4) (λ2 - λ1)) +
    1/((λ3 - λ1) (λ3 - λ2) (λ3 - λ4)) +
    1/((λ4 - λ1) (λ4 - λ2) (λ4 - λ3))]

Out[9]= 0
```

The deflection of the centerline for time $t < t_1$ can be expressed as

$$x(t) = A\sum_{i=1}^{4} \frac{1}{\prod_{j=1}^{4}(\lambda_i-\lambda_j)}\left[\frac{\sqrt{\pi}\, e^{t\lambda_i^2}(1+\operatorname{erf}(\lambda_i\sqrt{t}))-2\lambda_i\sqrt{t}}{\sqrt{\pi}\,\lambda_i^3} - \frac{t\lambda_i^2+1}{\lambda_i^3}\right],\quad i \neq j \ldots\ldots\ldots\ldots (17)$$

The deflection of the centerline for time $t > t_1$ can be expressed as

$$x(t) = B\sum_{i=1}^{4} \frac{1}{\prod_{j=1}^{4}(\lambda_i-\lambda_j)}\left[\frac{e^{t\lambda_i^2}(1+\operatorname{erf}(\lambda_i\sqrt{t}))-1}{\lambda_i}\right],\quad i \neq j, \qquad \ldots\ldots\ldots\ldots\ldots\ldots\ldots (18)$$

the deflection of the centerline using $\Delta p = C(1 - e^{-bt})$ as a better approximation to the physiological pressure gradient is

$$x(t) = C\sum_{i=1}^{4} \frac{e^{-bt}\left(\lambda_i^2 - b e^{bt} - \lambda_i^2 e^{bt} + e^{t(b+\lambda_i^2)}(1+\operatorname{erf}(\lambda_i\sqrt{t})) - \sqrt{b}\,\lambda_i \operatorname{erfi}\sqrt{bt}\right)}{\lambda_i(b+\lambda_i^2)\prod_{j=1}^{4}(\lambda_i-\lambda_j)},\quad i \neq j, \ \ldots\ \ldots\ldots (19)$$

The theoretical curve for the membrane vibration is plotted in [2,5,6]. It can be seen that the calculated curve has a shape similar to the curve calculated from experimental observations [5]. The solution of the modelling equation (2) is determined for representative values of valve stiffness (k=9.8 × 10^3 dynes cm^{-1}), damping force factor (d=2.8 × 10^3 dyn sec cm^{-2}), and effective mass (m=195g), with constant Δp = 150 mm Hg. per sec),. We can obtain Blik solution [2] directly as

```
In[10]:= c = 2.8 * 10^3;
         m = 195 ;
         k = 9.8 * 10^6;
         p = 150;
         a = 0.15;
         A = p π a^2 / m ;
```

```
In[16]:= Clear[F, x, t, s];
                        c             k
         diffeq = x''[t] + — x'[t] + — x[t] == A;
                        m             m
         initialconditions = {x[0] → 0, x'[0] → 0};
         transformed = LaplaceTransform[diffeq, t, s];
         F =
           Solve[transformed, LaplaceTransform[x[t], t, s]];
         F = F[[1, 1, 2]];
         F = F /. initialconditions;
         x[t_] = InverseLaplaceTransform[F, s, t] // Expand;
         Plot[10^6 x[t], {t, 0, 0.1}];
```

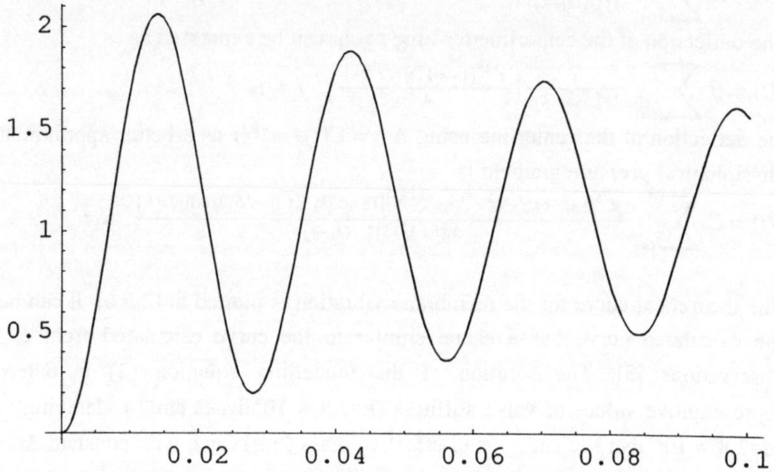

Since the fractional derivatives model approximates the physical situation more closely than other models [9], it would be expected that the results using the fractional derivatives model would be closer to the actual physiological values for deflection and rate of deflection of the centerline. The simplicity of the present solutions makes them ideal for testing the accuracy of numerical methods. The stiffness factor, *k*, in valve vibration and sound production explains the reason for an increased amplitude of the pulmonary component of the second sound relative to the aortic component in pulmonary hypertension. Another important factor is the effect of the viscosity factor, *c*, on the valve vibration. Since anemic patients have reduced viscosities, they have augmented heart sounds[6]. We hope that the above analysis helps to explain some previously unexplained clinical observations using factors that relate to valve vibration and resulting heart sounds. In his excellent paper, Gerd Baumann [3] demonstrated that *Mathematica* is able to handle fractional derivatives if the operations are defined in an appropriate way. He also implemented the Riemann–Liouville and the Weyl calculi with a minimal number of functions. In this paper we just use Baumann code [3] for the Riemann–Liouville fractional derivative that is able to do the Caputo fractional derivative, which is very important in real applications.

```
In[25]:= Caputo[f_, {x_, order_, a_: 0}] :=
            f Caputo[1, {x, order, a}] /;
             FreeQ[f, x] && Not[SameQ[f, 1]];

In[26]:= Caputo[c_ f_, {x_, order_, a_: 0}] :=
            c Caputo[f, {x, order, a}] /; FreeQ[c, x];

In[27]:= Caputo[f_ + g_, {x_, order_, a_: 0}] :=
          Caputo[f, {x, order, a}] + Caputo[g, {x, order, a}]

In[28]:= Caputo[ Caputo[f_, {x_, order1_, a_: 0}],
            {x_, order2_, a_: 0}] :=
          Caputo[f, {x, order1 + order2, a}] /; order1 < 0

In[29]:= Caputo[f_, {x_, 0, a_: 0}] := f

In[30]:= Caputo[f_, {x_, order_, a_: 0}] :=
          Module[{n, int, g, y, z, q},
           g = Function[{x}, f];
           If[NumericQ[order] && Simplify[order > 0],
            n = Ceiling[order]; q = order - n, q = order; n = 0];
           int = Simplify[Integrate[
                (x - y)^(-q-1) D[g[y], {y, n}], {y, a, z},
                GenerateConditions -> False], z < x] /. z -> x;
           int / Gamma[-q] /; FreeQ[int, y]
          ]
```

6 References

[1] Agrawal.O.P. A general Solution for a fourth-order fractional diffusion-wave equation defined in a Bounded Domain, Computers and Structures, vol 79 (2000) pp1497-1501.

[2] Blick. E. F, Sabbah. H.N and Stein. P.D. One-Dimensional Mode of Diastolic Semilunar Valve Vibrations Productive of Heart Sounds, Journal of Biomechanics, vol 12 (1979) pp232-227.

[3] Gerd Baumann. Knocking on Leibniz's Door, Mathematica in Education and Research, vol 8 (1999) pp58-65.

[4] Igor Podlubny. Fractional Differential Equations, Academic Press, New York, (1999).

[5] Mazumadar.J and Woodard, D. A Mathematical Study of Semilunar Valve Vibration,Journal of Biomechanics , vol 17 (1984) pp639-641.

[6] Mazumadar.J. An Introduction to Mathematical Physiology and Biology, Australian Mathematical Society,(1989).

[7] Sakakbira. Properties of Vibration with Fractional Derivative Damping of Order $\frac{1}{2}$, ASME International Journal ,vol 40 (1997) pp339-349.

[8] Suarez. L and Shokooh. L, Response of Systems with Damping Materials Modeled using Fractional Calculus, Appl Mech Rev,vol 48 (1995) pp118-S126.

[9] Suarez. L and Shokooh. L, An Eigenvector Expansion Method for the Solution of Motion Containing Fractional Derivatives, Journal of Applied Mechanics, vol 64 (1997) pp629-635.

[10] Thibodean, G. A and Pattom, K. T, Anatomy and Physiology, Mosby, (1995).

Designing a Distillation Column for Binary Mixtures

Akemi Gálvez, Andrés Iglesias
Dept. of Applied Mathematics and Computational Sciences, University of Cantabria, Santander, Spain
galveza@unican.es, iglesias@unican.es

1 Abstract

In this paper, we describe a *Mathematica* program for the design of distillation columns, clearly a very important topic in chemical industry. In particular, we consider McCabe-Thiele's procedure, which is a graphical method for designing distillation columns for binary mixtures. We take advantage of the symbolic, numerical and graphical *Mathematica* capabilities to create a general program that can be applied to any kind of binary distillation column. The performance of this program is shown through some illustrative examples.

2 Introduction

Chemical industry plays a basic role in our current technological and industrial development. One of the most typical operations in that industry is separation. Gasoline, kerosene, diesel fuel, heating oil, plastics, polymers, etc. are all made by distillation, the most common separation technique. This process is performed through distillation columns, which are designed to achieve this separation efficiently. They consume enormous amounts of energy, often contributing to more than 50% of plant operating costs [5]. Therefore, the correct design of distillation columns is a very important question in industry.

Traditionally, the design of distillation columns has been a numerical task and consequently many numerical programs have been developed during the last three decades. They are generally based on the idea of manipulating a relatively large number of look-up table numerical data obtained from experiments and then interpolating them to obtain the *vapor-liquid equilibrium* (VLE) curve, which is actually the starting point of the different distillation column design methods [1,4]. Determination of some column parameters (e.g. the number of trays or their location) is strongly affected by a great variety of errors that might imply additional costs due to an incorrect design. Symbolic computation programs, such as *Mathematica* or Maple, might help to overcome many of these error sources.

Here we describe a *Mathematica* program for distillation column design. For simplicity, we restrict ourselves to the case of continuous distillation columns for binary mixtures. Our approach is however very general in the sense that it can also be applied to any other kind of distillation columns by simply replacing our current assumptions by those of each specific case. In our examples we assume that the columns are designed through McCabe-Thiele's procedure. The program is general enough to analyze a number of different mixtures under different conditions as well as the role of many relevant parameters of this process. To this end, an adequate combination of symbolic and numerical calculations is achieved. From these calculations, both numerical and graphical outputs are obtained. Finally, the performance of the design method is shown through some illustrative examples.

3 Distillation: Basic Concepts and Terminology

According to [1], *distillation* is a process in which a liquid or vapor mixture of two or more substances is separated into its component fractions of desired purity, by the application and removal of heat. The main purpose of distillation is to separate a mixture of several components (water and alcohol, for instance) by taking advantage of their different volatilities. For example, the water boils at 100° C while alcohol boils at only 78.5° C. In this case, distillation consists of heating this water-alcohol mixture until its more volatile component (alcohol) vaporises. Consequently, the vapor that rises contains more alcohol than water than the liquid from which it came. After this heating stage, this vapor is cooled and condensed.

Distillation is performed through many different types of distillation columns. They can be classified into *batch columns* (the feed to the column is introduced batch-wise) and the most common *continuous columns* (no interruptions occur unless there is a problem with the column or surrounding process units). We only focus on binary columns, which are continuous and whose whose feed contains only two components.

The distillation column consists of several *trays*, which allow the simultaneous travel of liquid down the tray and vapor up the tray, allowing mixing of the two phases and therefore equilibrium [4]. The liquid mixture that is to be processed is known as the *feed* and it is introduced usually somewhere near the middle of the column to a tray known as the *feed tray*. The feed tray divides the column into a top section and a bottom section. The feed flows down the column where it is collected at the bottom in the *reboiler*. Heat from the reboiler vaporizes a portion of the liquid feed, and all vaporous feed travels up the column into top section, while all liquid feed travels down into the bottom section of the column. When the liquid reaches the bottom of the column, it is fed into the reboiler where heat generation causes a portion of liquid to vaporize. All vapor produced is returned to the bottom of the column, while any liquid remaining is removed as bottom product or *bottoms*.

A similar event occurs at the top of the column. The vapour moves up the column, and as it exits the top of the unit, it is cooled by a *condenser*. The condensed liquid is stored in a holding vessel known as the *reflux drum*. Some of this liquid is recycled back to the top of the column and this is called the *reflux*. The condensed liquid that is removed from the system is known as the top product or *distillate*. Thus, there are internal flows of vapour and liquid within the column as well as external flows of feeds and product streams, into and out of the column.

4 McCabe-Thiele's Method

One of the most typical methods to calculate the design parameters of a distillation column is the graphical scheme originally developed by McCabe and Thiele [3,4]. This is the simplest method to find the number of equilibrium stages for a given distillation process provided that the system contains only two components.

McCabe-Thiele's method requires additional assumptions. In particular: (1) the molar overflow is constant, (2) the molal heats of vaporization of the components are roughly the same and (3) heat effects are negligible. Under these conditions, McCabe-Thiele's method allows us to determine the theoretical number of stages for the separation of a binary mixture and hence, the number of trays required for the distillation column design. The method is also useful to analyze the effects of some column parameters on such a number.

The first step in McCabe-Thiele's method is to plot the vapor-liquid equilibrium data for a component of the binary mixture of composition x_f. The compositions of liquid and vapor are plotted on the x-axis and the y-axis respectively (see Fig. 1(left)) as a function of the geometric average relative volatility of the lighter component with respect to the heavier one, α (considered a constant here). This assumption holds well for the separation of similar components e.g. alcohols or hydrocarbons, but this is not true for non-ideal mixtures.

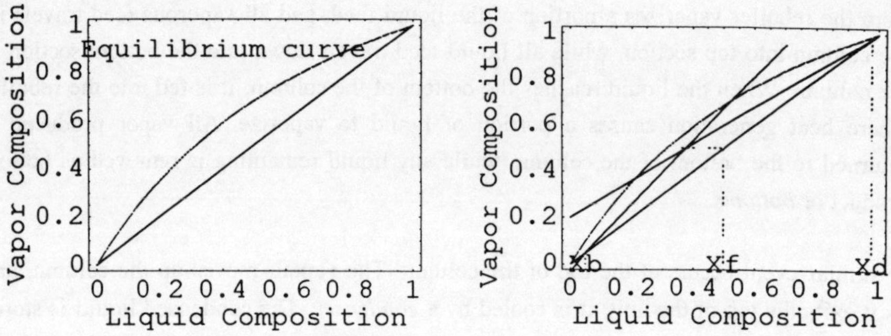

Figure 1: (left) vapor-liquid equilibrium curve; (right) operating lines.

Then, the operating lines defining the mass balance relationships between the liquid and vapor phases are drawn as follows. First, we set the desired compositions of the destillate x_d and the bottoms x_b. In the example of Figure 1 they are 0.85 and 0.05, respectively. The next step is to plot the operating line for the top section of the column as shown in Fig. 1(right). The slope of this operating line is given by $\frac{R}{R+1}$ where $R = \frac{L}{D}$ is called the reflux ratio and is a measure of how much of the material going up the top of the column is returned back to the column as reflux L, the rest being the distillate flow D. Then, we consider the feed line, plotted from the point $(x_d, 0)$ until the straight line $y=x$ and then, depending on the state of the feed, considering different slopes q. For example, $q=0$ corresponds to saturated vapor (see [3] for details). Finally, the operating line for the bottom section is obtained by tracing a vertical line from the point $(x_b, 0)$ until reaching the diagonal line, and then another line from such a point (x_b, x_b) until the intersection point between the operating line of the top section and the feed line, as shown in Figure 1(right).

It is important to remark that the previous scheme describes actually the case of total efficiency. However, in many situations it is more realistic to use of the so-called Murphree's efficiency $E_f \in [0, 1]$. The consequence of this factor is a modification of the original VLE curve (see Section 5 for details).

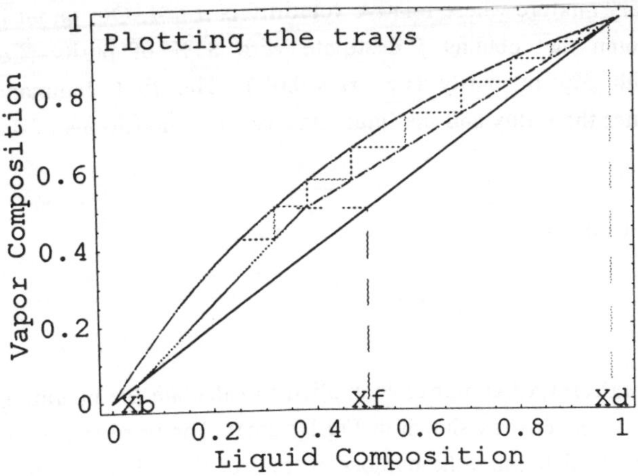

Figure 2: Construction of McCabe-Thiele diagram.

The next phase of McCabe-Thiele's graphical method is the determination of the number of stages in the distillation process. This is a sequence of similar steps giving rise to a number of "corner" sections, associated with the stages. To understand how they are constructed, let x and y denote the liquid and vapor compositions and the subscripts denote the origin of the flows or compositions. That is, $n-1$ will mean "from the stage below stage n" while $n+1$ will mean "from the stage above stage n". The liquid in stage n and the vapor above it are in equilibrium implying that x_n and y_n lie on the equilibrium curve. Since the vapour is carried to the tray above without changing composition, this is depicted as a horizontal line on the VLE plot. Its intersection with the operating line will give the composition of the liquid on tray $n+1$ as the operating line defines the material balance on the trays. The composition of the vapour above the $n+1$ tray is obtained from the intersection of the vertical line from this point to the equilibrium curve. Figure 2 shows this process for the ninth stage ($n=9$). Repeating this sequence back from point (x_d, x_d) until crossing the vertical line at x_b, the diagram of Figure 3(top) is obtained. The calculation of the number of stages reduces simply to count the number of "corners".

5 Some Illustrative Examples

The method described in the previous section has been implemented as a *Mathematica* package **MTBDC.m** (**M**cCabe-**T**hiele **B**inary **D**istillation **C**olumn). This section describes some examples of application of this program. The first one corresponds to a equimolar (i.e. $x_f = 0.5$) binary mixture whose relative volatility is $\alpha = 4$. Our target is to design a distillation column that obtains a destillate with 85% of purity (i.e. $x_d = 0.85$) and bottoms with 5% of purity (i.e. $x_b = 0.05$). The first command **Reflux[xd,xf,alpha]** determines the reflux and the liquid and vapor compositions of the more volatile component:

```
Reflux[0.85, 0.5, 4]
```

{1.75, 0.2, 0.5}

Then, the command **OperatingLines[xd,xf,alpha]** is applied to calculate the operating lines of the top and the bottom sections (as shown in Fig.1(right)). The third step is to apply McCabe-Thiele's method to determine the number of trays of the column as well as the location of the feed tray. For this we define the **MTPlot** command, which accepts five arguments, x_d, x_f, x_b, α and the efficiency E_f (see Section 4 for details) associated with different initial mixtures and/or final products. For example, **MTPlot[0.85,0.5,0.05,4,0.6]** returns a sequence of frames corresponding to the different steps of McCabe-Thiele's method. These frames can be saved as a QuickTime movie; in fact, the figures of this paper are simple frames of a movie generated by this way. This option is specially valuable for educational and training purposes, as we can obtain a virtually unlimited set of distillation columns in a fast, simple and error-free way. Some movies are included in the CD-ROM that contains the complementary material of this conference.

McCabe-Thiele's method itself does not allow us to determine all the relevant distillation column parameters and hence, some additional equations must also be considered. The discussion is beyond the scope of this paper and is not included here. See [1,4] for a detailed description on these topics. It is enough to say that a number of different *Mathematica* commands incorporating these additional equations have been implemented by the authors in the **MTBDC.m** package. For example, if we start with an equimolar mixture of heptane-octane with a relative volatility $\alpha = 2.17$ and efficiency $E_f = 1$ and we wish to obtain a destillate with 98% of heptane and bottoms with only a 5% of heptane, the column parameters can be determined as:

```
ColumnParameters[0.98, 0.05, 0.5, 2.17, 1]
```

Number of trays = 14 (rebolier included)   Number of Feed tray = 9
Distance between trays = 0.6 meters   Column height = 8.4 meters

Figure 3 shows the McCabe-Thiele's diagram of the column for total efficiency $E_f = 1$ (top) and when the efficiency of the process is only a 65%, that is $E_f = 0.65$ (bottom).

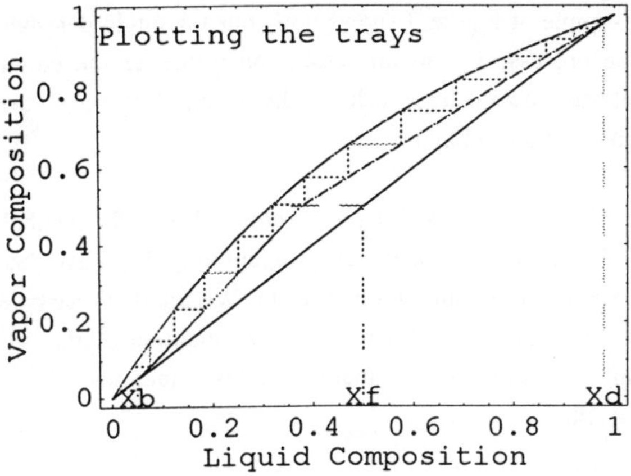

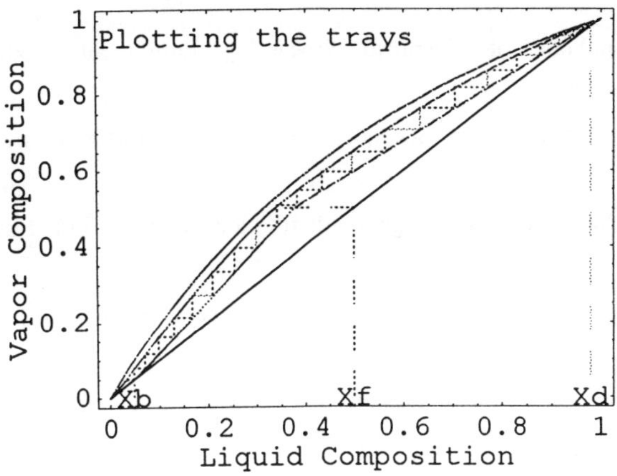

Figure 3: McCabe-Thiele diagram for a equimolar mixture of heptane-octane with $\alpha = 2.17$: (top) $E_f = 1$; (bottom) $E_f = 0.65$.

In the first case,the column has only 14 trays and its height is 8.4 m,while in the second case the number of trays increases up to 22 and the column height is now 13.2 m.

6 Conclusions and Further Remarks

In this paper, we describe a *Mathematica* program for the design of distillation columns for binary mixtures through McCabe-Thiele's method. This program is adequate to simulate the distillation columns and the different parameters involved in a very easy way. For instance, the simple example of Figure 3 shows how much a single parameter (the efficiency of the distillation process, E_f, in this case) can influence the column design, by modifying many relevant parameters such as the number of trays, their location and the height and diameter of the column.

All the commands have been implemented in *Mathematica* version 4 [6]. The symbolic capabilities and the powerful *Mathematica* functional programming [2] have been extensively used to make the program shorter and more efficient. We finally remark that in the case of multicomponent mixtures, the mathematical representation of the VLE becomes more complex, making necessary to use more complex equations. This is actually our proposal for future work.

7 References

[1] Cheremisinoff, N.P. Handbook of Chemical Processing Equipment (2000) Butterworth-Heinemann, Boston

[2] Maeder, R. Programming in Mathematica, Second Edition (1991) Addison-Wesley, Redwood City, CA

[3] McCabe, W.L., Thiele, E.W. Graphical design of distillation columns. Ind. Eng. Chem., 17 (1925) 605

[4] McCabe, W.L., Smith, J. C., Harriott, P. Unit Operations of Chemical Engineering, Sixth Edition (2001) McGraw-Hill, Boston

[5] Peters, M.S., Timmerhaus, K.D. Plant Design and Economics for Chemical Engineers, Fourth Edition (1991) McGraw-Hill, New York

[6] Wolfram, S. The Mathematica Book, Fourth Edition (1999) Wolfram Media, Champaign, IL and Cambridge University Press, Cambridge

A Symbolic and Graphical Gene Regulation Model of the lac Operon

Garret Suen and Christian Jacob
Dept. of Comp. Sci., University of Calgary,Canada

{sueng,jacob}@cpsc.ucalgary.ca

1 Abstract

We present a symbolic, grammar-based model for the classic *lac* operon gene regulation system implemented in Mathematica. This functional model focuses on the information processing aspect of gene regulation through pattern matching on symbolic expressions. Our *lac* operon notebook provides a viewer component for animated, 2D visualization of the simulated gene interaction processes, and is also connected to a 3D visualization engine.

2 Introduction

Biological research has changed drastically over the last decade. Nowadays, in an effort to minimize the time spent in the laboratory, data gathering and analysis is done primaily on computers. New interdisciplinary fields, biological computing and bioinformatics, seek a better understanding of biological phenomena through innovative programming techniques and algorithmic analysis. Classic,well understood biological models, can provide the basis for investigating larger, more complex models. For example, models of gene regulation in prokaryotic cells (without a nucleus) lead to a better understanding of gene regulation in more complex eukaryotic cells (with a nucleus). Here we present a symbolic model of the lactose (lac) operon, one of the simplest and best understood modes of gene regulation in the single-celled bacterium Escherichia coli (E.coli)[1,6]. We show how gene regulation mechanisms, which mainly rely on key-lock matching among cell components, can be immediately implemented through pattern matching on symbolic expressions. We give examples of the basic data structures used to encode an E.coli cell proteome, demonstrate selected rules that model interactions among proteins and other cell units, and show a simple way to visualize the dynamics of these interactions over time in 2D and 3D.

3 The Lactose Operon

E.coli is prokaryotic and resides in the human gut. E.coli DNA is free to interact with all other elements within the cell. The *lac* operon, in particular, is a group of genes found on the E.coli genome, the constituents of which represent a classic and intensively studied model for gene regulation [1,6,7].

3.1 Gene Regulation in the lac Operon: The Key Players

In a lactose-rich environment E.coli uses lactose as its primary food source. Lactose is converted into glucose, the bacterium's major energy source, and galactose. This process is controlled by the regulatory mechanisms of the *lac* operon structure, which encodes for proteins that facilitate the breakdown and conversion of lactose.The gene itself can regulate the creation of these molecules dependent on the amount of lactose present in the cell. More specifically, the *lac* operon consists of four genes (Fig.1): *lacI, lacZ, lacY* and *lacA*. *lacZ, lacY* and *lacA* are adjacent to one another on the operon. They are preceeded by a control complex comprising an operator region and a promoter region. This complex allows the protein RNA polymerase to bind, creating the enzyme β-galactosidase, which in turn breaks down lactose into glucose and galactose (Fig.1a).

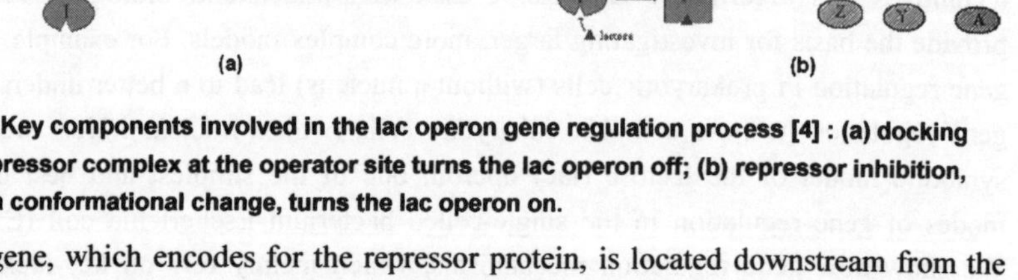

Figure 1. Key components involved in the lac operon gene regulation process [4] : (a) docking of the repressor complex at the operator site turns the lac operon off; (b) repressor inhibition, through a conformational change, turns the lac operon on.

The *lacI* gene, which encodes for the repressor protein, is located downstream from the main gene complex and is preceeded by its own control complex consisting of a single promoter region. The repressor serves as the basic control mechanism for the *lac* operon (Fig.1b).

3.2 *The lac Operon Model: DNA and Cell Cytoplasm*

In contrast to a probabilistic or logical network approach [2], our symbolic grammar-based model of the *lac* operon uses explicit representations of the components involved in the regulation process. Interaction and regulation mechanisms among these components are implemented as *Mathematica* rewrite rules [5]. A similar grammar-based approach is suggested in [3], but has never been implemented as a working, computational model. We model the DNA strand, the cytoplasm, the energy compounds (lactose, glucose etc.) and enzymes by explicit symbolic expressions. The following list data structure represents the two operon sections on the DNA and its surrounding cytoplasm (compare Fig.1):

```
Cytoplasm = {
  Operon[Promoter[], LacI[]],
  Operon[Promoter[], Operator[], LacZ[], LacY[]],
  RNAPolymerase[], Lactose, Glucose,
  LactosePermease[], Bgal[],
  RepressorTetramer[S1[], S2[], S3[], S4[]]}
```

We have simplified our model as follows: (1) as the Watson-Crick complementarity of the double-strand encoding does not have any influence on the functional aspect of gene regulation, we represent DNA as a single-stranded list of Operon[...]expressions; (2) we have not included explicit protein transcription and translation as this has no influence on the gene regulatory aspect. Hence, proteins are translated and transcribed immediately after RNA polymerase has docked onto and read the single DNA strand. (3) As we primarily focus on energy compounds and proteins that have a direct correlation to the operon, we do not consider galactose, a by-product of lactose cleaving, and thiogalactoside transacetylase, an enzyme encoded by the *lacA* gene. (4) As we primarily focus on the lactose-β-galactosidase interaction, we also do not consider the CAP catabolite repressor section, which regulates the production of β-galactosidase based on glucose concentrations.

4 *lac* Operon Gene Regulation: Step-by-Step

Here we give a detailed decription of the simulation steps required in our model,and say how the gene regulatory interactions are visualized.These iconic representations are automatically generated in the form of a frame-by-frame animation.Each frame represents the application of a single interaction rule among elements (e.g.proteins,enzymes) in the system.

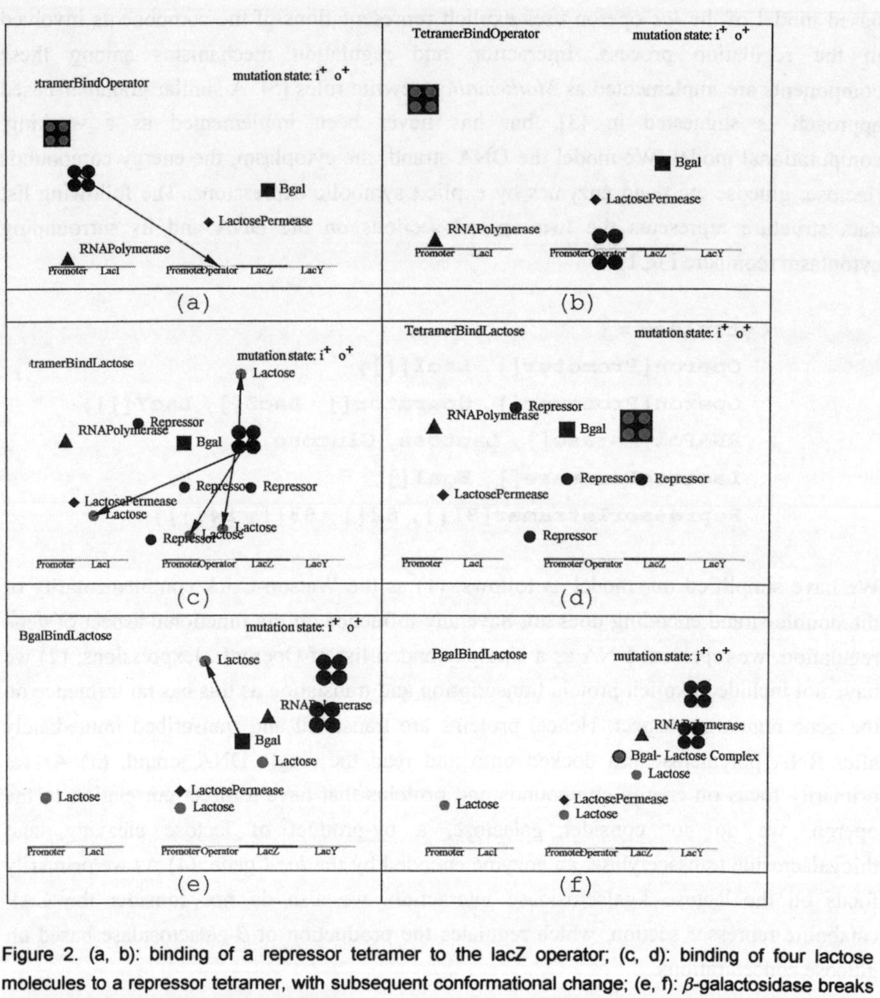

Figure 2. (a, b): binding of a repressor tetramer to the lacZ operator; (c, d): binding of four lactose molecules to a repressor tetramer, with subsequent conformational change; (e, f): β-galactosidase breaks down lactose into a β-gal-lactose complex.

Elements are represented as circles, squares or diamonds of different colours or graylevels (Fig.2). DNA elements are simple rectangle bands at the bottom of the display. All elements involved in the model are included in this visualization. In order to enhance the understanding of the dynamic interactions, arrows indicate which elements are interacting in each animation frame according to a particular rule.

4.1 Transcription and Translation of Structure Genes

RNA polymerase enzymes are present in large quantities in the cell. Therefore, we explicitly list them in the cytoplasm. RNA polymerase has a direct affinity for the promoter sites located on the operon. Once docking has occured, RNA polymerase moves along the operon and transcribes/translates the associated structure gene (section 6). Hence, the repressor is synthesized from *lacI*, Lactose Permease from *lacY* and *β*-galactosidase from *lacZ*. This model skips the more complicated process of *m*RNA (messenger RNA) creation and its subsequent conversion into a protein through the action of ribosomes. Here is an example of the rules we use to capture the docking of RNA polymerase onto a promoter site and its subsequent reading of *lacI*. Then RNA polymerase reads *lacI* and synthesizes a repressor:

```
Cytoplasm
/. {x___,Operon[Promoter[],LacI[]],y___,RNAPolymerase[],
z___}:>{x,Operon[Promoter[RNAPolymerase[]],LacI[]],y,z}

%/.Operon[Promoter[RNAPolymerase[]],LacI[]]:>
Operon[Promoter[],LacI[RNAPolymerase[]]]

%/.{x___,Operon[Promoter[],LacI[]]}:>
{x,Operon[Promoter[],LacI[]],y,z,Repressor[],RNAPolymerase[]
}
```

4.2 Binding of the Repressor Tetramer to the lacZ Operator

Four repressor molecules synthesized through *lacI* form a repressor tetramer, which has an affinity for binding to the operator region that preceeds *lacZ* (Fig.2a,2b). Once bound, transcription of *lacZ* cannot be accomplished, as the docking site for the RNA polymerase is blocked by the repressor. Hence, in the absence of lactose, this mechanism turns *lacZ* off in order to preserve cellular resources.

4.3 Lactose Entering the Cell

Lactose can enter the cell from the outside environment due to lactose permease, a protein encoded by *lacY*. In the presence of lactose, *lacZ* is turned on. This is accomplished through the binding action of four lactose molecules to a single repressor tetramer (Fig.2c,2d). This binding causes structural deformation of the repressor, such that it is no longer able to bind to the operator region (Fig.1b). RNA polymerase can then bind to the operator and synthesize β-galactosidase.

4.4 β-Galactosidase Breaks Down Lactose

The release of the repressor tetramer from the operator allows *lacZ* to be transcribed and translated again, resulting in an increase in the synthesis of β-galactosidase. This subsequently cleaves lactose into its constituent parts and reduces the concentration of lactose in the cell (Fig.2e,2f). Removal of lactose from the cell allows the repressor to bind to the operator region again and reduce the production of β-galactosidase, thus controlling gene regulation.

5 The *lac* Operon Notebook: Animations and Data Plots

Our *lac* operon gene regulation system iteratively applies the rewrite rules described above. We simulate interactions among the elements by repeated rule application. Each one allows rewrite rules to execute on the current state of the cell. The order of the rewrite rules applied in each round is chosen randomly, so as to model the randomness found in the *lac* operon system. Users of the *lac* operon notebook can interactively run the simulation by choosing the number of rounds the simulation runs. The output for each round consists of a list of rewrite rules and their associated graphics, which display the current state of the cell (Fig.2). Data plots are also generated for each simulation run. They chart the concentrations of lactose, glucose, repressor and β-galactosidase over the simulation time period.

6 Visualization in 3D

In order to enhance our functional,grammar-based model of gene regulation by a visualization component,we have designed a 3D visualization engine,based on the a library.To model the lac operon realistically,important coding regions are included into the DNA strand.The lacZ gene,with its associated control complex is incorporated into the DNA structure.In addition,the lacI gene along with its control complex is also part of the visualization.All other interactive elements including RNA polymerase,repressor molecules,and β-galactosidase molecules are rendered as spheres of different colours (Fig.3).

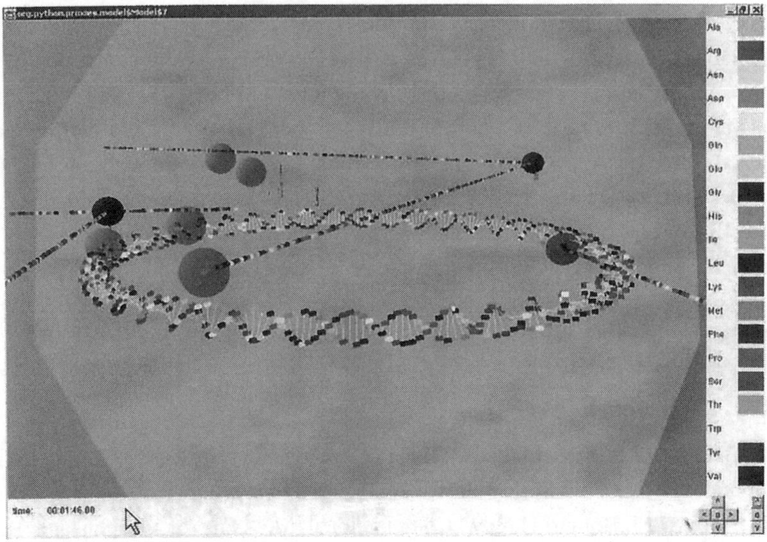

Figure 3. A snapshot during the synthesis of β-galactosidase:several RNA polymerases are attached to the circular double helix of the DNA.Codons, i.e., triplets of nucleotides (A,G,C,T) are represented by colour-coded spheres.Both mRNA strands and chains of amino acids are represented as long cylinders.

The primary processes transcription and translation are not yet considered in our model. One major objective of our 3D visualization is to model these processes, on an appropriate abstract level, as they occur in the cell. These intermediate processes ultimately are of interest in understanding gene regulation processes in general and the *lac* operon, in particular. Transcription is the process of converting a DNA template strand into an intermediate single strand of *mRNA*. RNA polymerase will read a given template strand of DNA and transcribe the codons into *mRNA*. Using colour-coded codons allows for ease of interpretation when viewing the various DNA and RNA structures. Translation is the process of converting a *mRNA* strand into the appropriate protein. It is facilitated by two enzymes, ribosomes and transfer RNA (*tRNA*). They affect the creation of proteins based on the *mRNA* encoding. The colour-coded amino acids directly relate to specific codons on both *mRNA* and DNA strands.

7 Conclusion and Future Work

We have presented a symbolic, grammar-based model of the classic *lac* operon gene regulation system. The model focuses on the information processing aspect of gene regulation. Symbolic expressions represent all structures and functions involved, so it is easy to combine it with a genetic programming engine. This will enable us to actually evolve gene regulation mechanisms and compare our results with other regulatory techniques evolved by nature. We are currently working on the integration of our model into our evolutionary computation environment Evolvica [5]. We will also include activators, such as CAP catabolites, into the model. They regulate the production of β-galactosidase based on glucose concentrations. Finally, we are working on an extension of the current 3D visualization engine, so that all aspects of the functional model are directly translated into animated, 3D visualizations which are automatically generated from the modeled genomic interaction processes. Further information is available at: http://www.cpsc.ucalgary.ca/~jacob/LacOperon. We would like to thank Julie Andreotti and Ian Burleigh for their help with the implementation of the 3D visualization.

8 References

[1] Beckwith,J.R.,Zipser,D.(eds.):The Lactose Operon.The Operon.Cold Spring Harbor Laboratory Press,Cold Spring Harbor,New York (1970).

[2] Bower,J.M.,Bolouri,H.(eds.):Computational Modeling of Genetic and Biochemical Networks.MIT Press,Cambridge, MA (2001).

[3] Collado-Vides,J.:Towards a grammatical paradigm for the study of the regulation of gene expression.In:Goodwin,B.,Saunders,P.(eds.):Theoretical Biology.Epigenetic and Evolutionary Order from Complex Systems.Johns Hopkins University Press,Baltimore,ML (1992):211-224.

[4] Crotty,S.,Basu,A.,Onufryk,C.,Ingram,V.:MIT Biology Hypertextbook.http://cyberbio.mit.edu/esgbio (1996).

[5] Jacob,C.:Illustrating Evolutionary Computation with Mathematica.Morgan Kaufmann Publishers,San Francisco (2001).

[6] Müller-Hill,B.:The lac Operon.A Short History of a Genetic Paradigm.de Gruyter,Berlin (1996).

[7] Ptashne,M.,Gann,A.:Genes& Signals.Cold Spring Harbor Laboratory Press,Cold Spring Harbor,New York (2002).

Algebraic Construction of Smooth Interpolants on Polygonal Domains

Elisabeth Anna Malsch, Gautam Dasgupta
Columbia University, USA

malsch@civil.columbia.edu, gd18@columbia.edu

1 Abstract

Given a set of vertices for a convex or concave polygon, a smooth and bounded interpolant (valid in the interior of the domain) may be constructed in an explicit algebraic form. Rational polynomial interpolants are suitable for *convex* polygons; the present treatment addresses *concave* domains. To resolve the concavity, square roots appear in the numerator and denominator of the rational interpolant. Finite element text books do not generally discuss tessellation into concave domains. Procedural programming languages, which are predominantly used in conventional finite elements, pose formidable challenges to the implementation of abstract geometrical constructs necessary to address convexity issues. Concepts of projective geometry are utilized within *Mathematica* in symbolic form to yield shape functions on concave polygons.

2 Introduction

A smooth and bounded approximation to a response function (e.g. temperature, potential) on a polygonal domain is constructed. Given the values of the response function $z = z(x, y)$ at the vertices (x_i, y_i) of a convex polygon, Wachspress' interpolation technique constructs an algebraic function that approximates z not only on the polygonal boundary, but also inside it. Here an analogous method applicable to *concave* polygons is formulated.

Approximate response functions are important in engineering applications. In computer graphics, smooth bounded functions depict a continuous surface. Other interpolation techniques (e.g. least square and spline fitting) require tessellation, which invariably results in unevenness. For example, a sharp color mismatch results across artificial interfaces when a polygonal domain is tessellated using triangles and finite element isoparametric quadrilaterals [6]. Such discontinuities are avoided using Wachspress' shape functions on convex domains [5]. The present formulation furnishes a similar smooth interpolant on a concave polygon. Other applications include: generating test functions for computational modeling[8]; computing shape functions for finite element computations[2,9].

Mathematica is indispensable in this development because it can combine geometrical computations (through list processing), polynomial manipulation and visualizations (in particular, colored contour plots). Maintaining expressions in symbolic form for as long as possible is important for avoiding round-off errors. It provides an environment for the development of concepts necessary for formulating C++ code at a later stage.

3 Input Data

The interpolation can be constructed on any two dimensional slice of any higher order dimensional polyhedron. Accordingly, the dimension of the data must be defined, and we then define nodal locations. In 2D:

```
xy = {x, y};
xyn = {xn, yn};
n = 4; (*number of nodes*)
pts = Array[Through[xyn[#]] &, n]
```

```
{{xn[1], yn[1]}, {xn[2], yn[2]},
 {xn[3], yn[3]}, {xn[4], yn[4]}}
```

The signed connection must also be defined. For example, for a triangle with three nodes and for a triangle with an quadrilateral inclusion and an interior point:

```
connect = {{1, 2, 3}};
connect = {{1, 2, 3}, {4}, {5, 6, 7, 8}};
```

For a simple interpolation the quantity which is going to be interpolated must be defined at every nodal point

```
dval = Array[dn, n]
```

4 Geometric measures and minimum functions

It is useful to construct the interpolation with respect to geometric measures as opposed to nodal location, the benefit being that it is then invariant under changes of coordinate system. The functions below define the distance between two points, the signed area of a triangle (positive or negative, depending on whether vertices are numbered clockwise or anti-clockwise) and the unsigned area of a triangle. The first and third are independent of dimension: the signed area function works in 2D only.

```
dist[a_, b_] := Sqrt[(a - b).(a - b)];
area[a_, b_, c_] :=
  (1/2) Det[Map[Append[#, 1] &, {a, b, c}]] /;
   Length[a] == Length[b] == Length[c] == 2
areaUS[a_, b_, c_] := Module[{lths =
      Apply[dist, Partition[{a, b, c}, 2, 1, {1}], {1}],
     s}, {s = (1/2) Apply[Plus, lths];
      Sqrt[s * Apply[Times, (s - lths)]]}[[1]]];
```

The geometric definitions can be used to define minimum functions. For example, the distance function **dist[xy, xyi]** is minimum at the point defined by the vector **xyi,** and the area function **area[xy, xyi, xyj]** is zero and minimum along the line connecting point **xyi** to point **xyj**. A function which is zero along the line segment connecting two points, and nowhere else, is defined in terms of the perimeter of the triangle:

```
perim[xy_, a_, b_] :=
  dist[xy, a] + dist[xy, b] - dist[a, b]
```

Contour plots of *dist, area, areaUS* and *perim*, for appropriate nodal points, are shown in order in Figure 1. Each case shows the distribution of the appropriate influence function.

4.1 Interpolation construction

Boundaries can be defined separately using functions which are minimum and zero at a line. Any functions which satisfy the requirements are valid interpolations. Accordingly, any given interpolation is not unique for any given set of nodal points. The code below is required to display the support functions graphically.

```
outline[pts_] := Line[Append[#, #[[1]]] &[pts]]
name[pts_] :=
  Table[Text[ToString[i], pts[[i]], {-2, 0},
    Background → RGBColor[1, 1, 1], TextStyle →
     {FontWeight -> "Bold"}] {i, Length[pts]}]
nodes[pts_] := Point /@ pts
mask[pts__] := Polygon[Flatten[Append[#, #[[1]]] & /@
     {{{-.1, -.1}, {-.1, .1}, {.1, .1}, {.1, -.1}} +
       ({#[[1]], #[[2]], #[[4]], #[[3]]} &[Flatten[
           Outer[List, ##] & @@ ({Min[#], Max[#]} & /@
             Thread[points]), 1]]), points}, 1]]
```

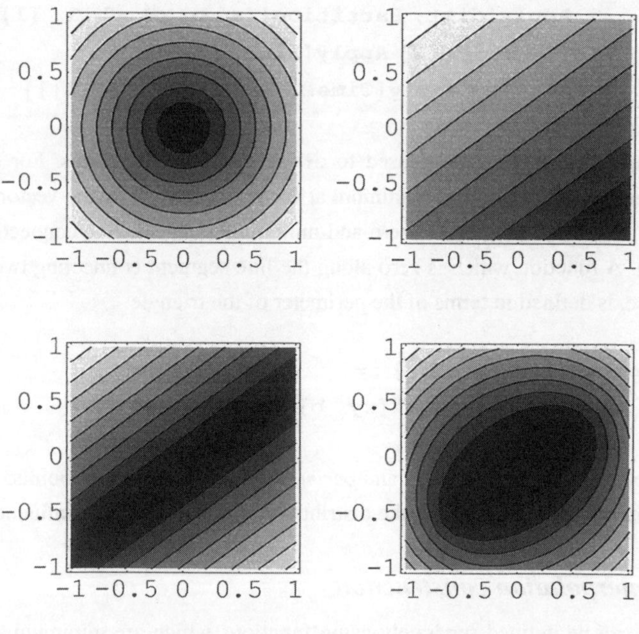

Figure 1

5 Convex Polygons

Using geometric measures the interpolation can be constructed simply. The following code does this for a triangle, and the resulting output forms the first panel in Figure 2.

The contours represent internal interpolations of the boundary conditions specified at the vertices. The theoretical development is presented in [4].

```
varArea[x_, y_] := Apply[area[##, {x, y}] &,
  Partition[pts, 2, 1, {1}], {1}]
zeroArea[x_, y_] :=
 Apply[Times, Partition[varArea[x, y],
   Length[pts] - 2, 1, {Length[pts]}], {1}]
convex[x_, y_] := (zeroArea[x, y] * areas) /
  (zeroArea[x, y].areas)
points = {{0, 0}, {1, 0}, {1, 1}};
connect = {{1, 2, 3}};
pts = Part[points, connect[[1]]];
areas = Apply[area, Partition[pts, 3, 1, {2}], {1}];
fc = Compile[{{x, _Real}, {y, _Real}},
   Evaluate[convex[x, y]]];
colorfunction = (RGBColor[1, (1 - #), (1 - #)] &);
i = 1;
pContour = ContourPlot[
   Check[fc[x, y][[i]], 0., CompiledFunction::"cfn"],
   {x, Min[Thread[points][[1]]],
    Max[Thread[points][[1]]]},
   {y, Min[Thread[points][[2]]],
    Max[Thread[points][[2]]]},
   DisplayFunction → Identity,
   ColorFunction → colorfunction, Contours → 20,
   PlotPoints → 40, PlotRange → {0, 1}];
 tri = Show[pContour,
   Graphics[
    Flatten[{{RGBColor[1, 1, 1]}, {mask[pts]}}]],
   Graphics[Flatten[{{PointSize[.02]},
      {outline[pts], name[pts], nodes[pts]}}]],
   PlotRange → All, Frame → False, AspectRatio → 1,
   DisplayFunction → $DisplayFunction]
```

Other convex polygons can be treated in the same way: a quadrilateral and a pentagon are also shown in Figure 2.

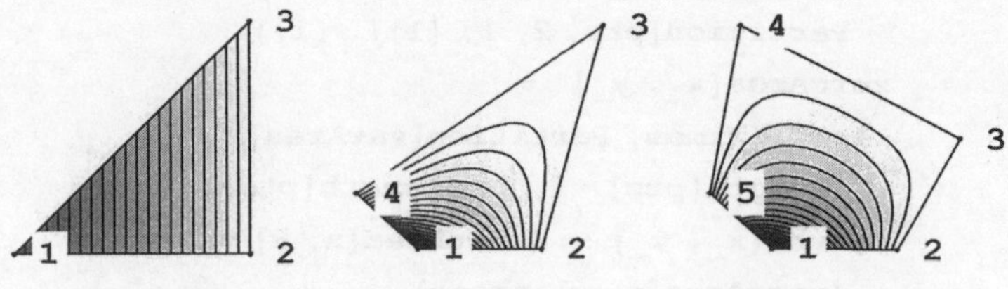

Figure 2

An algebraic example is presented:

```
points = {{0, 0}, {1, 0}, {3/2, 1}, {-1/2, 1/4}};
connect = {{1, 2, 3, 4}};
pts = Part[points, connect[[1]]];
areas = Apply[area, Partition[pts, 3, 1, {2}], {1}];
convex[x, y][[1]] // Simplify
```

$$-\frac{(7+6x-16y)(-2+2x-y)}{14+12x+59y}$$

6 Concave Polygons

Previous attempts at representing concave polygonal domains have required that the concave polygon be split into a set of convex sub-polygons. Assembling these convex sub-polygons does not serve the same purpose as assembling convex elements in finite element analysis. A distributed approximation to the original boundary conditions on the lines joining contiguous convex sub-polygons can be computed, subject to enforcing continuity of the shape functions across the boundary [3, 7]. Alternatively, a method which does not require any division can be formulated. The **perim** function explicitly accounts for concavity, and **zeroArea[a,b,c]** is replaced by **zeroPerim[a,b,c]**. Results for a quadrilateral and a pentagon are shown in Figure 3.

```
varPerim[x_, y_] := Apply[perim[{x, y}, ##] &,
  Partition[pts, 2, 1, {1}], {1}]
zeroPerim[x_, y_] :=
```

```
  Apply[Times, Partition[varPerim[x, y],
     Length[pts] - 2, 1, {Length[pts]}], {1}]
concave[x_, y_] := (zeroPerim[x, y]) /
   (Plus @@ zeroPerim[x, y])
```

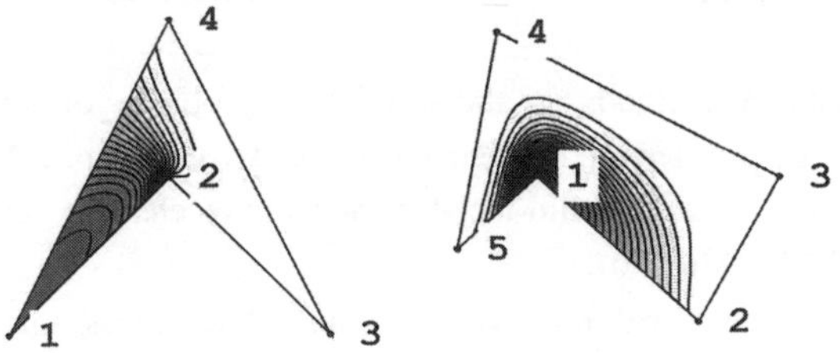

Figure 3

An algebraic example is presented:

```
points = {{-1, -1}, {0, 0}, {1, -1}, {0, 1}};
connect = {{1, 2, 3, 4}};
pts = Part[points, connect[[1]]];
areas = Apply[area, Partition[pts, 3, 1, {2}], {1}];
concave[x, y][[1]] // Simplify
```

$$\left(\left(-\sqrt{5}+\sqrt{x^2+(-1+y)^2}+\sqrt{(-1+x)^2+(1+y)^2}\right)\left(-\sqrt{2}+\sqrt{x^2+y^2}+\sqrt{(-1+x)^2+(1+y)^2}\right)\right)\Big/ \left(-\sqrt{2}-\sqrt{5}+\sqrt{x^2+(-1+y)^2}+\sqrt{x^2+y^2}+\sqrt{(-1+x)^2+(1+y)^2}+\sqrt{(1+x)^2+(1+y)^2}\right)^2$$

7 Conclusion

The interpolations in sections 5 and 6 result in contour plots. Evaluating the functions *concave* or *convex* results in the algebraic form of the interpolants. For a convex polygon, the closed algebraic form is a rational polynomial. For the concave polygon, the form includes an irrational polynomial component. Figures 2, 3 and 4 illustrate that the interpolations **concave** and **convex** defined in sections 5 and 6 are smooth and bounded within their respective domains.

Acknowledgements: This research was supported by a National Science Foundation Grant number CMS:0202232. We gratefully acknowledge Dr. Peter Mitic's editorial comments and his help in the preparation of the final version of this paper.

8 References

[1] Brebbia, C. A., *The Boundary Element Method for Engineers.* John Wiley and Sons, 1978.

[2] Courant, R., "Variational methods for the solution of problems of equilibrium and vibrations", in Bulletin of the American Mathematical Society, 1943. pp. 1-43

[3] Dasgupta, G. and Treil, J., "Maxillo-facial frame: Finite element shapes", in The Mathematica Journal, 8:235–246,2001.

[4] Dasgupta, G., " Interpolants within convex polygons: Wachspress' shape functions", in ASCE, February 2002.

[5] Dasgupta, G. and Malsch, E. A., "Boundary Element color interpolation for instrumentation, imaging and internet graphics industry", in Engineering Analysis with Boundary Elements, Elsevier Science, UK, vol.26,2002, pp.379--389.

[6] Foley, J. D., Van Dam, A., Feiner, S.K and Hughes, J. F., *Computer Graphics Principles and Practice.* Addison-Wesley, New York, 1996.

[7] Malsch, E.A. and Dasgupta, G., "Shape functions for concave quadrilaterals", in Proceedings of the First MIT Conference, K.J.Bathe ed.v.2 (Elsevier Science:June 2001). pp.1617--1619

[8] Wachspress, E. L., *A Rational Finite Element Basis.* Academic Press, 1975.

[9] Zienkiewicz O. C. and Taylor, R. L., *The Finite Element Method.* 4th Edition. McGraw-Hill, New York, 1989

The Method of Meshless Fundamental Solutions with Sources at Infinity

Peter Mitic
Positive Corporation Ltd., Hampshire, U.K.
peter.mitic@virgin.net
Youssef F. Rashed
Wessex Institute of Technology, Southampton, U.K.
youssef@wessex.ac.uk

1 Abstract

The method of external source collocation is used to solve a discretised boundary value problem, $\nabla^2 U = 0$, where U is the potential in a two-dimensional simply-connected region D, subject to a mixture of Neumann and Dirichlet boundary conditions. Numerical analysis has, to date, been hindered by an accumulation of round-off error, which has made it impossible to investigate accuracy of the Meshless Fundamental Solutions method unless sources are near the boundary. Symbolic analysis allows a full investigation of ill-conditioned systems in which sources can be placed "at infinity". This analysis provides an indication of how many sources must be used and where they should be placed.

2 Introduction

Solutions to the equation $\nabla^2 U = 0$, where U is the potential in a two-dimensional simply-connected region D, with U or $Q=\nabla U$ specified on the boundary, are often impractical or impossible to obtain analytically if D is not a simple shape, or if the boundary conditions are complex. In these cases, the usual approach is to discretise the boundary into segments and apply numerical techniques. Severe ill-conditioning results if many sources are used or if some of them are far from the boundary. In this paper we consider few sources with a simple boundary, and use symbolic analysis to avoid accumulating round-off error.

3 Outline and Analysis of the Method

Figure 1 shows a general case of such a region, D, with boundary ∂D and $\nabla^2 U = 0$ in D. The boundary is discretised into Z elements such as (B_{i1}, B_{i2}) with midpoint m_j. D is surrounded by W external sources, $S_1, S_2, \ldots S_W$, each with potential given by a fundamental solution:

$$U(n_{ij}) = \frac{\log(n_{ij})}{2\pi} \quad \text{......................(1)}$$

where n_{ij} is the distance from m_j to S_i. (See [1], [2] or [3])

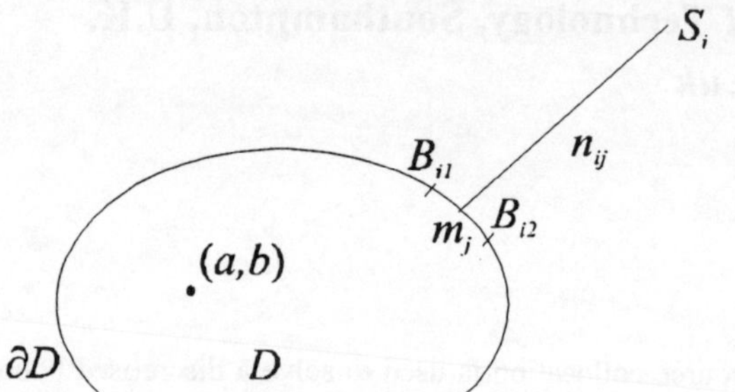

Figure 1

The justification for the Meshless Method may be found in any discussion of the Method of Fundamental Solutions.[4] is a detailed example.The main principles are:

- A source S, external to the boundary, gives rise to a potential at a given point X on or inside the boundary.
- The potential is given by the fundamental solution for the problem which depends on the problem geometry and the physical characteristics of the materials concerned.
- Several sources, $S_1 \ldots S_n$, external to the boundary, give rise to different potentials at the same point.
- The actual potential at X is a linear combination of the potentials due to the sources concerned, but boundary conditions must be satisfied.

The number of sources need not be the same as the number of boundary elements, but there must be enough to 'represent' all boundary conditions. If there are too few, the solutions obtained may be wrong. If there are too many, the correct linear combination of sources will be determinable by a least squares (or equivalent) process. The method of solution is described briefly in [5]. The details that appear here are relevant to the ensuing discussion of convergence and ill-conditioning, and are necessarily more detailed than described in [5]. The *Mathematica* implementation is coded in package `Meshless.m`.

4 Symbolic Analysis of a Simple Problem

Only the simplest of BEM problems may be analysed symbolically or with rational arithmetic. Computation in more complex problems is too slow. Hence, we use the *Brebbia Square* problem, described in [6] and [7], with four sources. Figure 2 shows the *Brebbia Square*, for which $\nabla^2 U = 0$ in the region D, and boundary values for U and Q are as shown. Each of four sources is a distance x from the boundary.

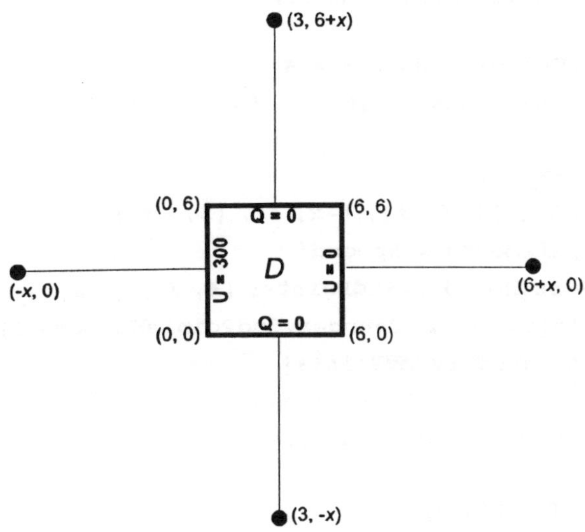

Figure 2

The Constant Element BE analysis in [6] and [7] yields the following values for the potential U, and fluxes Q_x and Q_y at the interior points P.

P	*U*	*Qx*	*Qy*
(2,2)	200.28	-50.303	-0.14976
(2,4)	200.28	-50.303	0.14976
(3,3)	150.01	-50.215	-0.00004
(4,2)	99.74	-50.306	0.14564
(4,4)	99.74	-50.306	-0.14564

Table 1

4.1 *Overall implementation in Mathematica*

The most important reason for using *Mathematica* is to allow analysis of extremely ill-conditioned systems. The source code is also particularly compact. Package `Meshless.m` exports the function `MeshlessBoundarySolutionMatrix`, which does the bulk of the computation. It implements a BEM and returns the computed BE matrix.

4.2 *Detailed implementation in Mathematica*

The *Mathematica* inputs, (collectively equation 2), for this problem are below. They define the boundary and sources, compute the BE midpoints, define the boundary conditions and their type and compute the required results.

```
BoundaryEndPointsNumber = 4;
BoundaryEndPoints = {{6, 0}, {6, 6}, {0, 6}, {0, 0}};

Sources2[x_] :=
  {{6 + x, 3}, {3, 6 + x}, {-x, 3}, {3, -x}}
BoundaryElements = Append[
   Map[Part[BoundaryEndPoints, {#, # + 1}] &,
     Table[i, {i, 1, BoundaryEndPointsNumber - 1}]],
    {Last[BoundaryEndPoints],
     First[BoundaryEndPoints]}];
BoundaryNodes = Map[MidPoint, BoundaryElements];

UQ = {0, 0, 300, 0};
UQType = {0, 1, 0, 1};
MboundarySolM =
MeshlessBoundarySolutionMatrix[BoundaryElements,
     UQ,   UQType, Sources2[x]] // Simplify // N

DetMBoundarySolM = Det[MBoundarySolM];
Table[DetMBoundarySolM, {x, 1, 100, 5}]
Plot[DetMBoundarySolM, {x, 1, 100}]
```

........................(2)

The output corresponding to MboundarySolM is a complicated 4-by-4 matrix. A plot of its determinant against x demonstrates progressive ill-conditioning of the matrix. When

$x \sim 100$, DetMBoundarySolM $\sim 10^{-8}$. The following inputs, (collectively equation 3), compute the potential, (solPab), and *x*- and *y*- fluxes, (solUab), at an interior point (*a*,*b*) of the square.

```
MBoundarySolution [x_] :=
   MeshlessBoundarySolution[
    BoundaryElements, UQ, UQType,  Sources2[x],
   "UseInverse"] // Simplify // N
MBoundarySolution[x];

InteriorPoints = {{a, b}};
solPab =
 First[intSolPotential[MBoundarySolution[x],
        InteriorPoints, x]]
solUab =
 intSolFlux[MBoundarySolution[x], InteriorPoints, x]
```

........................(3)

The unpalatable results solPab and solUab form the basis for further analysis.

4.3 Limiting expressions when $x \sim 0$ and $x \to \infty$

Taking the limit in the outputs from (3) as $y \to \infty$ should yield a simple function for the potential and flux at (*a*,*b*). The results are inconclusive because *Mathematica* displays a 'division by zero' error. Despite this, the correct potential, $U(a,b)=300-50a$, and fluxes, $Ux=-50$, $Uy=0$, are obtained. These results show that the limiting potential is independent of *b* and that the limiting flux is independent of *a* and *b*. This is confirmed by numerical examples.

Further numerical trials show that it is not, in general, safe to place the sources too close to the boundary. The result is grossly inaccurate for $x<20$, but can be improved by increasing the number of sources. If more sources are used, symbolic computations fail due to insufficient memory. Symbolic analysis is therefore quite limited.

4.4 Condition Number of the Boundary Element matrix

The Condition Number (CN) of the boundary element matrix, (MboundarySolM in equation 2), provides some insight into how ill-conditioning varies with the distance, *x*, of the sources from the boundary. *Mathematica* provides the function SingularValues, which implements Singular Value Decomposition. Using it to compute the CN is therefore easy:

```
ConditionNumber[n_] :=
Module[{mbs, usv},
   mbs = MBoundarySolutionMatrix[x]  /. {x -> n};
   usv =  SingularValues[mbs] [[2]];
   Max[usv] / Min[usv] ]
```

........................(4)

Figure 3 shows that the CN increases approximately quadratically as x increases. The largest CN, (x = 100), corresponds to a BE matrix determinant of the order of 10^{-9}, which is something that C++ or Fortran can cope with. Problems arise with more sources: the columns of the resulting BE matrices represent nearly-parallel vectors. For example, determinants of the order of 10^{-195} occur if 20 sources are used with x = 100. In such cases, obtaining equivalent results using C++ or Fortran is not possible due to severe round-off error.

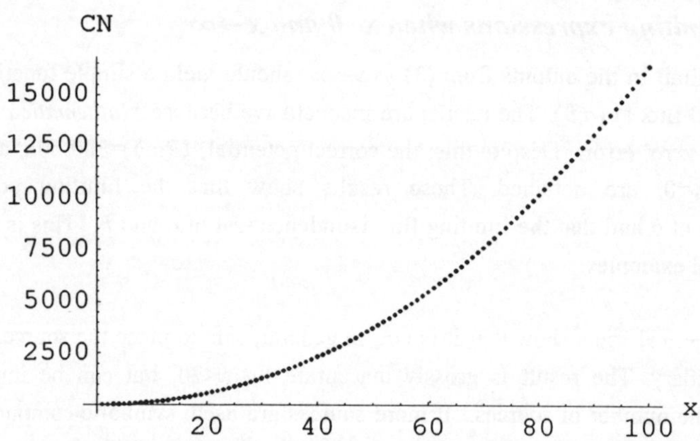

Figure 3

5 Numerical Investigations

The variables in a numerical analysis of this problem are the number of sources, the number of boundary elements and source placement. The latter comprises the distance of the sources from the boundary and their positions relative to each other. In order to mirror the symbolic analysis of the previous section, we use 4 sources, each a variable distance x from the boundary. Each side of the square has 4 boundary elements. Table 2 shows the results of calculations of potential and flux at (2,4) as x varies from 10 to 100 ('infinity'!) in steps of 10.

x	Potential	x-Flux	y-Flux
10	199.031	-49.0782	-0.558438
20	199.663	-49.7132	-0.186076
30	199.834	-49.8615	-0.0911756
40	199.901	-49.9186	-0.053867
50	199.935	-49.9465	-0.0355091
60	199.954	-49.9622	-0.0251509
70	199.966	-49.9718	-0.0187411
80	199.973	-49.9782	-0.0145017
90	199.979	-49.9826	-0.0115532
100	199.983	-49.9859	-0.00942013

Table 2

These results show that the computed solutions converge to the exact solutions, (200, -50 and 0), and that convergence is fast. Therefore, in order to compute a 'solution', it is sufficient to compute one solution using a sufficiently large value for x. The results in Table 2 indicate that 'sufficiently large' means, in practice, 5 to 10 times the dimension of the boundary. This gives a maximum error of 2.5% in the y-flux. For comparison, using x = 500 gives {199.999, -49.9994, -0.000395234} for the potential, x-flux and y-flux respectively. Although this appears to be an improvement, when more sources and/or boundary elements are used ill-conditioning limits further improvement in accuracy.

6 Discussion

The limiting process of the previous section shows that it is possible to obtain an accurate symbolic solution for the *Brebbia Square* problem using only a few sources. This is advantageous because more sources increase ill-conditioning effects. *Mathematica*'s Limit function does not work smoothly in all cases we have tried, but the results are easily backed up by numerical calculations. This method is also inherently efficient because the prerequisites for calculating the potential and flux at an interior point are determinable without knowing the co-ordinates of the interior point. 'Traditional' boundary element methods need to know target point co-ordinates in advance in order to compute results, and the computation must be reformulated to accommodate new target points. Numerical computations execute much faster than part symbolic computations, and it is sufficient to place sources sufficiently far from the boundary to achieve a given accuracy. Ill-conditioning effects can be reduced by using a minimal number of sources and boundary elements.

We have done many other numerical investigations, varying the number of sources, the number of boundary elements and source placement. The main outcome of these experiments is that the actual configuration of sources is immaterial: the shape profile of the sources need not be like the shape of the boundary.

7 Conclusion

The analysis in this paper demonstrates that 'sources at infinity' is a viable configuration for the method of Meshless Fundamental Solutions. The number of sources required will be discussed elsewhere, but it should be noted that the number of sources must reflect the degrees of freedom inherent in the boundary conditions of the problem. This determines the minimum number of sources required. Beyond that, extra sources produce more accurate results but ill-conditioning effects increase with the number of sources.

References

[1] Mathon, R. and Johnston, R. The Approximate Solutions of Elliptic Boundary Value problems by Fundamental Solutions, in SIAM J. Numerical Analysis, vol.14,#4,Sept 1977

[2] Karageorghos, A. and Fairweather, G. The method of Fundamental Solutions for the solution of Plane Potential Problems, in IMA J.of Numerical Analysis, vol.9,pp231-42,1989

[3] Cartwright, D.J. Underlying principles of the Boundary Element Method. WIT Press 2001

[4] Goldberg, M.A. and Chen, C.S. The method of Fundamental Solutions for potential, Helmholtz and diffusion problems, in Boundary Integral Methods - Numerical and Mathematical Aspects (ed. Goldberg, M.A.) pp103-76, CMP 1998

[5] Burgess, G. and Mahajerin, E. The Fundamental Collocation Method Applied to the Nonlinear Poisson Equation in two Dimensions, in Computers and Structures, vol. 17, #6, pp 763-767. 1987 Pergamon Journals

[6] Brebbia, C.A. The Boundary Element Method for Engineers. Pentech Press 1980

[7] Brebbia, C.A. and Dominguez, J. Boundary Elements: an Introductory Course. McGraw-Hill/CMP 1989

Mathematica Implementation in the Training Courses on Continuum Mechanics: Elasticity Theory, Hydromechanics, Theory of Filtration

Alexander N. Papusha (Continuum Mechanics&Offshore Exploration Department) Murmansk State Technical University

Alexander.Papusha@mstu.edu.ru

1 Abstract

The main aim of the presentation is to show how basic concepts of *Mathematica* can be applied to fundamental problems of continuum mechanics. In modern electronic courses covering this topic, symbolic solutions are used for:

- deriving general equations for the deformation and motion of the continuum (fluid) in any curvilinear system of coordinates by symbolic techniques,
- finding symbolic solutions, mainly corresponding to conservation laws,
- solving wave problems that appear in the theory of filtration.

Examples from the theory of filtration of the mixture "*oil* + *gas* + *water*", when waves of saturation take place in porous soil, are solved by computer algebra techniques. Such problems can now be solved by *Mathematica* in the class as a routine computer procedure.

2 Nonlinear Elasticity Theory: Symbolic calculations

It is well known that the key results of the classic theory of elasticity (linear theory) have been adequately confirmed by experiment. However, it is now possible and important to return to the original basic concepts of the field, using nonlinear techniques, because computer tools such as *Mathematica* are an excellent means for teaching fundamentals, and for research into difficult problems, in nonlinear cases.

2.1 *Theory of deformations: Traditional and non-traditional evaluations*

Although traditional (cylindrical and spherical) curvilinear coordinate systems are widely used in basic courses on the nonlinear theory of elasticity [1,2], the full set of nonlinear components of the deformation tensor cannot be found even in well-known textbooks (typically, only the linear elements are presented).

In order to obtain all the elements of the deformation tensor in any curvilinear coordinate system it is necessary to load the package **DeformationTensor.m** .

```
In[1]:= << Calculus`VectorAnalysis`; << DeformationTensor.m;
```

```
In[2]:= n = 3; SetCoordinates[Cartesian]; NewCoordinatesToCartesian =
          CoordinatesToCartesian[Coordinates[]];
        NewCoordinates = Coordinates[];
```

```
In[4]:= ϵ = TensorCurveLinerDeformation[
            NewCoordinatesToCartesian, NewCoordinates, n];
```

```
In[5]:= γ1,3 = ϵ[[1, 3]] /. {Xx -> x, Yy -> y, Zz -> z}
```

Out[5]= $u_1^{(0,0,1)}(x, y, z)\,(u_1^{(1,0,0)}(x, y, z) + 1) + u_2^{(0,0,1)}(x, y, z)\,u_2^{(1,0,0)}(x, y, z) + (u_3^{(0,0,1)}(x, y, z) + 1)\,u_3^{(1,0,0)}(x, y, z)$

The last output may be readily compared with the traditional formula derived by the hand (see, for example [2, pp. 29]). Verification of the output gives the same result.

The same package is used for different coordinate systems. Below, cylindrical coordinates are used for test purposes.

```
In[6]:= SetCoordinates[Cylindrical];
        NewCoordinatesToCartesian = CoordinatesToCartesian[Coordinates[]];
        NewCoordinates = Coordinates[];
```

The set of nonlinear elements forming the third row of the nonlinear tensor appear in the output from the following calculation:

```
In[7]:= TensorCurveLinerDeformation[
            NewCoordinatesToCartesian, NewCoordinates, n][[3]] /.
        {Rr → r, Ttheta → θ, Zz → z} // ExpandAll
```

Out[7]=
$$\Big\{u_1^{(1,0,0)}(r,\theta,z)\,u_1^{(0,0,1)}(r,\theta,z)+u_1^{(0,0,1)}(r,\theta,z)+u_2^{(0,0,1)}(r,\theta,z)\,u_2^{(1,0,0)}(r,\theta,z)+ u_3^{(0,0,1)}(r,\theta,z)\,u_3^{(1,0,0)}(r,\theta,z)+u_3^{(1,0,0)}(r,\theta,z),\; -\frac{u_2(r,\theta,z)\,u_1^{(0,0,1)}(r,\theta,z)}{r}+\frac{u_1^{(0,1,0)}(r,\theta,z)\,u_1^{(0,0,1)}(r,\theta,z)}{r}+\frac{u_1(r,\theta,z)\,u_2^{(0,0,1)}(r,\theta,z)}{r}+ u_2^{(0,0,1)}(r,\theta,z)+\frac{u_2^{(0,0,1)}(r,\theta,z)\,u_2^{(0,1,0)}(r,\theta,z)}{r}+ \frac{u_3^{(0,0,1)}(r,\theta,z)\,u_3^{(0,1,0)}(r,\theta,z)}{r}+\frac{u_3^{(0,1,0)}(r,\theta,z)}{r},\; \frac{1}{2}\,u_1^{(0,0,1)}(r,\theta,z)^2+\frac{1}{2}\,u_2^{(0,0,1)}(r,\theta,z)^2+\frac{1}{2}\,u_3^{(0,0,1)}(r,\theta,z)^2+u_3^{(0,0,1)}(r,\theta,z)\Big\}$$

In order to compare the last output with the linear elements of the deformation tensor presented in typical textbooks, let us first derive the corresponding nonlinear element by computer. Then, let us linearize the element $\epsilon_{3,2}$. The code needed for this procedure is presented below:

```
In[8]:= ϵ3,2 =
        Expand[TensorCurveLinerDeformation[NewCoordinatesToCartesian,
              NewCoordinates, n][[3, 2]]] /. {Rr → r, Ttheta → θ, Zz → z}
```

Out[8]=
$$-\frac{u_2(r,\theta,z)\,u_1^{(0,0,1)}(r,\theta,z)}{r}+\frac{u_1^{(0,1,0)}(r,\theta,z)\,u_1^{(0,0,1)}(r,\theta,z)}{r}+\frac{u_1(r,\theta,z)\,u_2^{(0,0,1)}(r,\theta,z)}{r}+u_2^{(0,0,1)}(r,\theta,z)+\frac{u_2^{(0,0,1)}(r,\theta,z)\,u_2^{(0,1,0)}(r,\theta,z)}{r}+\frac{u_3^{(0,0,1)}(r,\theta,z)\,u_3^{(0,1,0)}(r,\theta,z)}{r}+\frac{u_3^{(0,1,0)}(r,\theta,z)}{r}$$

```
In[9]:= γ3,2 = DeleteCases[ϵ3,2, (x_)*(y_)/(z_)] /. {Rr → r, Ttheta → θ, Zz → z}
```

Out[9]=
$$u_2^{(0,0,1)}(r,\theta,z)+\frac{u_3^{(0,1,0)}(r,\theta,z)}{r}$$

The last symbolic output and the element from the textbook are clearly identical. See, for instance, textbook [2, pp.232] and compare with the formula

$$e_{3,2} = \frac{1}{r}\frac{\partial u_3}{\partial \theta} + \frac{\partial u_2}{\partial z} \tag{1}$$

More powerful and impressive examples of the nonlinear deformation tensor can be derived in non-traditional coordinate systems. Let us consider *EllipticCylindrical* coordinates. An example, similar to the above of the nonlinear element $\gamma_{3,2}$ of the deformation tensor is presented below.

```
In[12]:= γ3,2 = Expand[
              TensorCurveLinerDeformation[NewCoordinatesToCartesian,
     NewCoordinates, n][[3, 2]]] /. {Uu → u, Vv → v, Zz → z} // Timing
```

Out[12]=

$$\Bigg\{43.01\ \text{Second},\ \frac{\sqrt{2}\,\sin(2v)\,u_1(u,v,z)\,u_1^{(0,0,1)}(u,v,z)}{\sqrt{\cosh(2u)-\cos(2v)}} - \frac{\sqrt{2}\,\sinh(2u)\,u_2(u,v,z)\,u_1^{(0,0,1)}(u,v,z)}{\sqrt{\cosh(2u)-\cos(2v)}} - \frac{\sqrt{2}\,\cos(2v)\,u_1^{(0,1,0)}(u,v,z)\,u_1^{(0,0,1)}(u,v,z)}{\sqrt{\cosh(2u)-\cos(2v)}} + \frac{\sqrt{2}\,\cosh(2u)\,u_1^{(0,1,0)}(u,v,z)\,u_1^{(0,0,1)}(u,v,z)}{\sqrt{\cosh(2u)-\cos(2v)}} + \frac{\sqrt{2}\,\sinh(2u)\,u_1(u,v,z)\,u_2^{(0,0,1)}(u,v,z)}{\sqrt{\cosh(2u)-\cos(2v)}} + \frac{\sqrt{2}\,\sin(2v)\,u_2(u,v,z)\,u_2^{(0,0,1)}(u,v,z)}{\sqrt{\cosh(2u)-\cos(2v)}} - \frac{\cos(2v)\,u_2^{(0,0,1)}(u,v,z)}{\sqrt{\cosh(2u)-\cos(2v)}} + \frac{\cosh(2u)\,u_2^{(0,0,1)}(u,v,z)}{\sqrt{\cosh(2u)-\cos(2v)}} - \frac{\sqrt{2}\,\cos(2v)\,u_2^{(0,0,1)}(u,v,z)\,u_2^{(0,1,0)}(u,v,z)}{\sqrt{\cosh(2u)-\cos(2v)}} + \frac{\sqrt{2}\,\cosh(2u)\,u_2^{(0,0,1)}(u,v,z)\,u_2^{(0,1,0)}(u,v,z)}{\sqrt{\cosh(2u)-\cos(2v)}} + \frac{\sqrt{2}\,u_3^{(0,0,1)}(u,v,z)\,u_3^{(0,1,0)}(u,v,z)}{\sqrt{\cosh(2u)-\cos(2v)}} + \frac{\sqrt{2}\,u_3^{(0,1,0)}(u,v,z)}{\sqrt{\cosh(2u)-\cos(2v)}}\Bigg\}$$

It seems that in the last output a non trivial result has been obtained by *Mathematica*. We can apply other coordinate systems in the same way.

2.2 *Nonlinear dynamic equations in non-traditional coordinate systems*

Just as it is very difficult to find, published, all the nonlinear elements of the deformation tensor in non-traditional coordinate systems, the study of associated nonlinear dynamical equations in such systems looks like an impossible operation. But the computer technique created here is able to derive dynamical equations in any curvilinear coordinate system.

A special package **<<ContinueDynamicEquations.m** was created for realizing the main idea of this presentation.

```
In[13]:= << ContinueDynamicEquations.m
```

In the first step **Paraboloidal** coordinates (implemented natively in *Mathematica*) are applied.

```
In[14]:= SetCoordinates[Paraboloidal];
         NewCoordinatesToCartesian= CoordinatesToCartesian[Coordinates[]];
         NewCoordinates = Coordinates[];
```

The short form of the first equation of the motion of an individual particle of the continuum is presented in *Paraboloidal* coordinates below.

```
In[15]:= Short[ContinueDynamicEquations[NewCoordinatesToCartesian,
              NewCoordinates, n][[1]] /. {Pphi -> φ, Uu -> u, Vv -> v}, 15]
```

Out[15]//Short=

$$\frac{\rho\cos(\varphi)\,u_2^{(0,0,0,2)}(u,v,\varphi,t)\,u_1^{(0,0,1,0)}(u,v,\varphi,t)\,u^5}{\sqrt{u^2+v^2}\,v^3+u^2\sqrt{u^2+v^2}\,v}+\frac{\rho\sin(\varphi)\,u_2^{(0,0,0,2)}(u,v,\varphi,t)\,u_2^{(0,0,1,0)}(u,v,\varphi,t)\,u^5}{\sqrt{u^2+v^2}\,v^3+u^2\sqrt{u^2+v^2}\,v}+$$
$$\frac{\rho\sin(\varphi)\,u_2^{(0,0,0,2)}(u,v,\varphi,t)\,u_1^{(0,0,1,0)}(u,v,\varphi,t)\,u_1^{(0,1,0,0)}(u,v,\varphi,t)\,u^5}{\sqrt{u^2+v^2}\,v^3+u^2\sqrt{u^2+v^2}\,v}+\frac{\rho\sin(\varphi)\,u_2^{(0,0,0,2)}(u,v,\varphi,t)\,u_2^{(0,0,1,0)}(u,v,\varphi,t)\,u_2^{(0,1,0,0)}(u,v,\varphi,t)\,u^5}{\sqrt{u^2+v^2}\,v^3+u^2\sqrt{u^2+v^2}\,v}+$$
$$\ll 1741 \gg+\frac{2v^2\,\sigma_{3,1}(u,v,\varphi,t)\,u_1^{(0,0,1,0)}(u,v,\varphi,t)\,u_1^{(0,0,2,0)}(u,v,\varphi,t)}{(v\,u^3+v^3\,u)\,u^2}+\frac{2v^2\,\sigma_{3,1}(u,v,\varphi,t)\,u_2^{(0,0,1,0)}(u,v,\varphi,t)\,u_2^{(0,0,2,0)}(u,v,\varphi,t)}{(v\,u^3+v^3\,u)\,u^2}+$$
$$\frac{2v\,u_1(u,v,\varphi,t)\,\sigma_{3,1}(u,v,\varphi,t)\,u_3^{(0,0,2,0)}(u,v,\varphi,t)}{(v\,u^3+v^3\,u)\,u^2}+\frac{2\,\sigma_{3,1}(u,v,\varphi,t)\,u_3^{(0,0,1,0)}(u,v,\varphi,t)\,u_3^{(0,0,2,0)}(u,v,\varphi,t)}{(v\,u^3+v^3\,u)\,u^2}==0$$

Clearly, the size of this symbolic expression is a real obstacle for the full solution of this problem using *Mathematica*.

3 Nonlinear theory of filtration: symbolic and numerical calculations

The last part of this paper is devoted to the application of symbolic and numerical techniques to the nonlinear theory of filtration with an application in the form of an oil technology simulation. The study focuses on the motion of the fluid mixture: "*gas+water+oil*"; let us consider a well-known model of filtration due to Leveret andMascot [see [3], pp. 287]. The mathematical model of the motion of the mixture "*gas+water+oil*" is described by the following system consisting of a PDE with boundary and initial conditions.

$$\frac{\partial s_1}{\partial\tau}+\frac{\partial f_1}{\partial\zeta}=0,\;\; \frac{\partial s_2}{\partial\tau}+\frac{\partial f_2}{\partial\zeta}=0,\;\; s_1+s_2+s_3=1;$$
$$s_i=S_i(\zeta,0);\;\; s_i=S_{i0}(0,\tau);\;\; (i=1,2). \tag{2}$$

Here s_i- are the saturations of the water, gas and oil in the mixture, f_i-are the Leveret-Mascot functions. Clear forms of the all the components of the model are presented below.

In[1]:= $s_1 = S_1(\zeta, \tau);\ s_2 = S_2(\zeta, \tau);\ f_1(s_1, s_2) = \dfrac{s_1^2 + s_2^2}{s_1^2 + \eta_{10}\,(1 - s_1)^2};$

$f_2(s_1, s_2) = \dfrac{s_2\, s_1}{s_2^2 + \eta_{20}\,(1 - s_2)^2};$

In[2]:= $g[s_1, s_2] = \dfrac{\partial s_1}{\partial \tau} + \partial_{s_1} f_1(s_1, s_2)\,\dfrac{\partial s_1}{\partial \zeta} + \partial_{s_2} f_1(s_1, s_2)\,\dfrac{\partial s_2}{\partial \zeta} == 0;$

$h[s_1, s_2] = \dfrac{\partial s_2}{\partial \tau} + \partial_{s_1} f_2(s_1, s_2)\,\dfrac{\partial s_1}{\partial \zeta} + \partial_{s_2} f_2(s_1, s_2)\,\dfrac{\partial s_2}{\partial \zeta} == 0;$

In[3]:= **Off[NDSolve::"eerr"]**

In[4]:= **solution =** $\{S_1, S_2\}$ **/.**
First[NDSolve[$\{g[s_1, s_2]$ /. $\eta_{10} \to .0001$, $h[s_1, s_2]$ /. $\eta_{20} \to .5$,
$S_1[\zeta, 0] == \zeta\,(1 - \zeta^2)$, $S_2[\zeta, 0] == \zeta\,(1 - \zeta^4)$, $S_1[0, \tau] ==$
$1 - e^{-\tau}$, $S_2[0, \tau] == 1 - e^{-\tau}\}$, $\{S_1, S_2\}$, $\{\zeta, 0, 1\}$, $\{\tau, 0, 3\}$**]]**;

In[5]:= **con1 = ContourPlot[solution[[1]][**ζ, τ**], {**ζ**, 0, 1}, {**τ**, 0, 1},**
ColorFunction → Hue, DisplayFunction → Identity];
con2 = ContourPlot[solution[[2]][ζ, τ**], {**ζ**, 0, 1}, {**τ**, 0, 1},**
ColorFunction → Hue, DisplayFunction → Identity];
con3 = ContourPlot[1 − solution[[2]][ζ, τ**] − solution[[1]][**ζ, τ**], {**ζ**, 0, 1},**
{τ**, 0, 1}, ColorFunction → Hue, DisplayFunction → Identity];**

Based only on the concept of the mobility (in the broad sense: for more details see in [3, pp.pp. 285-288]) of the different components of the mixture, two non-similar types of fluid motion can be observed in the model: *wave* propagations of the saturations in the mixture and *smooth* replacement of petroleum by water. Pictures of distributions of the wave saturations (first water, second gas, third oil) are presented below, and below that, smooth replacements.

```
In[12]:= Show[GraphicsArray[{{con1, con2, con3}, {con4, con5, con6}}],
         DisplayFunction → $DisplayFunction];
```

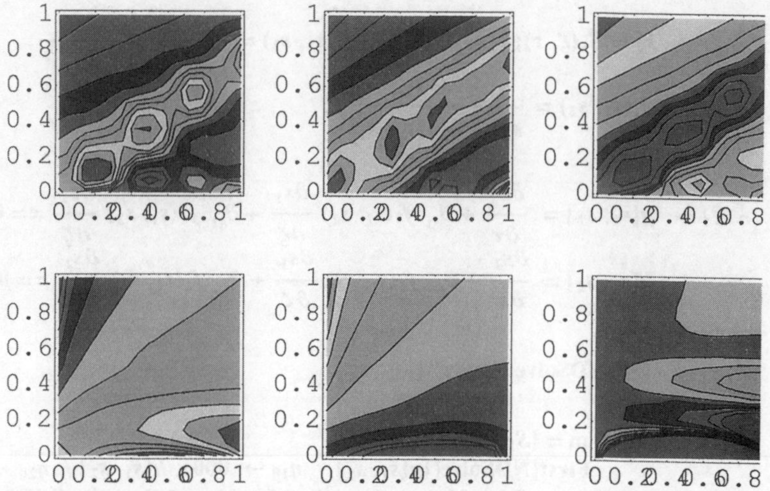

It is impossible to foresee the motion of a mixture beforehand.

4 Conclusion

In the paper several themes are discussed in detail which deal with traditional courses used for the education of engineers under the general heading of Continuum Mechanics. During last five years in MSTU training courses on mathematical models in continuum mechanics were established at two departments of the University: "Computer Sciences and Programming" and "Continuum Mechanics and Offshore Exploration". Symbolic techniques are especially suitable in the University classes for these departments.

5 References

[1]. Lurie A.I. Teoria uprugosty. (In Russian). Moscow, "Nauka", 1970, 939 p.

[2]. Sedov L.I. Mechanika splozhnoy sredy. vol.**1,2** (In Russian).- Moscow, "Nauka", 1973, 536 p.

[3]. Basniev K.C., Kochina I.N., Maksimov V.M. Podzemnay hydromechanika.(In Russian) - Moscow, "Vyshay shkola" 1993, 414 p.

Studying stability of the equilibrium solutions in the restricted many-body problems

Alexander N. Prokopenya
Brest State Technical University

prokopenya@brest.by

1 Abstract

We study the stability of the equilibrium solutions in the elliptic restricted many-body problems using *Mathematica*. The calculations done in the case of five interacting bodies have shown that the radial equilibrium solutions are unstable for any value of the mass m_0 of the central body, while some bisector equilibrium solutions are stable in linear approximation if m_0 is sufficiently large. However, there is a domain of instability of the bisector solutions in the vicinity of the resonant point. Its boundaries are calculated with the method of infinite determinants.

2 Introduction

The restricted problem of three bodies is well known and has a lot of applications; it has been discussed in detail by V.Szebehely [1]. Many eminent mathematicians have investigated it in the last two centuries and many important results have been obtained. This problem stimulated development of new qualitative, analytical and numerical methods for studying nonlinear hamiltonian systems. Nevertheless, development of the stability theory of hamiltonian systems has not yet been completed, and investigations in the field are very topical. In recent work of E.A.Grebenikov [2], two new dynamical models, called the restricted many-body problem,s were proposed. In one, called the noninertial model, a point particle P of infinitesimal mass moves in the gravitational field generated by the particles $P_0, P_1, \ldots, P_n$. The particle P_0, having a mass m_0, is at rest at the centre of a regular n-gon, while the others, having equal mass m, are at the vertices of the polygon and move in circular or elliptic orbits about the particle P_0. The corresponding configuration for $n = 4$ is shown on Figure 1. E. A. Grebenikov, D. Kozak-Skovorodkin and M. Jakubiak [3] investigated the particular case of the problem, when the particles $P_1, \ldots, P_n$ move uniformly in a circular orbit and the test particle P is constrained to move in their orbital plane. The aim of the present paper is to find all equilibrium solutions of the equations of motion within the framework of the elliptic restricted many-body problem and to study their stability. Both of these problems are very complicated andrequire the use of computer software. Here we use *Mathematica* [4] for doing both numerical and symbolic calculations and for visualization .

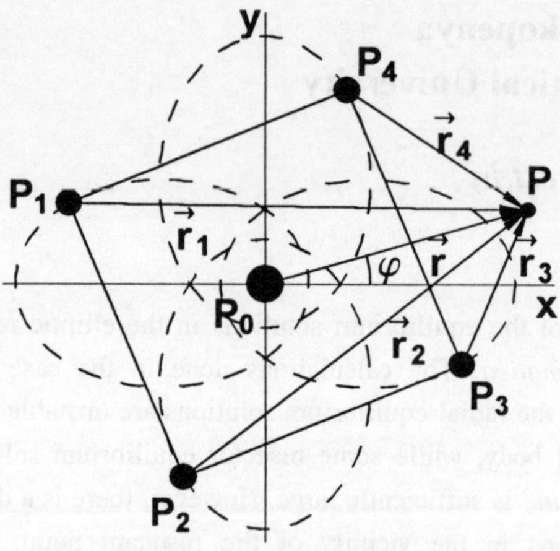

Figure 1

3 Equilibrium solutions of the equations of motion

Let us consider the motion of the particle P of infinitesimal mass m_1 in the gravitational field generated by the particles $P_0, P_1, \ldots, P_n$. Denoting its cylindrical coordinates as ρ, φ, z and using the notations shown on Figure 1 we can write the Lagrangian of the system in the form

$$L = \frac{1}{2} m_1 (\dot{\rho}^2 + \rho^2 \dot{\varphi}^2 + \dot{z}^2) + G m_1 \left(\frac{m_0}{r} + \sum_{j=1}^{n} \frac{m}{r_j} \right) \tag{1}$$

where G is a constant of gravitation, the dot denotes the derivative $\frac{d}{dt}$,

$$r = \sqrt{\rho^2 + z^2}\,,\; r_j = \sqrt{\rho^2 + \rho_j{}^2 - 2\,\rho\,\rho_j \cos(\varphi - \varphi_j) + z^2} \quad (j = 1, 2, \ldots, n)$$

and cylindrical coordinates of the particles $P_1, \ldots, P_n$ are given by

$$\rho_j = \frac{p}{1 + e \cos \nu}\,,\; \varphi_j = \nu + \frac{2\pi}{n}\,,\; \rho_j{}^2\,\dot{\nu} = \sqrt{G\,p\,M_n} \equiv c. \tag{2}$$

Here p and e are the parameter and eccentricity of the elliptic orbits of the particles and

$$M_n = m_0 + \frac{m}{4} \sum_{j=1}^{n-1} \left(\sin\left(\frac{\pi}{n} j\right)\right)^{-1}$$

Taking into account relationships (2) it is expedient to make a substitution in (1) according to the rule

$$\rho_j(t) \to \frac{p}{1 + e\cos\nu},\ \rho(t) \to \frac{p}{1 + e\cos\nu}\,\rho(\nu)$$

$$z(t) \to \frac{p}{1 + e\cos\nu}\,z(\nu),\ \varphi(t) \to \nu + \varphi(\nu)$$

In a new configurational space known as the Nechvil space [3], the particles $P_1, \ldots, P_n$ rest at the vertices of a regular polygon inscribed into a unit circle. Using the polar angle ν as a new independent variable we can rewrite the Lagrangian (1) in the form

$$\begin{aligned} L = \frac{1}{2} m_1\, c \Bigg(& \dot{\rho}^2 + \rho^2(\dot{\varphi} + 1)^2 + \dot{z}^2 \\ & - \frac{e\cos\nu}{1 + e\cos\nu}(\rho^2 + z^2) + \frac{2Gp}{c^2(1 + e\cos\nu)}\left(\frac{m_0}{r} + \sum_{j=1}^{n} \frac{m}{r_j}\right)\Bigg) \end{aligned} \tag{3}$$

where $r = \sqrt{\rho^2 + z^2}$, $r_j = (1 + \rho^2 - 2\rho\cos(\varphi - 2\pi j/n) + z^2)^{1/2}$. With the Lagrangian (3) the equations of motion may be written as

$$\begin{aligned} & \ddot{\rho} - \rho(\dot{\varphi} + 1)^2 + \frac{e\cos\nu}{1 + e\cos\nu}\rho = \\ & \frac{1}{M_n(1 + e\cos\nu)}\left(\frac{m_0}{r^3}\rho + m\sum_{j=1}^{n} \frac{\rho - \cos(\varphi - 2\pi j/n)}{r_j{}^3}\right), \\ & \rho\ddot{\varphi} + 2\dot{\rho}(\dot{\varphi} + 1) = -\frac{m}{M_n(1 + e\cos\nu)} \sum_{j=1}^{n} \frac{\sin(\varphi - 2\pi j/n)}{r_j{}^3}, \\ & \ddot{z} + \frac{e\cos\nu}{1 + e\cos\nu} z = -\frac{z}{M_n(1 + e\cos\nu)}\left(\frac{m_0}{r^3} + m\sum_{j=1}^{n} \frac{1}{r_j{}^3}\right). \end{aligned} \tag{4}$$

The equilibrium solutions of the system (4) may be defined as $\rho \equiv R = \text{const}$, $\varphi \equiv \beta = \text{const}$, $z = \text{const}$. Then the third equation in (4) is satisfied only if $z = 0$. In this case the second equation in (4) may be written in the form

$$\sum_{j=1}^{n} \frac{\sin(\beta - 2\pi j/n)}{(1 + R^2 - 2R\cos(\beta - 2\pi j/n))^{3/2}} = 0 \tag{5}$$

The function in the left-hand side of (5) is periodic in β with period $\frac{2\pi}{n}$. Therefore, it is sufficient to analyse this equation on the interval $0 \le \beta < \frac{2\pi}{n}$. Taking into account a symmetry of the system we can conclude that equation (5) is satisfied for any R if $\beta = 0$ or $\beta = \frac{\pi}{n}$. According to the terminology of E.A.Grebenikov [3] the equilibrium positions on the ray $\beta = 0$ are called the radial equilibrium solutions. The corresponding values of R are obtained as roots of the equation $f_1 = 0$ where

```
f1[μ_, R_, n_] :=
  μ (R - 1 / R^2) + R / 4 Sum[Sin[π j / n]^-1, {j, n - 1}] -
   Sum[(R - Cos[2 π j / n])
     (1 + R^2 - 2 R Cos[2 π j / n])^(-3/2), {j, n}];
```

and $\mu = m_0/m$. The function f_1 has a point of discontinuity $R = 1$. Its typical behaviour as a function of R is seen from the following graphs for $n = 3$ and different values of μ.

```
In[11]:= Plot[{f1[0, R, 3], f1[0.1, R, 3], f1[1, R, 3]},
           {R, 0, 3}, PlotRange → {-1, 1}, PlotStyle →
            {{}, {Dashing[{0.04}]}, {Dashing[{0.02}]}}];
```

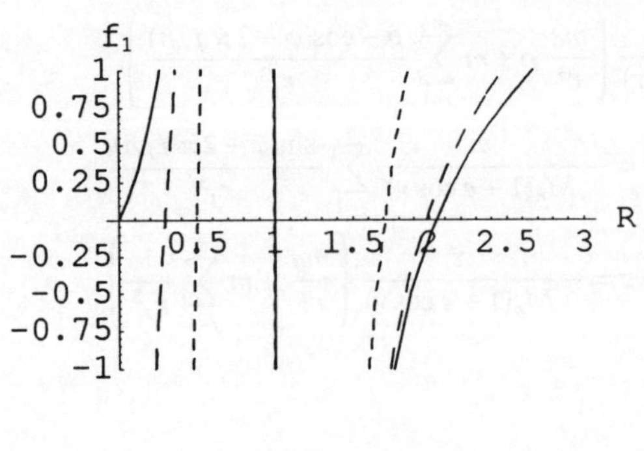

We see that there exist two roots of the equation $f_1 = 0$ for any μ, and both of them tend to a limit $R = 1$ as $\mu \to \infty$. The equilibrium positions on the ray $\beta = \frac{\pi}{n}$ are called the bisector equilibrium solutions. They are determined from the condition $f_2 = 0$ where

```
In[12]:= f2[μ_, R_, n_] :=
      μ (R - 1 / R^2) + R / 4 Sum[Sin[π j / n]^-1, {j, n - 1}] -
       Sum[(R - Cos[(2 j - 1) π / n])
         (1 + R^2 - 2 R Cos[(2 j - 1) π / n])^(-3/2), {j, n}];
```

The behaviour of f_2 as a function of R is shown on the following graphs for $n = 3$ and different values of μ.

```
In[14]:= Plot[{f2[0, R, 3], f2[0.005, R, 3], f2[0.1, R, 3]},
      {R, 0, 2.5}, PlotRange → {-1, 1}, PlotStyle →
       {{}, {Dashing[{0.04}]}, {Dashing[{0.02}]}}];
```

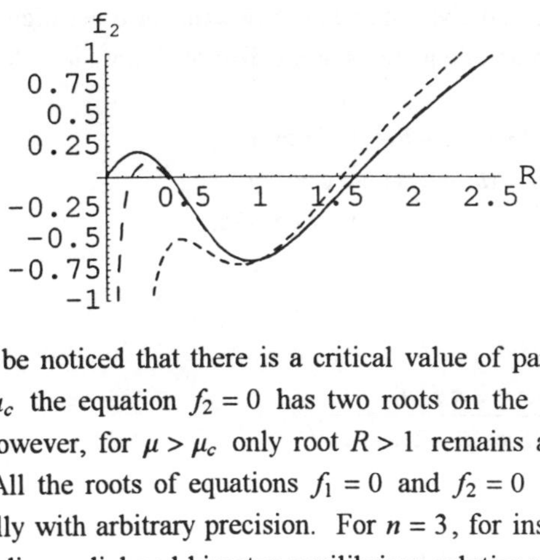

It should be noticed that there is a critical value of parameter $\mu = \mu_c$ depending on n. For $\mu < \mu_c$ the equation $f_2 = 0$ has two roots on the interval $0 \le R < 1$ and one root $R > 1$. However, for $\mu > \mu_c$ only root $R > 1$ remains and it tends to a limit $R = 1$ as $\mu \to \infty$. All the roots of equations $f_1 = 0$ and $f_2 = 0$ and parameter μ_c can be found numerically with arbitrary precision. For $n = 3$, for instance, we have $\mu_c = 0.014$. The corresponding radial and bisector equilibrium solutions for $\mu < \mu_c$ are shown on Figure 2 as the points N_1, N_2 and S_1, S_2, S_3 respectively.

It should be emphasized that although the equilibrium solutions are some points in the Nechvil space, they determine the elliptic trajectories of the particle P with the parameter pR and eccentricity e in the barycentric inertial frame of reference.

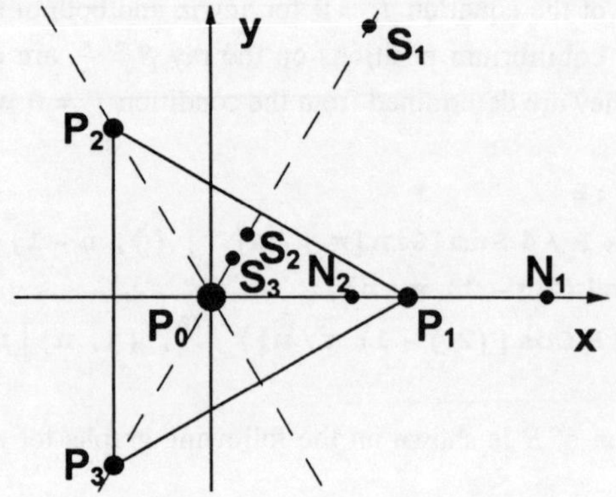

Figure 2

4 Stability of the equilibrium solutions

To study stability of the equilibrium solutions let us make the following substitution in the equations of motion (4): $\rho(\nu) \to R + u(\nu)$, $\varphi(\nu) \to \beta + \gamma(\nu)/R$. Expanding the equations in powers of $u(\nu)$, $\gamma(\nu)$ and $z(\nu)$, and neglecting terms of order higher or equal to 2, we obtain the linearised system of equations of the disturbed motion in the form

$$\begin{aligned} \ddot{u} - 2\,\dot{\gamma} &= (3 + 2\,a - b)/(1 + e\cos\nu)\,u, \\ \ddot{\gamma} + 2\,\dot{u} &= (b - a)/(1 + e\cos\nu)\,\gamma, \\ \ddot{z} &= -(1 + a + e\cos\nu)/(1 + e\cos\nu)\,z, \end{aligned} \tag{6}$$

where

$$a = \frac{m}{M_n R} \sum_{j=1}^{n} \frac{\cos(\beta - 2\pi j/n)}{(1 + R^2 - 2R\cos(\beta - 2\pi j/n))^{3/2}},$$

$$b = \frac{3m}{M_n} \sum_{j=1}^{n} \frac{\sin^2(\beta - 2\pi j/n)}{(1 + R^2 - 2R\cos(\beta - 2\pi j/n))^{5/2}}.$$

The third equation in (6) is independent on the first and the second ones and determines the disturbed motion of the particle P along the Oz axis. It is just the Hill's equation that was investigated in detail in [5]. It was shown that there are the domains of instability of this equation in the $\mu - e$ plane in the vicinity of the points $a = (2k-1)^2/4 - 1$ ($k = 1, 2, \ldots$) and the boundaries of these domains were found in the form of power series in e.

The first two equations in (6) are linear second order differential equations with periodic coefficients which are analytic functions of parameter e in the domain $|e| < 1$. According to the general theory [6], behaviour of their solutions for sufficiently small e is determined by their characteristic exponents λ_k calculated for $e = 0$. They are

$$\lambda_k = \pm\left(-1 + a \pm \left(\sqrt{1 + 10\,a + 9\,a^2 - 12\,b - 12\,a\,b + 4\,b^2}\,\right)^{1/2} \quad (k = 1, 2, 3, 4) \qquad (7)$$

Numerical calculations carried out for $n = 3$ show that for each radial equilibrium solution there are four characteristic exponents (7) but that at least one of them is a positive real number for any μ from the interval $0 \le \mu < \infty$. As the characteristic exponents are continuous functions of e, there will exist at least one characteristic exponent with a positive real part for sufficiently small e. So, according to Liapunov's theorem on linearized stability, we can conclude that radial equilibrium solutions of the elliptic restricted problem of five bodies are unstable. The same is true for the bisector equilibrium solutions if parameter μ is small enough. But if $\mu > 43.1811$ for $n = 3$ then all characteristic exponents (7) for equations (6) determining the disturbed motion of the particle P in the vicinity of the bisector equilibrium solution S_1 (see Figure 2) are distinct pure imaginary numbers. It means that this solution is stable in linear approximation for $\mu > 43.1811$ and sufficiently small e. Nevertheless, numerical analysis of the characteristic exponents (7) as functions of parameter μ shows that for $\mu = 57.1053$ the resonance condition

$$\lambda_k \pm \lambda_l = i\,N (k, l = 1, 2, 3, 4; N = 0, \pm 1, \pm 2, \ldots)$$

is fulfilled when $\lambda_1 = -\lambda_2 = i/2$. This means that in the vicinity of this point in the $\mu - e$ plane the domain of instability of the bisector solution S_1 may exist. The boundaries of this domain are some curves $\mu = \mu(e)$ which are characterized by the presence of periodic solutions with period 4π. Hence, we can attempt to determine the boundaries directly by seeking a solution of the system (6) in the form of Fourier series

$$u(v) = c_0 + \sum_{k=1}^{\infty} (c_k \cos(k\,v/2) + d_k \sin(k\,v/2)). \qquad (8)$$

$$\gamma(v) = c_0 + \sum_{k=1}^{\infty} (p_k \cos(k\,v/2) + q_k \sin(k\,v/2)). \qquad (9)$$

Substituting (8), (9) into equations (6) and setting coefficients of $\cos(k\nu/2)$ and $\sin(k\nu/2)$ to zero we obtain the infinite sequence of equations determining coefficients of the Fourier series. For a solution to exist, the corresponding determinant of the infinite system must vanish, thus determining the stability boundaries. For $n=3$, for instance, this method gives the stability boundaries in the form

$$\mu = 57.1053 \pm 114.067e + 196.30e^2 \pm 282.29e^3 + 386.5e^4. \qquad (10)$$

5 Conclusion

Using *Mathematica* we have obtained equations determining the equilibrium solutions in the elliptic restricted many-body problems in the Nechvil configurational space, and solved them in the case of five interacting bodies. We have proved that for $n=3$ the radial equilibrium solutions are unstable for sufficiently small e and any value of the parameter μ. The bisector equilibrium solutions S_2, S_3 are also unstable, while the solution S_1 is linearly stable if μ is sufficiently large (namely $\mu > 43.1811$) and the values of the parameters μ and e do not belong to the domain of instability bounded by the curves (10). All symbolic calculations were carried out for an arbitrary number of interacting bodies, so the numerical calculations can be easily repeated for arbitrary n.

Acknowledgments

The author is very grateful to Prof. Evgeni A. Grebenikov for the interesting and useful discussion of the stability problem.

6 References

[1] Szebehely V. Theory of orbits. The restricted problem of three bodies, New York, London, Academic Press, 1967.

[2] Grebenikov E.A. Two new dynamical models in celestial mechanics, in Romanian Astronomical J., Vol. 8, No.1, p. 13-19, Bucharest 1998.

[3] Grebenikov E.A., Kozak-Skovorodkin D., Jakubiak M. The methods of computer algebra in the many-body problem, Moscow, Ed. RUDN, 2002 (in Russian).

[4] Wolfram S. The *Mathematica* book, Cambridge, University Press, 1996.

[5] Prokopenya A.N. Calculation of the characteristic exponents for a Hill's equation, in Proc. 8th Rhine Workshop on Computer Algebra, Mannheim, Germany, March 2002. Eds. Kredel H. and Seiler W.K., University of Mannheim, 2002, p. 275-278.

[6] Yakubovich V.A., Starzhinskii V.M. Linear differential equations with periodic coefficients, John Wiley, New York, 1975.

webMathematica as a Core Service for the Calculation of the Drought Indicator for South Africa

Dusan Sakulski
dsakulski@yahoo.com
Prof. David Stephenson
steph@civil.wits.ac.za
Prof. Prvoslav Marjanovic
banem@civil.wits.ac.za
University of the Witwatersrand, School of Civil and Environmental Engineering, Johannesburg, South Africa

1 Abstract

The increase in the frequency of disasters and their associated damage in the SADC region is part of a worldwide trend, which results from growing vulnerability and may reflect changing climate patterns. Global risks seem to be increasing. These trends make it all more necessary for South Africa to initiate the development and implementation of the national Disaster Hazard and Vulnerability Atlas. This is a database-driven, web-enabled interactive "virtual book". It consists of various "chapters", such as drought, flood, cyclones, storms.etc. *Mathematica* is in use as a core service (engine) for the calculation of the Standardized Precipitation Index (SPI) as a drought indicator. web*Mathematica* scripts have been written to perform complex probabilistic rainfall data analysis.

Special thanks to the National Disaster Management Centre of the Republic of South Africa (http://sandmc.pwv.gov.za) for their support in this project, and for hosting the Atlas (http://sandmc.pwv.gov.za/atlas/).

2 National Disaster Hazard and Vulnerability Atlas

Worldwide hazard, vulnerability and risk-increasing trends have initiated a need for the development and implementation of the National Disaster Hazard and Vulnerability

Atlas for South Africa. The main idea was to design and develop a database-driven, web-enabled interactive "virtual book" (Atlas). It consists of various "chapters", such as drought, flood, cyclones, storms, ... It enables users, using just a web browser, to search and select various data, images, maps and graphs; to perform various calculations; to run models on-the-fly and copy-paste results to the local computer; and to print their own pages of the Atlas.

Key elements of the Atlas include:
- Development of National disaster related hazard, vulnerability and risk assessment tools, able to report periodically on the National exposure to natural and man-made hazards, and on patterns and trends or changes in that exposure, and to guide priorities in natural disaster vulnerability reduction efforts.
- Development of an integrated National disaster hazard and vulnerability information network to provide the tools needed by National, provincial and local governments, the private sector, and the general public. The network will also facilitate much-needed augmentation of education and training.

In general, here are some of the main elements of the Atlas:
- Comprehensive database driven, web-enabled tools to identify hazard, vulnerability and risk-prone areas.
- Understanding and addressing hazard, risk and vulnerability.
- Information exchange tools.

In particular, the Atlas attempts to:
- Carry out research on factors contributing to environmental hazards, risks and vulnerabilities and measures to alleviate it.
- Develop methodologies for the analysis of environmental related hazard and vulnerability indicators, and to improve environmental / disaster management.

3 Drought and Standardized Precipitation Index (SPI)

Drought is a normal, recurrent feature of climate, although many erroneously consider it a rare and random event. It occurs in virtually all climatic zones, although its characteristics vary significantly from one region to another. Drought is a temporary aberration and differs from aridity since the latter is restricted to low rainfall regions and is a permanent feature of climate [1].

Drought is an insidious hazard of nature. Although it has scores of definitions, it originates from a deficiency of precipitation over an extended period of time, usually a season or more. This deficiency results in a water shortage for some activity, group, or environmental sector. Drought should be considered relative to some long-term average condition of balance between precipitation and evapotranspiration (i.e., evaporation + transpiration) in a particular area, a condition often perceived as "normal". It is also related to the timing (i.e., principal season of occurrence, delays in the start of the rainy season, occurrence of rains in relation to principal crop growth stages) and the effectiveness of the rains (i.e., rainfall intensity, number of rainfall events). Other climatic factors such as high temperature, high wind and low relative humidity are often associated with it in many regions of the world and can significantly aggravate its severity [2].

The understanding that a deficit of precipitation has different impacts on the ground water, reservoir storage, soil moisture, snowpack, and streamflow led McKee et al. [3] to develop the Standardized Precipitation Index (SPI). The SPI was designed to quantify the precipitation deficit for multiple time scales. These time scales reflect the impact of drought on the availability of the different water resources. Soil moisture conditions respond to precipitation anomalies on a relatively short scale, while ground water, streamflow, and reservoir storage reflect the longer-term precipitation anomalies. For these reasons, McKee et al. [3] originally calculated the SPI for 3-, 6-,12-, 24-, and 48-month time scales.

The SPI calculation is based on the long-term precipitation record for a desired period. This long-term record is fitted to a probability distribution, which is then transformed into a normal distribution so that the mean SPI for the location and desired period is zero [4]. Positive SPI values indicate greater than median precipitation, while negative values indicate less than median precipitation. Because the SPI is normalized, wetter and drier climates can be represented in the same way, and wet periods can also be monitored using the SPI.

4 Why web*Mathematica* and a new integrated database-driven web-enabled technique?

The core of every IT related application is data. Many IT applications developers do not take this fact seriously enough. At the beginning of the IT application life cycle, it does not seem to be important. As a result, in the later phase, when applications become more complex, and especially in the process of application maintenance, life becomes more and more difficult. The usual result is the loss of the integrity and coherence of the application. Amongst other time related data, rainfall data belong to the time series data category. This is a special data type, which always requires detailed data management planning and implementation. Also, time series data are highly consuming of storage space and they usually require different storage algorithms. This is why a database approach towards time series data is extremely important.

Web-enabled technology (Internet, Intranet) has bridged a large gap in IT implementation. It is not important any more what computer or operating system is in use. To be able to use a web-enabled application, all the user needs is a web browser and Internet/Intranet connection. In a few years, the use of web technology will not necessary require a computer. It will be an integral part of many communication devices (TV, cell phones).

The majority of web-based applications are developed using languages like VB Script (ASP), Perl, Java or Java Script. None of them are suitable for performing mathematical and/or statistical analysis or optimizations. IT developers will, very probably, use Excel or C++, and spend a lot of time debugging and library linking. All in all, there is no one technical environment that enables the seamless integration of symbolic and numeric abilities like *Mathematica* (or, even better, web*Mathematica*, which makes the power of *Mathematica* available via Internet / Intranet).

5 All together step-by-step

Step 1: The user submits input by selecting either rainfall station or quaternary catchment (quad), water deficit period (i.e. 3 months) and starting month (e.g. January) . The application will pass the values of two variables from the web server to the web*Mathematica* server and generate an SQL query on-the-fly: "SELECT RAIN FROM RAIN WHERE QUAD = 'A10A' AND INTERVAL = 3 AND STARTMONTH = 1". Using *Mathematica* DB Module, this string is submitted to database server running an MS SQL (ORACLE in future) database. The result of the query is returned to the web*Mathematica* server as a list (set of numbers).

Step 2: *Mathematica* (using Statistical Add-ons) calculates the empirical cumulative probability (the dots on the graph below). The empirical cumulative probability was found optimally as suggested by Panofsky and Bier [5] where precipitation data is sorted in increasing order of magnitude so that the kth value is $k-1$ values from the lowest and where n is the sample size (Figure 1). The x-axis value represents 3-month cumulative rainfall [inch].

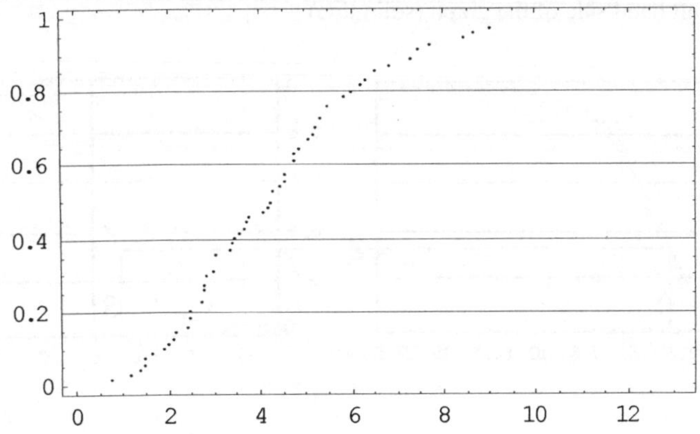

Figure 1

Step 3: Using historical rainfall time series data, the α and β parameters of the Gamma(α, β) distribution are calculated, and a Gamma distribution CPD is fitted (solid line, Figure 2). The x-axis value represents 3-month cumulative rainfall [inch]:

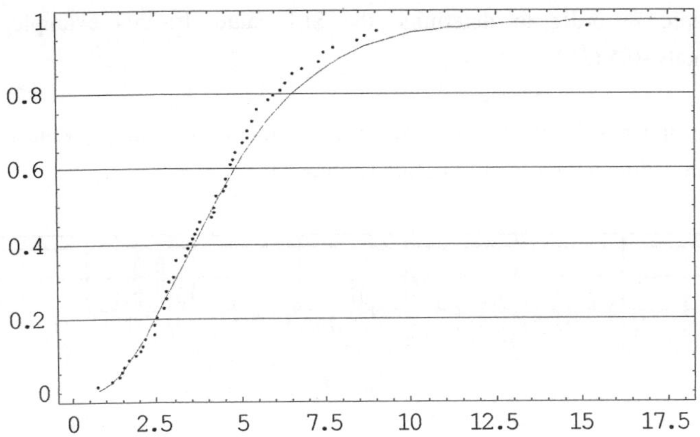

Figure 2

Step 4: The smooth curve on the right hand side of Figure 3 denotes the cumulative probability distribution of the Standard Normal random variable Z, using the same cumulative probability scale of the empirical distribution and fitted Gamma distribution, as on the left hand side of the graph (solid line).

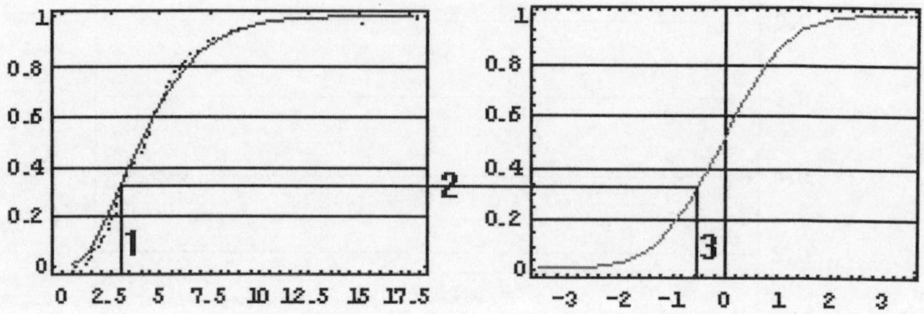

Figure 3

For example, to find the SPI for the 3.5 inch total (Jan + Feb + Mar) rainfall, simply go vertically upwards from the 3.5 inch mark on the *x*-axis on the left-hand side of Figure 3, until the fitted Gamma cumulative probability (solid line) curve is intersected (1). After that, go horizontally (maintaining equal cumulative probability) to the right until the curve of the Standard Normal cumulative probability distribution (solid line) is intersected (2). Then proceed vertically downward to the *x*-axis on the right side of the above graph, in order to determine the SPI value. In this example, the SPI is approximately -0.5 (3).

Step 5: To plot a SPI time series graph for the whole period, simply repeat steps 1-4 for each month ending a three-month period (Figure 4):

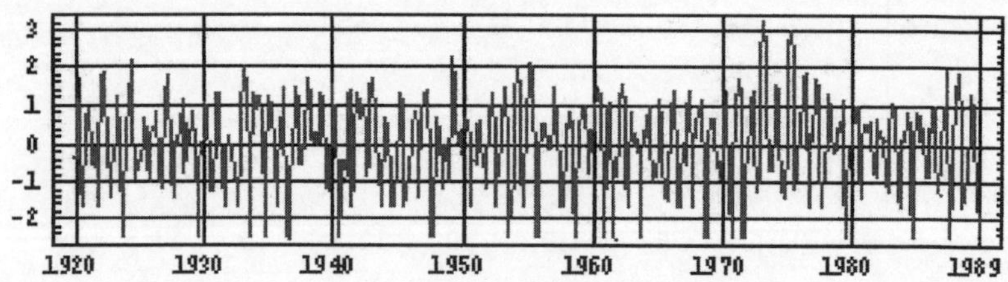

Figure 4

6 Present and future works

At present, web*Mathematica* is in use to enable on-the-fly calculation and analysis of various other hazard, vulnerability and risk indicators. Analysis of the worldwide monthly rainfall data is enabled using web*Mathematica* (http://sandmc.pwv.gov.za/safari2000/nasa/rain/cntrlist.asp). The UN ISDR (International Secretariat for Disaster Reduction) has expressed an intention to utilise Atlas as a hazard, vulnerability and risk monitoring tool (Global Atlas).

7 References

[1] Dracup, J.A., K.S. Lee, E.G. Paulson. On the Definition of Droughts. *Water Resources Research*, 1980, Vol. 16, No. 2, pp. 297-302.

[2] Wilhite, D. A. Understanding and Defining Drought. National Drought Mitigation Center, USA, 1995.

[3] McKee, T. B., N. J. Doesken, and J. Kleist. The relationship of drought frequency and duration to time scales. Preprints, 8th Conference on Applied Climatology, 17-22 January, Anaheim, CA, pp. 179-184, 1993.

[4] Edwards, D.C. and T.B. McKee. Characteristics of the 20th Century drought in the United States at multiple time scale. Climatology Report No. 97-2, Colorado State University, Fort Collins, Colorado, 1997.

[5] Panofsky, H. A., and G. W. Brier. Some Applications of Statistics to Meteorology. Earth and Mineral Sciences Continuing Education, College of Earth and Mineral Sciences, The Pennsylvania State University, University Park, Pennsylvania, 224 pp, 1958.

Evaluating the Efficiency of the Carnot Cycle with a van der Waals Gas

Haiduke Sarafian
Pennsylvania State University
has2@psu.edu

1 Abstract

The efficiency of a perfect heat engine running with an ideal gas operating between two heat reservoirs is optimized by Sadi Carnot. It is known that the efficiency is independent of the working body and depends on the temperatures of the two heat reservoirs only. The detail of the proof, however, in most of the approaches relies implicitly on the assumption that the gas is ideal. The author revisits the concept and has generalized the proof assuming the working body to obey the nonideal gas law. Pedagogically, with this realistic explicit model it is shown that the efficiency of the cycle stays intact. *Mathematica*'s symbolic calculation power is deployed throughout the work. A *Mathematica* package is crafted to automate and evaluate the efficiency of the cycle. This package can prove useful in teaching thermo physics and engineering college courses.

2 Introduction

In introductory physics and chemistry courses, gases are assumed ideal e.g. Halliday, Resnick and Walker [1]. By introducing three macroscopic quantities, (volume, pressure and temperature), one describes analytically the general features of the gases. In terms of these three variables, one investigates the characteristics of special processes such as isochoric, isobaric and isothermal processes. By introducing heat, entropy and auxiliary quantities such as specific heats, one quantifies the behavior of the ideal gas undergoing processes such as adiabatic and isentropic processes. In intermediate college thermodynamic physics courses, by replacing the ideal gas with a nonideal gas, one routinely revisits the characteristics of the first three aforementioned processes. However, to the author's surprise, neither the popular intermediate Zemansky [2] nor the graduate Pathria [3] physics texts address the quantifications of the last two processes. Furthermore, the recent published survey of theoretical concepts in physics by Longair [4] and the specialized thermo physics monograph by Landsberg [5] have both ignored addressing the issue.

We considered a nonideal gas and have studied its characteristics when undergoing an adiabatic process. We have then applied the derived characterization to analyze a comprehensive practical problem, the Carnot cycle. In order to relate the body of the work to pedagogy of physics and chemistry, among the possible choices, the vastly used and familiar van der Waals gas law is chosen to describe the properties of the nonideal gases. To further its pedagogical objectives, *Mathematica*'s symbolic calculation power is deployed throughout the work. *Mathematica's* superb graphics are used to display for the first time the Carnot cycle working with a real gas. Furthermore, a *Mathematica* package is prepared to automate and evaluate the efficiency of the cycle.

3 An adiabatic process and the van der Waals equation

Here we consider a nonideal gas whose state is subject to van der Waals gas law:

$$\left(P+\frac{a}{V_m^2}\right)(V_m-b)=\mathrm{RT} \qquad \text{.....................(1)}$$

R is the gas constant and the pair (a, b) is specific to the gas and assumed constant. V_m is the molar volume and P is its associated pressure.
According to the first law of thermodynamics, $\mathrm{dQ} = \mathrm{dU} + \mathrm{dW}$. Assuming the internal energy of the gas is a function of volume and temperature, dU becomes $\mathrm{dU}(V, T) = \left(\frac{\partial U(V,T)}{\partial V}\right)_T \mathrm{dV} + \left(\frac{\partial U(V,T)}{\partial T}\right)_V \mathrm{dT}$. On the other hand, $\left(\frac{\partial U(V,T)}{\partial T}\right)_V = c_v$ where c_v is the specific heat at constant volume, and $\left(\frac{\partial U(V,T)}{\partial V}\right)_T = T\left(\frac{\partial P}{\partial T}\right)_V - P$, Zemansky [6]. Therefore, by solving eq(1) for P and substituting the solution into the right hand side of the last equation in the text, and substituting the element of work with $\mathrm{dW} = \mathrm{PdV}$, then deploying *Mathematica*, symbolically the first law simplifies to:

```
In[1]:= vdw[P_, V_, T_] = (P + a/V^2) (V - b)
          == R T;
        solP = Solve[vdw[P, V, T], P] // FullSimplify;
        dU = Dt[U[V, T]];
        dU = dU /. {dU[[1, 2]] -> cv,
                 dU[[2, 2]] -> T D[P /. solP[[1]], T] - P /. solP[[1]]} /.
              dU[[1, 1]] -> dT /. dU[[2, 1]] -> dV;
        dQ = dU + (P /. solP[[1]]) dV // FullSimplify
```

.....................(2)

```
Out[5]= dV R T/(-b + V) + dT cv
```

$$\text{Out[5]=}\ \frac{\mathrm{dV\,R\,T}}{-\mathrm{b}+\mathrm{V}}+\mathrm{dT}\,\mathrm{c_V}$$

.....................(3)

For an adiabatic process there is no heat exchange, $dQ = 0$, and eq(3) becomes $0 = c_v\,dT + \frac{RT}{v-b}\,dv$. Substituting the solution of this equation for T in eq(1) gives an equation describing the adiabatic process. Deploying *Mathematica* to do this:

```
In[7]:= eqTV = ∫ 1/T dT == - R/cv ∫ 1/(V - b) dV + Log[c1] /.
          {x_ Log[y_] :> Log[y^x]} /.
         {Log[x_] + Log[y_] :> Log[x y]};
       solT = Solve[eqTV, T];
       (P /. solP[[1]]) /. T -> T /. solT[[1]];
       adiabatic[P_, V_] = (P + a/V^2) (V - b)^(1+R/cv) ==
         C1
```

.....................(4)

$$\text{Out[11]=}\ \left(\mathrm{P}+\frac{\mathrm{a}}{\mathrm{V}^2}\right)(-\mathrm{b}+\mathrm{V})^{1+\frac{\mathrm{R}}{\mathrm{c_V}}} == \mathrm{C}_1$$

.....................(5)

Eq(5) is arranged so that $C_1 = Rc_1$, where c_1 is the integration constant. This equation closely resembles the original van der Waals equation and hence it is easy to memorize. For an ideal gas the value of the pair $(a, b) \to 0$ and the exponent $1 + \frac{R}{c_v} \to \gamma$, where $\gamma = \frac{c_p}{c_v}$. In this limit eq(5) reduces to the well known adiabatic equation of an ideal gas, $pv^\gamma = C$. Theoretically, based on the modeled structure of the gas constituents, the value of γ can be evaluated.

4 Efficiency of the Carnot cycle and the van der Waals equation

In the previous section we have studied the implication of the state of a nonideal gas in describing an adiabatic process. However, for an isothermal process, i.e. a process with sustained temperature, the van der Waals gas law is the relationship between the relevant macroscopic variables V and P. We have focused on the characteristics of these two processes only because they are the building blocks of the Carnot cycle. One might recall that the Carnot cycle is composed of two alternative pairs of *i*sothermal and *a*diabatic processes (*iaia*). It is also worth recalling that the efficiency of any reversible cycle is correlated with the efficiency of the Carnot cycle. In Halliday, Resnick and Walker [7] and Mandl [8] however, the ideal gas law is implicitly assumed to graphically describe the cycle. This leaves an impression that the implication of the ideal gas law is crucial to the proof. Therefore, among other objectives, the pedagogical goals of this work are to display on a P-V diagram a Carnot cycle operating with a nonideal gas, and to apply *Mathematica* to evaluate its efficiency. Here, formally, we present the analyses; then we will close the work with a numerical example.

Figure 1 (last page) shows a Carnot cycle running with a nonideal gas. Its P-V diagram is composed of: an isotherm $\overline{12}$ followed by an adiabatic $\overline{23}$ branch, an isotherm $\overline{34}$ and, closing in the cycle, an adiabatic $\overline{41}$. In this reversible cycle, since there is no change in the internal energy, according to the first law, the work done is equal to the net exchanged heat. Therefore, the ratio of the net work done to the input work is the same as the ratio of the net exchanged heat to the intake heat. Hence, the efficiency is $\epsilon = \frac{\text{net exchanged heat}}{\text{intake heat}}$. The net exchanged heat is $Q_{12} - Q_{34}$ and hence $\epsilon = \frac{Q_{12}-Q_{34}}{Q_{12}} = 1 - \frac{Q_{34}}{Q_{12}}$. It is standard practice to deploy the ideal gas law and deduce the equality $\frac{Q_{34}}{Q_{12}} = \frac{T_{\text{low}}}{T_{\text{high}}}$, yielding $\epsilon = 1 - \frac{T_{\text{low}}}{T_{\text{high}}}$.

Here we show that this holds true even if we apply the van der Waals gas law. The calculation, however, is done by deploying *Mathematica's* symbolic calculation power.

For an isothermal process dT = 0 and eq(3) gives $dQ = \frac{dV\,RT}{-b+V}$. Integrating both sides of this equation yields the value of the exchanged heat.

```
In[12]:= eqQ = ∫_{Vi}^{Vf} (R T)/(V - b) dV == ∫_{Qi}^{Qf} dQ /.
         {Log[x_] - Log[y_] :> Log[x/y],
          Qf - Qi -> Qif}
```

....................(6)

```
Out[13]= R T Log[(-b + Vf)/(-b + Vi)] == Qif
```

....................(7)

For the two isotherms $\overline{12}$ and $\overline{34}$, this gives:

```
In[14]:= eqQ12 = eqQ /. {vf -> v2, vi -> v1, Qif -> Q12, T -> Thigh};
         eqQ34 =
            eqQ /. {vf -> v3, vi -> v4, Qif -> Q34, T -> Tlow};
```

....................(8)

and the heat-ratio becomes:

```
In[16]:= heatratio = eqQ34[[1]] /
            eqQ12[[1]]
```

....................(9)

$$\text{Out[17]=}\ \frac{\text{Log}\left[\frac{-b+v_3}{-b+v_4}\right] T_{low}}{\text{Log}\left[\frac{-b+v_2}{-b+v_1}\right] T_{high}}$$

....................(10)

Now let us form the connecting equations. Vertices 1 and 2 are connected by an isotherm, as are vertices 3 and 4. Equating the isothermal equations connecting these vertices yields:

```
In[18]:= isoteq12 = vdw[p1, v1, T1][[1]] == vdw[p2, v2, Thigh][[1]];
         isoteq34 = vdw[p3, v3, T3][[1]] ==
            vdw[p4, v4, Tlow][[1]];
```

....................(11)

Similarly, vertices (2,3) and (4,1) are related by adiabatic processes. We get:

```
In[20]:= adiaeq23 =
           adiabatic[p2, v2][[1]] == adiabatic[p3, v3][[1]];
         adiaeq41 = adiabatic[p4, v4][[1]] ==
            adiabatic[p1, v1][[1]];
```

....................(12)

Now let us form the denominator of the heat-ratio.

```
In[22]:= adiaeq23[[2]]/adiaeq41[[1]] == adiaeq23[[1]]/adiaeq41[[2]];
```

....................(13)

On the other hand,

```
In[23]:= isoteq12[[2, 1]]/isoteq12[[1, 1]] == isoteq12[[1, 2]]/isoteq12[[2, 2]];
```

....................(14)

Similarly,

```
In[24]:= isoteq34[[1, 1]]/isoteq34[[2, 1]] == isoteq34[[2, 2]]/isoteq34[[1, 2]];
```

....................(15)

Applying these identities, the left-hand-side, *lhs*, and the right-hand-side, *rhs*, of eq(13) are:

```
In[25]:= lhs = adiaeq23[[2]]/adiaeq41[[1]] /.
           {isoteq34[[1, 1]]/isoteq34[[2, 1]] :> isoteq34[[2, 2]]/isoteq34[[1, 2]]};
```

....................(16)

Similarly,

```
In[26]:= rhs = adiaeq23[[1]] / adiaeq41[[2]] /.
           {isoteq12[[2, 1]] / isoteq12[[1, 1]] :> isoteq12[[1, 2]] / isoteq12[[2, 2]]};
```

......................(17)

Raising both sides of the output of eqs(16) and (17) to the power of $\frac{C_v}{R}$ gives:

```
In[27]:= qed = PowerExpand[lhs^(Cv/R)] == PowerExpand[rhs^(Cv/R)];
```

......................(18)

and hence the heat-ratio becomes:

```
In[28]:= heatratio /. qed[[1]] :> qed[[2]]
```

......................(19)

$$\text{Out[29]=}\ \frac{T_{low}}{T_{high}}$$

......................(20)

This concludes the proof of the Carnot theorem with explicit usage of the nonideal gas law.

For a chosen set of parameters, (a, b) and (c_1, c_2), a *Mathematica* package is crafted to automate the plotting of Figure 1. The package also applies two different methods to evaluate the efficiency of the cycle. In the first method, which is a geometrical approach, the program integrates the analytic isotherm and adiabatic equations by setting the vertices of the cycle within the boundaries. This gives the area of the shaded piece of Figure 1, corresponding to the value of the net work. In the second method, a physical approach, the program applies $eqQ_{12} - eqQ_{34}$, given by eq(8), and evaluates the same quantity. The program then applies this value and forms the ratio needed to evaluate the efficiency.

For demonstration purposes only, and in order to display the features of a nonideal gas and to emphasise its characteristic differences from an ideal gas, a set of suitable parameters is chosen. These are: $(a, b) \to (5.46\,, 0.03\,)$, $(c_1, c_2) \to (44.8, 50)$ and $1 + \frac{R}{c_v} = 1.4$. The area of the shaded piece is evaluated as 1.76 bar.ml, yielding a 10% efficiency.

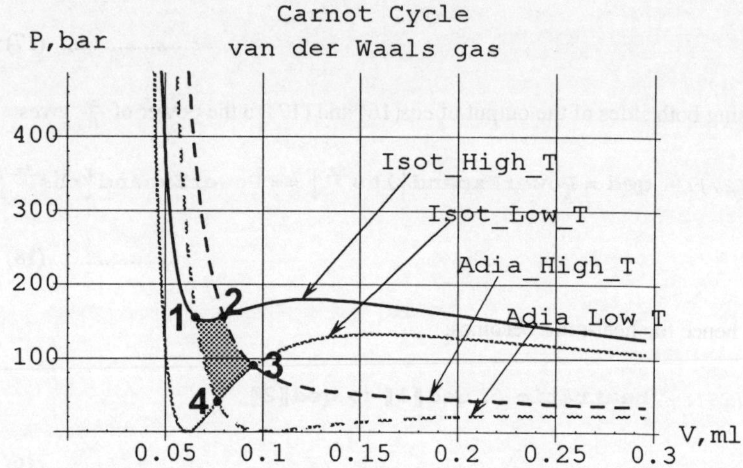

Figure 1

5 References

[1] Halliday, D., Resnick, R. and Walker, J. Fundamentals of Physics (chapter 19-21), 2001, John Wiley and Sons, Inc., Vol 2, 6th edition.

[2] Zemansky, M.W. Heat and Thermodynamics, 1968, McGraw-Hill.

[3] Pathria, R.K. Statistical Mechanics, 1978, Pergamon Press.

[4] Longair, M.S. Theoretical concepts in physics, 1994, Cambridge.

[5] Landsberg, P.T. Problems in thermodynamics and statistical physics, 1971, Pion-Limited.

[6] pg 291, reference [2].

[7] pg 489, reference [1].

[8] Mandl, F. Statistical physics, (chapter 5, pg 122), 1976, John Wiley and Sons, Inc.

Stepwise maths exercises on the web

Giovannina Albano
University of Salerno, Italy
albano@diima.unisa.it
Bernardo D'Auria
University of Salerno, Italy
dauria@diima.unisa.it
Saverio Salerno
University of Salerno, Italy
salerno@unisa.it

1 Abstract

In this paper we want to show an e-learning application, MoMAM@th, to support teaching/learning of mathematics on the web. MoMAM@th exposes to the user an infinite set of interactive exercises, randomly generated on-the-fly, referring to specific topics. The innovation of this application is in using exercises that are split into simple steps. At each step the software guides the student towards the solution of the problem and the student is required to interact, filling some blank boxes. In this way MoMAM@th can follow the reasoning of the student and at the same time helps him/her in each single step. WebMathematica is the engine used to develop MoMAM@th.

2 MoMAM@th overview

The learning process produces competencies, which can be classified as follows: know-what, know-how, know-being. Many studies (Duffy et al, 1993; Minsky, 1986; Bateson, 1973) have shown the importance of the active and constructive role of the learner in the learning process; this is particularly true in acquiring know-how, because the necessity to put oneself into play for the learner becomes stronger. The process of acquiring competencies can be split into three phases: the acquisition of the information, the personal elaboration and the verification.

MoMAM@th can be classified as a cognitive tutor: it aims to acquire competencies in basic mathematics, with emphasis on the know-how, even if it impacts all the previous phases in different ways, as we will see in the following. A set of rules about domain knowledge allows the system to perform the task required of the student, so that the student solution is compared with the system solution step-by-step. Errors are promptly detected and suitable feedback is given.

MoMAM@th generates exercises on the fly any time s/he requires. The underlying algorithm are constructed so that the exercises generated are consistent from a didactical viewpoint and are different each time.

The main idea of MoMAM@th is to allow the students to perform exercises split into "elementary" steps. At each step the software suggests elementary operations to do and provides the possibility of consulting theory references. The student is required to interact with the software, completing some blank boxes in order to do computations or to take into account theoretical results. By *elementary step* we mean a task which is considered well-known (that is a pre-requisite) or it is an exercise presented in MoMAM@th (that is it has been split into simple parts). In the latter case, if the student fails that step, s/he is sent back to a suitable exercise in order to perform that step in detail again. This decomposition is not formalized in the application, but it could be easily represented by means of an ontology and the process of solving the elementary single steps could be automatized, letting the user open the linked exercise instantiated with the data of the studied example. This may be seen as an upside-down procedure for solving the exercise that the student may use in case of difficulty.

MoMAM@th is accessible through a common web browser. The student accesses a personal working area via a login and password where the available topics and various types of related exercises are presented by means of a user-friendly interface with menu and buttons. The exercises are classified by topic and subject. Students can simulate an exam, which is a session with exercises randomly generated from chosen topics. Students are required to give only the final answer and, if s/he fails, s/he can access the step-by-step learning path.

Finally MoMAM@th currently offers on-line teacher support in an asynchronous way (by email), and synchronous tools (e.g. chat) are planned.

3 Learning Features

In this section we want to illustrate how MoMAM@th works from didactical viewpoint. Even if MoMAM@th was not originally intended for self-evaluation, it offers self-evaluation features (Trinchero, 2002), due to its structure. In fact, at each step the student is required to involve him/herself in completing some blank boxes and

MoMAM@th gives automatic correction and a prompt feedback of the correctness of his/her answers. Moreover, if the student fails, the software helps him/her to identify his/her difficulties or gaps precisely. It suggests recovery paths, refering the user back to suitable theoretical results or previous exercises. Actually each step of an exercise, because of its interactivity, can be thought of as questions that monitor competencies involving computational and logic abilities, rather than testing simple knowledge. Indeed, in this way, mastery of the knowledge is indirectly checked, so competencies of higher level are detected. In order to make clear what we mean and how it has been realised in MoMAM@th, we will show some examples.

Consider the following exercise.

"Determine the dimension and a basis of the kernel of the homomorphism"

```
f (x1, x2, x3, x4) = (-4 x1 - 4 x2 + 2 x3 + 2 x4,
  2 x1 - x2 - x3 - 5 x4, -5 x2 - 3 x3 - 2 x4)
```

At the first step the software requires a computation of the kernel of the linear system defined by f. Suppose that the student gives the following system:

```
-4 x1 - 4 x2 + 2 x3 + 2 x4 = 0
2 x1 - x2 - x3 - 5 x4 = 1
```

The software detects two theoretical mistakes: the definition of the kernel and the representation of a vector subspace. So the given warnings are: "Pay attention: the number of the inserted equations is not sufficient to represent the kernel of f. Moreover the linear system is not homogeneous. You should recover what is the kernel of a homomorphism and how to represent it. Then try again to construct the system."

At any step, the first time the student fails, s/he is obliged to try again, and is not able to see the solution. We chose this way in order to get the student to involve him/herself in solving the problems. In fact it is well known through teaching experience that, if solutions are available, the students are not much motivated to try to solve.

Back to the above exercise, at the second step, the student is required to compute the dimension and a basis of the kernel. Suppose the student gives the answer:

```
dim Ker f = 1
B_Ker f = {{1, -2, 0, 5}, {-1, 1, 0, 2}}
```

Here the system detects two computational errors. The dimension of the kernel is wrong and the elements in the basis are not in the kernel of f. Such information makes it clear that knowledge of the relationship between the rank of matrix and solving a linear system is missing. Also, the inconsistency between the given dimension of the kernel and the number of vectors in the given basis shows a theoretical gap in knowledge about the definition of the dimension of a vector space! The software gives the following computational and theoretical warnings.

"The dimension of Ker f is not correct: see how to compute the rank of a matrix and recall that the dimension of the NullSpace of a linear system is equal to the difference between the number of variables and the rank of the coefficients matrix. The inserted vectors are not in the kernel of f: see how to solve a homogeneous linear system. The inserted dimension is not equal to the length of the given basis: the dimension of a vector space is exactly the number of vectors of any basis of the space. You should also see better the related theory."

Note that, in this exercise, solving a linear system is considered an elementary step. This is because in a previous session such a procedure was explained step by step.

Let us see another exercise.

"Determine the dimension and a basis of the Image of the homomorphism"

```
f (x1, x2, x3, x4) = (-5 x1 - x2 - 5 x3 - 2 x4,
    2 x1 + 3 x2 + 4 x4, -21 x1 + x2 - 2 x3 - 2 x4)
```

At the first step the student is required to give a set of generators of Im f. The student has to insert the dimension and a basis of Im f. Suppose the answer is:

```
dim Im f = 3
B_Im f = {{-5, 0, -2}, {-3, 0, -2, 1}, {10, 0, 4}}
```

With respect to these answers, MoMAM@th establishes that the given dimension is correct and it is consistent with the given number of vectors, but it also detects conceptual and theoretical mistakes due to the fact that the vectors are linearly dependent and one of them is not in the environment vector space. This establishes that the student does not know the definition of basis and moreover s/he does not reflect on the context in which s/he actually works. Here is the feedback from the software. "The dimension of the space Im f is correct, but the given basis is not. The length of the inserted vectors is not always correct because Im f is a vector subspace of R^3. The inserted vectors are linearly dependent while a basis is constituted by linearly independent vectors. See better the theory and insert the missing values."

4 MoMAM@th Architecture

This section deals with the technical and architectural details of MoMAM@th, and gives an overview of the applications we integrated and in which way they are used. MoMAM@th essentially may be divided in two parts. The first is an accounting and delivery part, which uses the application uPortal, developed by JA-SIG. The second is an algorithmic part, which uses *Mathematica* for the execution and Web-*Mathematica* for deployment to the web.

uPortal is a free and sharable application used to develop web-portals. It is particularly suited for didactical purposes and so is often used by universities. It allows the web-page to be divided into different areas, each one independent of the other, by means of a dedicated channel. More channels can be collected together in folders.

MoMAM@th actually uses uPortal only in a simplified way. We used it together with Microsoft IIS (the web-server application), Tomcat, (the servlet engine necessary to execute uPortal servlets) and Microsoft SQL Server, which contains all the accounting information. The folders are used to manage the different sections. Each one is associated with a particular topic, e.g. Limits, Derivatives, Matrices, etc. After the user has selected one of them, MoMAM@th coherently activates the only channel we developed. It comprises a menu area and an exercise area. The menu area shows the full list of available exercises, and each item is linked to a specific WebMathematica MSP page or to a set of MSP pages. In the latter case the selected MSP page is randomly chosen. In this way we are able to develop different kinds of exercises, in terms of algorithms, that all relate to the same subject, e.g. exercises about Gauss and Rouché-Capelli methods for the solutions of a linear system.

After selecting a section the exercise area is filled with the last exercise in the section that the user accessed. The selected exercise appears in the same state that it was left in when it was last accessed.

The exercises appear in the exercise area and consist of a sequence of pages generated by Web-*Mathematica,* containing the text, graphs, formulae and tools necessary to solve the specific problem. As the Figure 1 shows, a single exercise is made up of a set of files. Some of them, the MSP page and some *Mathematica* Packages are fixed, and are identical for all the exercises. The only file different for each exercise is "Template.m", which contains the logic for the specific exercise. This choice was adopted in order to make the task of content development easier. The MSP page always calls the same functions in the "Hidden.m" package, which redirects them to the package "Template.m". In order to manage the state and the type of the exercise, the function called receives the parameter $$exercise that contains the exercise type and subtype numbers, and the current step number.

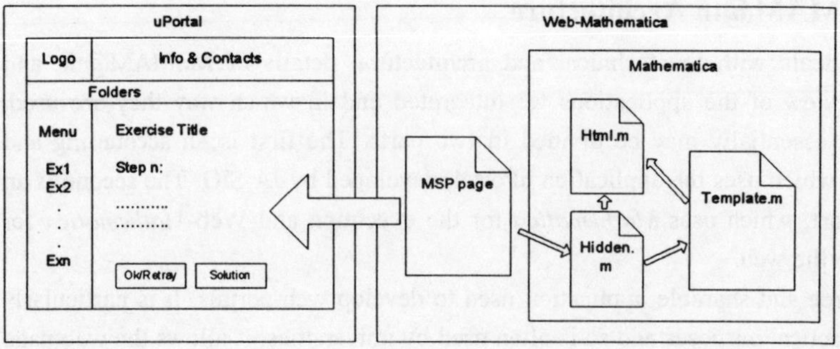

Figure 1

The MSP page initially calls the function Exercise, which updates the value of the variable $$exercise. Then the MSP page calls the function StepNumbers, which reports how many steps there are in the exercise. In this way MoMAM@th can manage exercises with a variable number of steps and/or with a number of steps that depends on the content of the exercise.

It's important to note that the contents of the Exercise, i.e. all the information needed to execute it, are randomly generated before the exercise is presented, and are stored statically in the variable $$param for the duration of the Exercise. This choice was taken in order to avoid passing data between steps and so to avoid difficult serialization of data. Indeed, in order to store an exercise with all its history it is necessary to store only the last value of $$exercise (which gives the step number), the constant value of $$param (which essentially is the exercise) and the list of the $$string values (which are the user's answers at each step). In the next section we will explain the contents of the $$param variable in more in detail, and give some examples of random generation. At the moment the mathematical formulae are shown by gif images and user input is in textual *Mathematica* syntax using input text boxes. In order to simplify the input, text boxes substitute only part of formulae. For example they can be used to supply the numerator of a fraction but not the denominator, or they can be arranged in an array to simulate a matrix. The content developer can add input boxes everywhere in the page using standard functions in the "Html.m" package. Next, the values of the text boxes are collected in the variable $$string by standard javascript. When the uses clicks a button on the page, these values are passed to the functions of the "Template.m" package as parameter, in order to evaluate the user's answers.

5 Exercise Logic

As we explained in the previous section, the entire logic of an exercise is contained in the package "Template.m". The content developer has only to respect a calling interface in order to create the logic for a given set of functions. In the following we report the set of functions called by "Hidden.m".

```
Exercise[t, tt]                   PrintStep[t, tt, s][p]
StepNumbers[t, tt][p]             Print Solution[t, tt, s][p]
Title[t, tt][p]                   PrintRightAnswer[t, tt, s][p]
ExerciseStatement[t, tt][p]  PrintSummary[t, tt, s][p]
                                  PrintVerification[t, tt, s][p
```

where the parameter list is:

```
t   Exercise type ($$exercise)
tt  Exercise subtype ($$exercise)
s   Number of the step ($$exercise)
p   Exercise parameters ($$param)
a   User answer ($$string)
```

The most important function is Exercise. Together with StepNumbers, it defines the exercise statement. As an example we give exercise [9,3], which concerns matrix diagonalization. The code shown was translated into English and adapted for this paper. Exercise is called before the exercise starts and creates random data for the exercise.

```
Exercise[9, 3] := Block[{A, P, b, c},
  A = If[Random[] < 0.25, b = Random[Integer, {-5, 5}];
    c = Random[Integer, {-5, 5}];
    {{0, b, 0}, {-b, 0, 0}, {0, 0, c}},
    DiagonalMatrix[
     Table[Random[Integer, {-5, 5}], {i, 1, 3}]]];
  P = Table[Random[Integer, {-2, 2}],
    {i, 1, 3}, {j, 1, 3}];
  While[Det[P] == 0, P = Table[Random[Integer, {-2, 2}],
     {i, 1, 3}, {j, 1, 3}]];
  Dot[Inverse[P], A, P]]
```

It is important to note that we usually generate the answer to the exercise in order to formulate the problem. In the example above we start by creating the eigenvalues of the matrix. In the 75% of the cases they are all real integers in the interval [-5,5]. In the remaining cases one is real and two are imaginary. The so generated matrix A is then post-multiplied by a matrix P and pre-multiplied by P^{-1} in order to get a matrix in a non-trivial form. This method gives us the opportunity to simplify the computation for the user. In this case, s/he will always find integer solutions.

The parameters of the function *Exercise* are the type (9) and subtype (3), and the returned value is simply the resulting matrix. We can get all the needed information to identify and execute the exercise using the *StepNumber* function, as below.

```
StepNumbers[9, 3][param_] :=
  Block[{l = Eigenvalues[param]},
    If[Select[l, (Im[#] =!= 0) &] === {},
      Length[Union[l, l]], 2]]
```

This uses the *param* parameter to find the exercise, i.e. the matrix. After checking for imaginary eigenvalues it answers 2 if there are any. Otherwise it gives a number of steps equal to the number of distinct real eigenvalues.

6 Future work

As a conclusion we want to outline what we plan for the future. Beyond the update of content, we plan to personalise the user working area by providing collaborative tools, an agenda, bookmarks, and a notepad, and to manage user profiles. Moreover, as most of the difficulties encountered by users are with mathematical input, we are experimenting with a new version using WebEQ to help with editing formulae by means of a graphical syntax. Formulae will be coded in a standard way using MathML.

7 References

Albano G., D'Apice C., D'Auria B., Salerno S. (1999) Statistical interpretation of mechanical models by using simulation - Proceedings ICSEE 2000 - pp. 215-220

Bateson G, (1973) Stes to an Ecology of Mind. Paladin Books

Duffy T., Jonassen D., Lowyck J. (Eds). (1993) Designing constructivist learning environments. Heidelberg, FRG: Springer-Verlag.

Minsky M. (1986) The society of Mind. New York: Simon and Schuster.

Trinchero R. (2002) Valutazione dell'apprendimento e formazione assistita dalla rete. In Docimologia, Coggi and Notti, PENSA Eds.

Design Science, Inc. (MathType, WebEQ, MathPlayer), http://www.dessci.com

MathML, http://www.w3.org/Math/

uPortal, http://mis105.mis.udel.edu/ja-sig/uportal/

Wolfram Research (Mathematica, WebMathematica), http://www.wolfram.com/

Enhancing Mathematical Teaching-Learning Process by *Mathematica*

C. D'Apice, R. Manzo, V. Tibullo
Department of Information Engineering and Applied Mathematics, Uniersity of Salerno, Italy
dapice@diima.unisa.it, manzo@diima.unisa.it, tibullo@diima.unisa.it

1 Abstract

In this paper we aim to show a *Mathematica* notebook based on a package built to demonstrate the working of *Linear Time Independent Systems* (LTI Systems).

This notebook lets the user play with a virtual LTI system, where he/she can choose things such as the input signal and the system transformation given as the system response to a pulse signal, and see in real time the shape of the output signal. The user can also see the response of the LTI system to canonical signals, or also the signal in the frequency domain, showing its frequency response in Bode or Polar diagrams. The pedagogical assumption we followed in the development of the package is that a student learns better and in a more stimulating way using the strategy of learning by doing. Manipulating data and observing the results of their manipulation a learner can reach a deeper knowledge of the subject. We well explain in more detail how the package works in the following sections.

2 Linear Time Independent Systems concepts

In the Theory of Signals, a *system* is a mathematical model of a process that produces an output signal in response to an input signal, where a *signal* is a real valued continuous or discrete function representing some quantity of the observed real world modeled in the system. We will represent a system as a black box, because we are not interested in the internal intricacies of the system. Many real world objects can be adapted to this model with a suitable definition of signal. In our case, we define a signal as a real valued function of time. Linear Time Independent systems are systems that have the further properties of linearity and time independence. A *linear system* is an additive and homogeneous system, where *additive* means the output signal of the sum of two input signals is the same of the sum of the two output signal, and *homogeneous* means that if the input signal is multiplied by a given factor, the output is multiplied by the same factor. *Time independent* means a translation in time of the input signal corresponds to an equal translation of the output signal.

3 Notebook Overview

In this section we will describe the realized notebook. When opened in *Mathematica*, it will show an initial frame with three buttons: the first to load the core package providing the required features, the second to begin the time analysis of a signal, and the third to perform the frequency analysis. After loading the package, we can go on with the time analysis, which shows another series of buttons, with the following functions:

Enter Input Signal

Enter Pulse Response

View Output Signal

Plot Output Signal

Common Signals Response

The last one gives the user a chance to see the output corresponding to commonly used input signals, among others the unit step and the sine.

If we choose to do a frequency analysis, the notebook will show another series of buttons, with the following functions:

Enter Transfer Function

View Frequency Response

Plot Polar Diagram

Plot Bode Diagram

Let now examine the notebook functions in more detail.

4 An Interactive Time Domain Analysis Session

When we open the notebook, an initial frame will be shown, giving the user the choice of performing a Time Domain analysis or a Frequency Domain analysis. The initial frame is shown in the following figure:

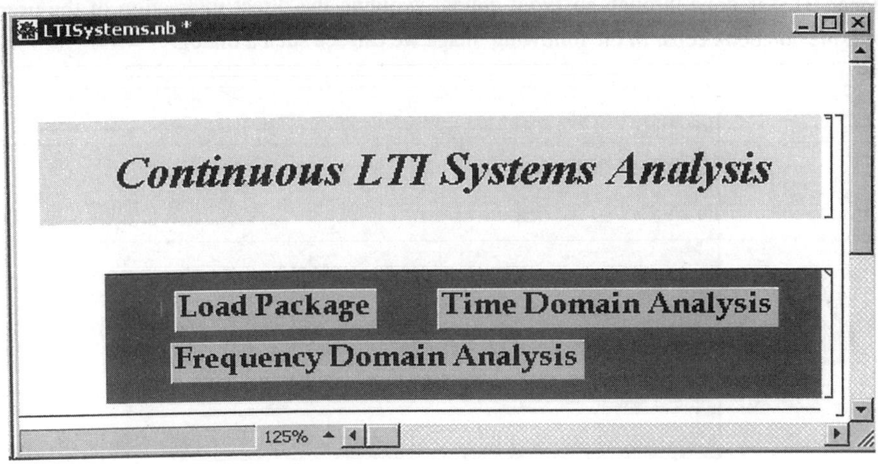

Figure 1

The user is presented with a choice. The first thing he/she should do is to load the package; next he/she can choose the button *Time Domain Analysis*, and the notebook will show the following frame:

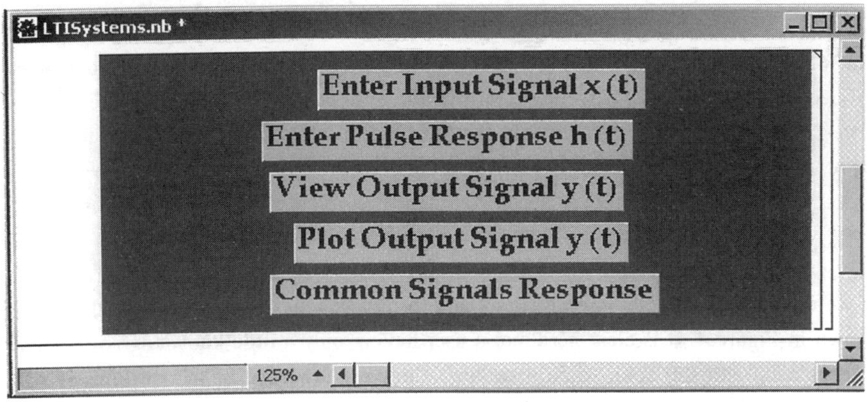

Figure 2

Now the first two buttons have to be selected, to instruct the notebook what type of system we are concerned with and what type of signal we want to transform. If we are interested to see the system response to a given input signal, we have to enter the input signal. This is done through an *Input* dialog, avoiding the direct interaction of the user with the notebook cells. In the following image we can see such a dialog.

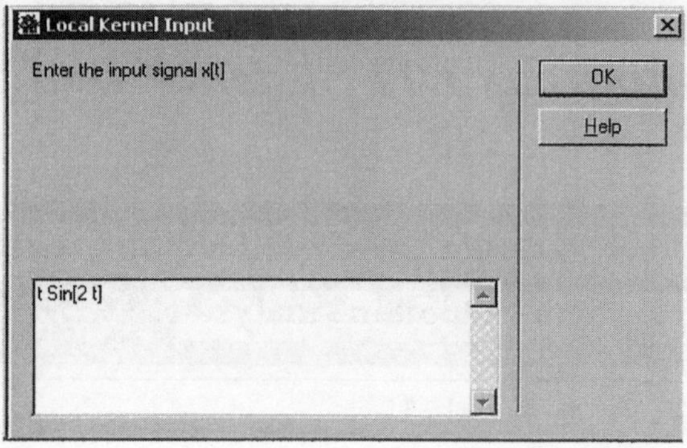

Figure 3

As an example, in this sample session we want to see the output corresponding to the following input signal

```
t Sin[2 t]
```

We follow a standard procedure to specify to the package the system we want to analyze. To accomplish such a task, we have to give the system pulse response, by pressing the *Enter Pulse Response* button. In this sample session, we choose to give the following input:

```
t^2
```

If we next ask to view the output signal and to plot this signal, we will have the output shown in Figure 4. It is clear here that a user having only a little knowledge of Systems Theory can, with this tool, reach a certain intuitive knowledge of the working of such systems, and by trying to put in different input signals and studying the pulse responses, he can see directly and in real time what the output will be. This tool can also be useful for a technician who already has a working knowledge of the phenomenon we are studying, because he can play with the system parameters both in the time and in the frequency domain, as we will see in the following of this paper.

The following figure shows what type of output we can expect from the notebook corresponding to our input.

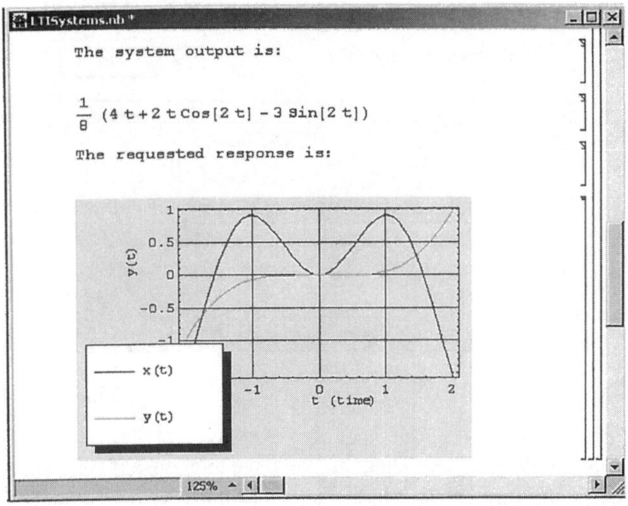

Figure 4

5 A Frequency Domain Analysis Session

If in the initial frame we select the button *Frequency Domain Analysis*, we are faced with the following choice of buttons:

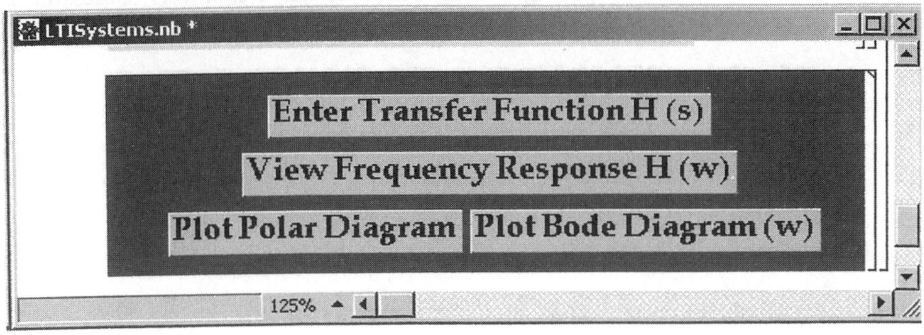

Figure 5

Let us choose the first button, so we can tell the notebook what the System Transfer Function is.

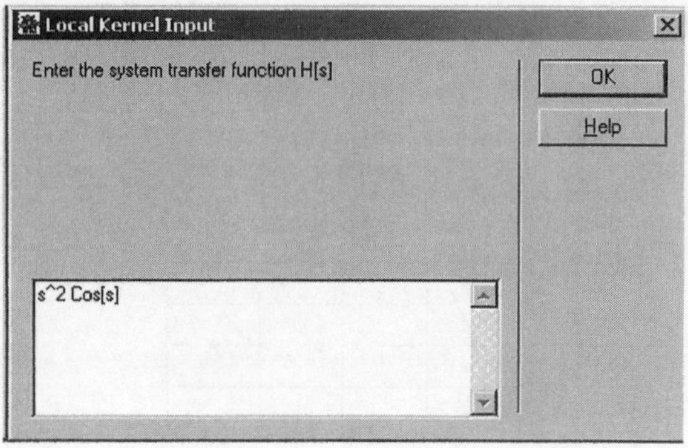

Figure 6

When we click on the button Bode Diagram, we have to choose if we want to see the plot of the Modulus or the plot of the Phase. Selecting the first one, Figure 7 (see below) shows the output we will see.

6 Conclusions

Mathematica offers great possibilities for producing efficient didactical packages. Its use together with traditional didactical lectures allows the learner to reach a deep knowledge of the subject of study developing student intuition and curiosity. Our experience of the utilization of this *Mathematica* package during University lectures is very successful; in particular students appreciate its potentialities for visualization.

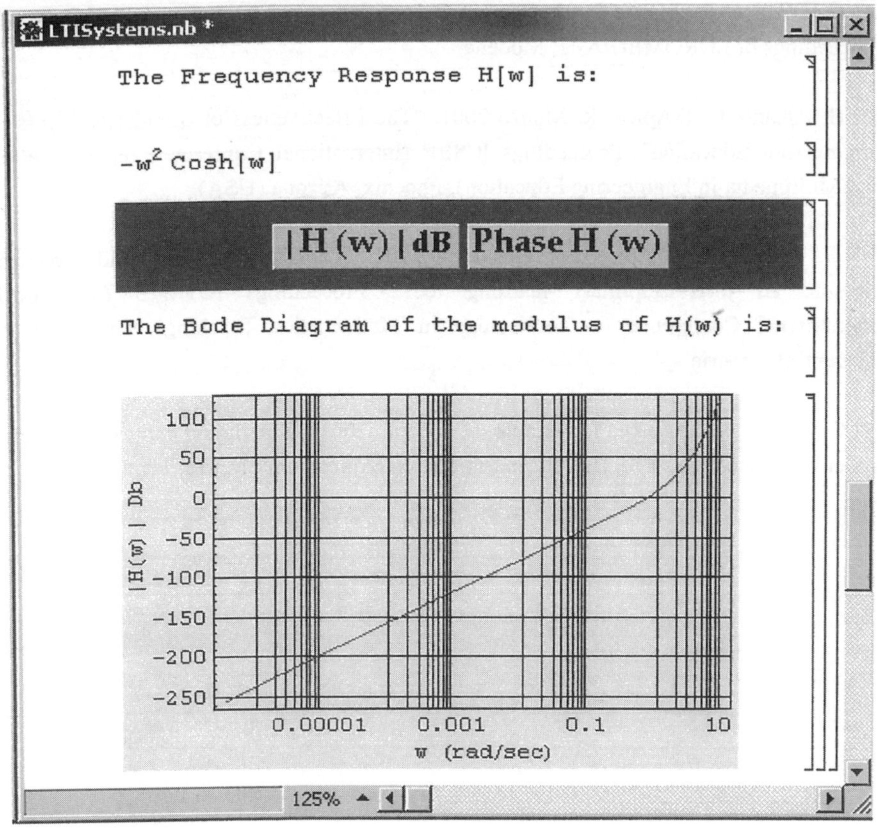

Figure 7

7 References

[1] A. A. Gokale 1996. "Effectiveness of Computer Simulation for Enhancing higher order thinking". Journal of Industrial Teacher Education, 33 (4), 36-46.

[2] P. Goodyear, M. Njoo, H. Hijne, Van Berkham 1991. "Learning processes. Students attributes and simulation". Education and Computing.

[3] H. Ruokano, S. Pohjolanein 1999. "Pedagogical principles for evaluation of hypermedia-based learning environments in mathematics". J. UCS, Vol. 4, n. 3.

[4] C. D'Apice, R. Manzo 2002. "To reach confidence in Maths *Learning by doing*". Proceedings of EUROMEDIA'02, Modena.

[5] G. Albano, C. D'Apice, R. Manzo 2001. "The Effectiveness of Simulation Tools in Engineering Education". Proceedings ICSEE (International Conference on Simulation and Multimedia in Engineering Education), Phoenix, Arizona (USA).

[6] G. Albano, C. D'Apice, M. Desiderio 2001. "Laplace Transform and electrical circuits: an interdisciplinary learning tool". Proceedings ICTMT5 (The Fifth International Conference on Technology in Mathematics Teaching), University of Klagenfurt, Austria.

[7] G. Albano, C. D'Apice, T. Durante, S. Salerno 2001. "Simulation games: theory and systems". Proceedings ESS (European Simulation Symposium), Marseilles, Francia.

Is small small enough? Conceptualisation of the continuous by means of the discrete

Ivy Kidron
Jerusalem College of Technology
Jerusalem, Israel
ivy@jct.ac.il

1 Abstract

In this paper, we analyze the influence of applying the double strand discrete-continuous on College students' conceptual understanding of central notions in analysis such as limit and derivative. The research is done in the context of a course on differential equations. Our aim is to analyze how using *Mathematica*, while iterating the simplest mathematical expression such as quadratic functions, the students are introduced to the great beauty of the field of dynamical systems and to a better understanding of convergence problems.

2. Introduction

This study illustrates how the discrete-continuous analysis sheds new light on the solution of differential equations (the logistic equation). In some cases, as a consequence of applying discrete approximations, the nature of a problem changes significantly. In such cases, there may be a big difference between the analytical solution to the problem, and the computational solution obtained by means of discrete methods. Moreover, applying numerical discrete methods to solve the logistic equation may lead to unexpected results such as the following: the fact that a small change in a parameter causes only a small effect, does not necessarily imply that a further small change in the parameter will cause only a further small change in the effect. Once a specific value of the parameter, called the bifurcation value, is passed, the effect could be a significant one! This is contrary to the belief that gradual causes have gradual effects. This study investigates how to take advantage of such situations of conflict to help students understand central concepts in analysis such as limit and derivative. We analyze how students' awareness of the limitation of discrete numerical methods, combined with error analysis, lead to a better understanding of the continuous methods. We also investigate the role played by *Mathematica* in the students' conceptualisation of the continuous by means of the discrete.

3 Conceptualisation of the continuous and "treasured" intuitions

In previous studies concerning the way students conceived irrational numbers, Kidron & Vinner [3] observed that the infinite decimal is conceived as one of its finite approximations ("*three digits after the decimal point are sufficient, otherwise it is not practical*"), or as a dynamic creature which is in an unending (a *potentially infinite*) process: in each next stage we improve the precision with one more digit after the decimal point. Courant [2] wrote that if the concept of limit yielded nothing more than the recognition that certain known numbers can be approximated to as closely as we like by certain sequences of other known numbers, we should have gained very little from it. The fruitfulness of the concept of limit in analysis rests essentially on the fact that limits of sequences of known numbers provide a means of dealing with **other** numbers which are not directly known or expressible. Thus, the limit concept should lead to a new entity and not just to one more digit after the decimal point.

The derivative function is defined as $f'(x) = \lim_{h\to 0} (f(x+h) - f(x))/h$. By means of animation and elementary programming in *Mathematica*, students can visualize the process of $(f(x+h) - f(x))/h$ approaching $f'(x)$ for decreasing h. The dynamic picture might reinforce the misconception that one can replace the limit by $\Delta y/\Delta x$ for Δx very small. How small? If we will choose $\Delta x = 0.016$ instead of 0.017, what will be the difference? There is a belief that **gradual causes have gradual effects.** and that **small changes in a cause should produce small changes in its effect** (Ian Stewart, [6]). This "treasured" intuition might explain the misconception that a change of, say, 0.001 in Δx will not produce a big change in its effect.

4 The evolution of the concept of derivative

Kleiner [4] explains that the derivative (as tangent) was *used* by Fermat and others in the first half of the 17th century, was *discovered* by Newton and Leibniz (as fluxion and differential, respectively) in the later part of that century, was rigorously *explored* and *developed* in the 18th century, and was *defined* in the 19th. The definition of the derivative, too, was given in stages: by Lagrange in the 1790s (an algebraic definition), by Cauchy in the 1820s (in terms of limits and infinitesimals) and finally by Weierstrass in the 1870s (in terms of ϵ and δ). Kleiner concludes that before introducing rigorous definitions we have to demonstrate the need for higher standards of rigor. This could be done by introducing counterexamples to plausible and widely held notions. Thus, we were looking for a counterexample that will help to overcome the belief that gradual causes have gradual effects. We found the counterexample in the field of dynamical systems. A dynamical system is any process that evolves in time. The mathematical model is a differential equation $dy/dt = y' = f(t, y)$, and we encounter again the derivative $y' = \lim_{\Delta t \to 0} \Delta y / \Delta t$. In a dynamical process that changes with time, time is a continuous variable. In the numerical method, there is a discretization of the variable "time". Our aim is that the students will realize that in some differential equations the passage to a discrete time model changes totally the nature of the solution. We aim to help students realize that gradual causes do not necessarily have gradual effects, and that a difference of 0.001 in Δt might produce a significant effect.

5 Discrete numerical methods: Old methods, new tools

Numerical methods are not new. The founders of the mathematical theory developed numerical discrete methods to solve problems that relate to dynamical continuous processes. The old masters pointed out the importance of analyzing the error that arises from applying numerical processes. However, historically, the approximation procedures were mainly a theoretical tool. With *Mathematica,* the students can apply the numerical algorithms developed by the old masters, and analyze the accuracy of the discrete methods.

6 Analytical vs. numerical solutions of differential equations

We will describe some cases in which the analytical solution obtained by means of continuous calculus is totally different from the numerical solution obtained by means of discrete numerical methods.

Numerical methods - the method of Euler. The students were given the following task: a point (t_0, y_0) is given and the derivative of the function $dy/dt = f(t, y)$ is given. Plot the function $y(t)$. The students were asked to find the next point (t_1, y_1) by means of $\frac{y_1 - y_0}{t_1 - t_0} = f(t_0, y_0)$. As t increases by the small constant step $t_1 - t_0 = \Delta t$, the students realized that they are moving along the tangent line in the direction of the slope $f(t_0, y_0)$. The students generalized and wrote the algorithm: $y_{n+1} = y_n + \Delta t \; f(t_n, y_n)$ for Euler's method. They were asked how to better approach the solution. They proposed to choose a smaller step Δt. The students worked on some examples in which the differential equations have analytical solutions, and in those examples, indeed, when Δt was a small step the plot of the numerical solution and the plot of the analytical solution "coincided".

The logistic equation: do gradual causes have gradual effects? The logistic equation $dy/dt = r\, y(t)\,(1 - y(t))$, $y(0) = y_0$ was introduced as the dynamics of the growth of a population. An analytical solution exists for all values of the parameter r. Now, we consider the Euler's method numerical solution of the logistic equation. The relation $(y_{n+1} - y_n)/\Delta t = r\, y_n(1 - y_n)$ can be written $y_{n+1} = y_n + r\Delta t\, y_n(1 - y_n)$. In other words, we iterate the expression $y + R\,y(1 - y)$, where $R = r\Delta t$. To perform the iterations we use *Mathematica*'s `NestList` command. As a first example, we consider the numerical solution of the equation $\dot{y} = 18\,y(1 - y)$, $y(0) = 1.3$ with $\Delta t = 0.1$. (Here, $R = 1.8$.) We will observe what happens when Δt is increaed slightly. We give below the command for the first plot `p1`; the others are obtained by similar means.

```
In[1]:= step[{t_, y_}] := {t+1, y+R y (1-y)}
        euler[n_] = NestList[step, {0, 1.3}, n]

In[3]:= R = 1.8

In[4]:= p1 =
         ListPlot[euler[25], PlotJoined -> True, PlotStyle ->
            {PointSize[0.02], RGBColor[0, 0, 0.996109]},
          PlotLabel -> "Δt = 0.1"]

In[38]:= Show[GraphicsArray[
            {{p1, p2, p3}, {p4, p5, p6}, {p7, p8, p9}}] /.
          (Ticks -> Automatic) -> (Ticks -> None),
         Frame -> True]
```

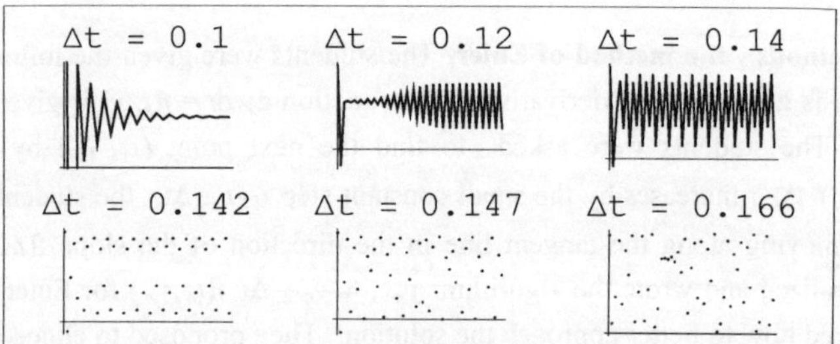


```
In[58]:= f[y_] := y + 18 * 0.15825 y (1 - y)
```

```
In[59]:= newdigits = NestList[f, Nest[f, 0.6, 1000], 30]

Out[59]= {1.29317, 0.213261, 0.691186, 1.29919,
  0.191956, 0.633784, 1.29493, 0.207062,
  0.674751, 1.29989, 0.189481, 0.626947,
  1.29317, 0.213261, 0.691186, 1.29919, 0.191956,
  0.633784, 1.29493, 0.207062, 0.674751,
  1.29989, 0.189481, 0.626947, 1.29317, 0.213261,
  0.691186, 1.29919, 0.191956, 0.633784, 1.29493}
```

In the first plot, the solution tends to 1 and looks like the analytical solution. In the second, third and fourth plot, the process becomes a periodic oscillation between two, four and eight levels. In the fourth plot, we did not join the points, in order that this period-doubling will be clearer. In the fifth and sixth plot, the logistic mapping becomes chaotic. The period-doubling is all over! Can we conclude that as the step Δt increases the dynamic behaviour of the mapping will be steady (one fixed point), periodic (with period-doubling 2, 4, 8, ...) and chaotic? We slightly decrease Δt in the seventh plot. For the first 40 iterations, the logistic map appears chaotic. Then, period 3 appears. As we increase Δt very gradually we get, in the eighth plot, period 6 and, in the ninth, dynamics that seems periodic, but it is not clear with which period. If we iterate for 1000 steps without writing or plotting any points and then continue for another 30 steps, we see clearly that the period is12 and that **the belief that gradual causes have gradual effects is false!** Sometimes, a small change in a cause can produce a big change in its effect.

7 Students' reactions

The students (N = 33, first year College students in a differential equations' course) were asked the following question: *In Euler's method, if we attribute a very small value to the step Δt, for example - 0.02, can we be sure that we have a good approximation to the solution?*This question was presented to the students before the introduction of the logistic equation. The students worked on some examples in which the differential equations had analytical solutions and in those examples, indeed, when Δt was reasonably small, the plot of the numerical solution and the plot of the analytical solution "coincided". Nevertheless, 88% of the students' answers expressed the claim that small may be not small enough. We observed the following categories of answers:

- Students' awareness of the limitations of the numerical methods ("T*he error accumulates*") was expressed in 39% of the answers.

- the claim "A small change in *t* could produce a big change in *y*" was expressed in 22% of the answers (" *Maybe an approximation with* $\Delta t = 0.02$ *will not be enough and we will miss some significant oscillations in the function*").
- discrete - continuous considerations and the claim that Δt must tend to 0 was expressed in 22% of the answers ("*The time, progresses all the time in a continuous way without jumps or pauses and we, in our ways, describe it only with little jumps which approach the continuous time but which are not equal to it*").

The students were also asked what is their opinion about the following statement: *If in Euler's method, using a step size* $\Delta t = 0.017$ *we get a solution very far from the real solution, then a step size* $\Delta t = 0.016$ *will not produce a big improvement, maybe some digits after the decimal point and no more.* This question was also presented to the students before the introduction of the logistic equation. The belief that "gradual causes have gradual effects" was expressed in 50% of the students' answers ("*It seems to me that if with* $\Delta t = 0.017$ *we didn't get a good solution, then* $\Delta t = 0.016$ *will not produce a big improvement*"). Among these students, was a student that wrote in the previous question "*A small change in t could produce a big change in y*". 25% claim that *gradual causes have not necessarily gradual effects.*

After introducing the students to the logistic equation, the students were asked to characterize the source of the error in Euler's method. The claim "*In the numerical method we omit the lim in the definition of the derivative*" was expressed explicitly by 30% of the students ("*The source of the error is in the hypothesis of the method that the derivative is equal to* $(y_{n+1} - y_n)/\Delta t$", "*The main problem is that we defined the derivative as the rate of change of the function. The rate of change is something continuous. However, we take a point and try to find a second point while ignoring the fact that the time between these two points is continuous*"). Other answers (30%) relate to the fact that the error accumulates in the numerical method or that the error depends on the initial point. 15% claim that the method is not suitable for certain kind of differential equations, supplying no explanation. We noted that for some students it might not be sufficient simply to introduce counterexamples to widely held notions. A theoretical discourse is necessary to help them reach the conclusion that in the definition of the derivative one cannot simply replace $\lim_{\Delta x \to 0} \Delta y / \Delta x$ by $\Delta y / \Delta x$ for very small Δx and that *passing to limits - or passing to discrete approximations- may change the nature of a problem significantly* [5].

8 The role played by *Mathematica* in the students' conceptualisation of the continuous by means of the discrete

Supporting different approaches to the same problem: *Mathematica* enables students to apply both the continuous analytical approach and the discrete numerical approach to the same problem. In both approaches the numerical, graphical and symbolical capabilities of the software were used.

Following Euler's algorithmic thinking: The students had to make Euler's algorithmic thinking very explicit in order to understand the syntax for Euler's numerical method.

Supporting qualitative analysis: *Mathematica* enables not only to solve numerically the differential equation, but also to analyze qualitatively the behaviour of the solution. For example, the students were helped by *Mathematica* to answer the question: What will happen in the long run? *Mathematica* was used to perform a very big number of iterations and to look at the results graphically and numerically. For example, we cite the possibility to take as an initial point the result of applying 1000 iterations on an original initial point.

Students' interaction with the software: *Mathematica* adds an experimental component to mathematics and helps to communicate the vitality of contemporary mathematics to the students (the field of dynamical systems), and this was done in our study while working on old numerical methods. Solving numerically the logistic equation, the students discover unexpected effects in an experimental way. This experimental component stimulates students to interact with the results obtained on the screen, changing parameters, reflecting and revising.

Realizing in a concrete way that some intuitions are false: *Mathematica* enables to change **very very gently** the value of a parameter and to help students test in a concrete way the fact that in mathematics, gradual changes in a cause do not necessarily produce gradual changes in its effect.

This study is about the conceptualisation of the continuous by means of the discrete. Experimenting with *Mathematica,* the students realized that in continuous methods, too, small changes in a cause can produce large changes in its effect. For example, in some differenitail equations, a change of 10^{-6} in the initial condition might cause a large change in the analytical solution. Applying the discrete continuous strand to solve the same problem added depth to the students' understanding.

differenitail equations, a change of 10^{-6} in the initial condition might cause a large change in the analytical solution. Applying the discrete continuous strand to solve the same problem added depth to the students' understanding.

We conclude with the words of Burger & Starbird [1]: *"If something is an anomaly in your world view, expand your world view to make it fit"*

9 References

[1] Burger, E. B. & Starbird, M. The Heart of Mathematics (chapter 6, p. 475), 2000, Key College Publishing

[2] Courant, R. Differential and Integral Calculus (Chapter 1, p. 39), 1937, Blackie & Son

[3] Kidron, I. and Vinner, S. Rational numbers and decimals at the senior high level-Density and comparison, in Proceedings of the 7th International Conference for the Psychology of *Mathematical* Education, Israel, 1983.

[4] Kleiner, I. Infinitely small and large in Calculus, in Educational Studies in Mathematics Vol.48 Nos 2-3 (Ed.Tall, D. & Tirosh, D.) 2001, Kluwer publishers

[5] Peitgen, H.O., Jurgens, H. & Saupe, D. Fractals for the classroom, (Part 2, pp. 295 - 301), January 1992, Springer - Verlag

[6] Stewart, I. What shape is a snowflake (Chapter 12, p. 148), 2001, W.H. Freeman

Editors' note:

We invite comment on the following:

1. The period doubling and chaotic effects described are really a function of the parameter R (see section 6) in the logistic map, and it is unfortunate that a numerical approximation by Euler's method casts R into an expression involving the step size. The n-th iterate of the logistic map for given y_0 and r depends on the stepsize in a wholly continuous and differentiable way: the gradient of the derivative is just steep in some cases.

2. The above does not weaken the fact that an approximation $\frac{\delta y}{\delta t} \sim \frac{dy}{dt}$ is still valid for small δt.

3. How widespread is the (student) belief that "a small change should produce a small effect"?

Visualization for Math-Education using Mathematica

Hyang Sook Kim
School of Computer Aided Science
And
Mathematica Technical and Training Center
Inje University, South Korea
mathkim@ijnc.inje.ac.kr
Young Mi Kim
School of Computer Aided Science
Inje University, South Korea
ymkim91@hanmail.net

1 Abstract

In this paper, we give examples of models we have created for use in university mathematics courses. We visualize the area of the unit circle, estimate the area of a figure by probabilistic simulation and consider the relation between sound and trigonometry. We give interesting, visual, meaningful and effective models for teaching the above subjects, which are obtained by the powerful functions of Mathematica such as animation of graphics, variety of visualization and speed of computation. This paper illustrates how one can use Mathematica to visualize abstract mathematical concepts, thus enabling students' effective understanding in the mathematics classroom. Development of these kinds of teaching and learning models can stimulate the students' curiosity about Mathematics and increase their interest.

2 Introduction

When one studies a mathematical problem nowadays, tools such as calculators and computers are available for students and teachers. Students can be actively engaged in reasoning, communicating, problem solving, and making connections with Mathematics and other disciplines [3]. The learning of mathematical properties and principles can be enhanced through visualization using computer graphics. Complicated calculations that are time-consuming by paper and pencil can be carried out instantaneously using Mathematica or some other mathematical software package. Some writers have asserted that the computational medium is affecting not only the means by which Mathematics is transmitted and learned, but also the content and context in which it is important [2].

The eminent mathematician and educator Gail S. Young [13] has compared the advent of the computer to the introduction of Arabic numerals into Europe or the invention of the calculus in its impact. He points out similarities with those earlier "revolutions" by means of which "hard problems became easy and solvable not only by an intellectual elite but by a multitude of people without special mathematical talent; problems arose that had not been previously visualized, and their solutions changed the entire level of the field". In the National Research Council (USA) report *Everybody Counts* [9] we also read " … computers have posed new problems for research [in Mathematics], supplied new tools to solve old problems, and introduced new research strategies".

For some years now people have been developing software for use in the mathematics classroom. Packages have varied greatly in their ease of use, their motivational appeal, their level of intellectual demand and their cross-platform accessibility [7]. *Everybody Counts*, taking a problem-solving view of mathematics, tells us that "research on learning shows that most students cannot learn mathematics effectively by only listening and imitating" [9].

Technology brings to students and teachers the opportunity to individualize learning: to generate illustrative examples, to follow interesting topics to the desired depth, to choose their own problems and then gather the tools for solving them. Information technologies have transformed the workplace, but not yet the schools. The various teaching-learning models which involve mathematical software need to be developed further.

The *Mathematica* code described in this paper is available from the authors. For the sake of brevity and readability, most of the code and graphics have been omitted from the paper, but all of them are included in the appendix CD-ROM of IMS2003.

3 Model 1 : Computing the area of the unit circle

In this section, we show how Mathematica can be used to illustrate several different computations of the area of the unit circle.

3.1 Using the inscribed regular polygon

One familiar way to define the area of a circle is as the limit of areas of inscribed regular polygons (*m*-gons) as the number of sides *m* tends to infinity. We begin with $m = 3$.

To begin with, we draw a unit circle and its inscribed triangle

Next, we consider the inscribed *m*-gon ($m = 3, 4, \ldots, 32$) of the unit circle.

Finally, we show that the area and the perimeter of an m-gon approach the area of the unit circle (π) and the perimeter of unit circle (2π) as m increases.

3.2. Using Riemann sums

A second method of computing the area is to treat the semicircle as the graph of a function. We begin by drawing a semi-unit circle.

First, consider the total area of the inscribed rectangles in the unit semi-circle using 2^m subdivisions.

Next, consider the sum of outer rectangles covering the unit semi-circle with the same number of subdivisions.

Finally we illustrate that the total areas of inscribed rectangles approaches the area of unit semi-circle ($\pi/2$) as m increases.

4 Model 2 : Estimating the area of a figure by probabilistic simulation[1]

4.1 Estimating the area of the unit circle

We draw a square of side 2 and its inscribed unit circle. Then we design Mathematica code which scatters points randomly (uniformly distributed) in the square and count the number of points within the circle. We can estimate the area of unit circle by computing (the area of the square)×(numbers of points in unit circle)÷(numbers of points in square).

(1) Estimating the area by a single sample of various sizes.

We try five cases using 500, 1000, 2000, 10000 and 100000 sample points respectively. Note that in these displays "data numbers" means sample size and "repeating numbers" means number of repetitions (samples).

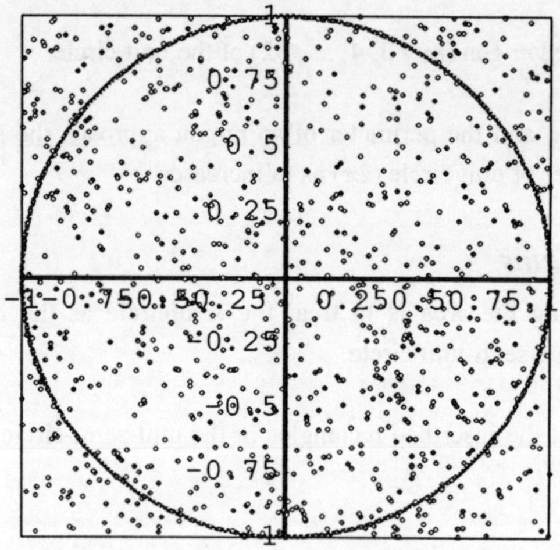

Data numbers is 1000

Numbers inside circle is 768

Since estimation of circlr area is

4 : π = 1000 :768

, thus we have Pi is $\frac{96}{125}\times 4$

Estimation of area is 3.07200000000000000000

Since the error is $\pi-\frac{96}{125}\times 4$

The error of area is 0.0695926535897932384 63

(2) Estimating the area by repeated samples

Next, we choose ten samples and record the average of the results. We use the same sequence of sample sizes. Then we can see that the error decreases as the sample size increases.

4.2 Estimating the area under a graph using probabilistic simulation.

We consider a function $f(x) = -x^2 + 1/4$, and estimate the area of the figure enclosed by f(x) and the x-axis. Again, we illustrate the effect of repeated sampling and of sample size.

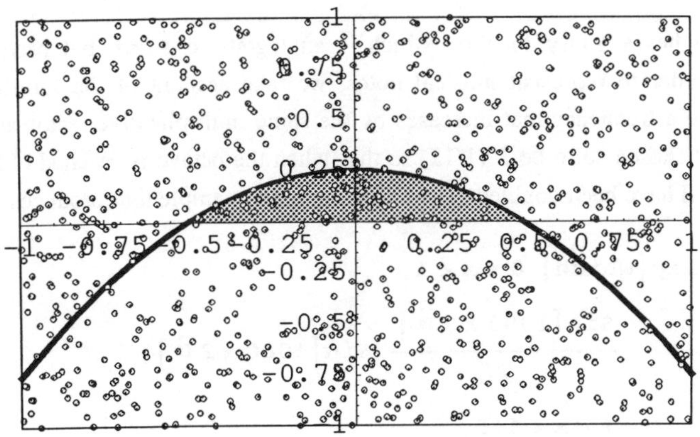


```
Data numbers=1000

Numbers inside Figure=55

Definite Integral=0.166667

Estimation of area =0.22000000000000000000

Error=-0.0533333
```

4.3 Computing the area of a triangle

In this section, show how the user can be allowed to specify a triangle, compute the area using the familiar formula, and then estimate it using probabilistic simulation. First, a coordinate grid is presented and we choose three points by clicking. The coordinates must then be copied to the list `data1`. Mathematica then draws the triangle. Beyond this, we calculate the area of the above triangle. By the same method as used in the preceding two sections we can estimate the area of the triangle and we compare the results with and without repeating samples. This idea raises the problem of estimating the area of any polygon.

5 Model 3 : Music

In this section we give various animations of trigonometric functions and show how musical notes can be produced using, for example, the sine and exponential functions. First, we define the function `SinPlot` which can be used to illustrate the relationship between the sine graph and the unit circle. Then we can animate the drawing of the sine graph as its argument runs from 0 to 2π.

Mathematica has a `Play` function which is analogous to `Plot` but produces sound instead. In order to represent musical notes, we take account of the familiar 12 tone scale and use a frequency that increases by the same multiplicative factor at each step. The factor is chosen so to be 2^(1/12) so that when the octave is reached (12 steps) the frequency will have been doubled. The following is an example that produces 6 notes.

$$
\begin{aligned}
&\mathbf{Play}\Big[\mathbf{Which}\Big[\ 0 < x < 1,\\
&\qquad \frac{\mathrm{Sin}\left[1000\, x\, 2^{\frac{-4}{12}}\right]}{e^{x\, 2^{\frac{-4}{12}}}}\ \mathrm{Sin}\left[1000\, x\, 2^{\frac{-4}{12}}\right],\\
&\qquad 1 < x < 1.5,\\
&\qquad \frac{\mathrm{Sin}\left[1000\,(x-1)\, 2^{\frac{3}{12}}\right]}{e^{(x-1)\, 2^{\frac{3}{12}}}}\ \mathrm{Sin}\left[1000\,(x-1)\, 2^{\frac{3}{12}}\right],\\
&\qquad 1.5 < x < 2,\\
&\qquad \frac{\mathrm{Sin}\left[1000\,(x-1.5)\, 2^{\frac{0}{12}}\right]}{e^{(x-1.5)\, 2^{\frac{0}{12}}}}\ \mathrm{Sin}\left[1000\,(x-1.5)\, 2^{\frac{0}{12}}\right],\\
&\qquad 2 < x < 4,\\
&\qquad \frac{\mathrm{Sin}\left[1000\,(x-2)\, 2^{\frac{3}{12}}\right]}{e^{(x-2)\, 2^{\frac{3}{12}}}}\ \mathrm{Sin}\left[1000\,(x-2)\, 2^{\frac{3}{12}}\right],\\
&\qquad 4 < x < 5,\\
&\qquad \frac{\mathrm{Sin}\left[1000\,(x-4)\, 2^{\frac{8}{12}}\right]}{e^{(x-4)\, 2^{\frac{8}{12}}}}\ \mathrm{Sin}\left[1000\,(x-4)\, 2^{\frac{8}{12}}\right]\Big],\\
&\quad \{x,\ 0,\ 5\}\Big]
\end{aligned}
$$

Using the same method, we have prepared a function that plays the song "Twinkle twinkle little star".

6 Conclusion

It is clear that computer systems are presenting new challenges and opportunities in the mathematics classroom. Their ubiquity has the potential to affect curriculum and teaching-learning methods both in high schools and universities. As Muller[8] observes, new educational technologies "will erect new barriers for some people, while[freeing others] to explore the world of Mathematics in a very different environment."

In this paper, we have presented three models designed to encourage effective teaching and learning in the mathematics classroom. Since abstract mathematical definitions and concepts need to be represented easily in order to facilitate the student's understanding, mathematical software and other technologies may stimulate better mathematics education. The more novel the environment is, the better the chance to arouse the students' curiosity. It may well be that these experiments raise more questions than they answer.

7 References

[1] Barrodale, Ian, Roberts, Frank D.K. and Ehle, Byron L(1971), Elementary Computer Applications, New York, Wiley

[2] Kaput, James J.(1998), Mixing new technologies, new curricula aid new pedagogies to obtain extraordinary performance from ordinary people in the next century, ICMI-EARCOME 1, 141-155.

[3] Kim, H.S.(2001), Teaching-learning method for plane transformation geometry with Mathematica, J.Korean Math.Edu.(ser.A) 40-1, 93-102.

[4] Kim, H.S.(2001), The visualization of the figure represented by parametric equations, J.Korean Math.Edu.(ser.A) 40-2, 317-333.

[5] Kim, H.S. and Kim, H.G.(2001) A-ha! Mathematica, Kyowoosa(Seoul).

[6] Lawrence, J. Dennis (1972), A Catalog of Special Plane Curves, New York, Dover.

[7] Maeder, Roman E.(1996), The Mathematica Programmer II, Academic Press.

[8] Muller, Eric R.(1998), Maple in the learning and teaching of Mathematics, Proc.of Fields-Nortel Workshop, 93-99.

[9] National Research Council.(1989), Everybody counts : A report to the nation on the future of mathematics education, Washington, DC : National Academy Press.

[10] Watson, Jane M.(1998), Technology for the professional development of teachers, ICMI-EARCOME 1, 171-190.

[11] Wellin, Paul R.(2000), Exploring Mathematics and science with Mathematica, Proc.of MTTC symp.1, 45-7

[12] Wolfram, S.(2000), Mathematica(4th ed.), Cambridge University Press.

[13] Young, Gail S.(1986), Epilogue in R.Ewing, K.Gross and C.Martin(Eds.), The Merging of Disciplines : New Directions in Pure, Applied and Computational

[Editor's note: this paper is a shortened version of a talk,

"Teaching and Learning Models for Mathematics using *Mathematica*" by Hyang Sook Kim and P. J. Ryan, given at NA-MKM 2002 (North American Workshop on Mathematical Knowledge Management , McMaster University, Hamilton, Ontario, Canada, June 2002

For details, see http://imps.mcmaster.ca/na-mkm-2002/talks/kim-ryan.pdf]

How high school students could present original math research using Mathematica

R. Miyadera, K. Miyabe, D. Kitajima, N. Fujii and K. Fujii
Kwansei Gakuin High School, Japan

Miyadera127@aol.com

1 Abstract

I have been teaching mathematics using *Mathematica* in high school for more than 4 years, and found out that it is possible for high school students to do very interesting research of mathematics if they can use good software and get proper advice. I am going to introduce the results of our research. Some of them are mathematically new, and others are not new but very inspiring since they show promising possibilities to students. In our research the use of *Mathematica* is essential. In this papar "we" sometimes means a group of students, and in other times it means students plus the instructor R.Miyadera.

2 An interesting pattern which emerges when we calculate numbers of the type $9999999999999999999999^{99}$ and the animation made by these patterns.

We played with large numbers. Two of us expanded $9999999999999999999999^{99}$, and found that there are very many 0s and 9s. We could not find why there are more 0s and 9s than numbers 1 and 8. I omitted the output of this expansion, because it takes too much space. It is essential to see the output to understand the following argument, so please calculate this on your own computer. It is easy to count the occurrences of each number.

```
Table[DigitCount[9999999999999999999999^227, 10, n],
 {n, 0, 9}]

{595, 492, 462, 480, 451, 473, 461, 479, 472, 629}
```

This proves that there are far more 0 and 9 in the expansion of the number. It took for us several days to realize that we could explain the findings by applying the Binomial theorem to $(9999\cdots999)^n = (100000\cdots000 - 1)^n$. We also found that n needs not to be of the type $99\cdots99$. We are very excited because high school math could explain this phenomenon. We found that in the case 99999^{999} 0 and 9 occur fewer times than 4 and 8. We realized that the size of the numbers cannot explain the difference of the outcomes.

```
Table[
 DigitCount[99999^999, 10, n], {n, 0, 9}]

{531, 479, 496, 503, 532, 463, 475, 501, 541, 474}
```

In short when ${}_nC_k$ is relatively small compared to $(100\cdots00)^k$, then we have a lot of 0 and 9 in the expansion. Therefore, if we increase the size of $100\cdots00$ and at the same time reduce the size of n, thus maintaining the size of the number $(100\cdots00 - 1)^n$ almost constant, we see 0 and 9 increase. To see this phenomenon with our own eyes, we made an animation. Here we just present the program, since the animation takes a lot of space.

```
int2[n_] := Block[{y, a, b, c},
   y = 9 (Sum[10^k, {k, 0, n}]);
   a = IntegerPart[20700 / (Log[y] // N)];
   b = y^a;
   c = Length[Part[RealDigits[b], 1]];
   Return[b]];
p = Apply[Join, {Table[n, {n, 1, 100}]}];
k = Length[p];
Do[m = Part[p, n];
  b = Part[RealDigits[int2[m]], 1];
  Show[Graphics[Evaluate[
     Table[Table[{Hue[Part[b, (80 y + x)] / 9.7],
        Disk[{x, y}, 0.6]},
       {y, 0, 79}], {x, 1, 80}]]], AspectRatio → 1/0.9],
  {n, 1, k}]
```

3 How Many Games Does it Take to Win the World Series?

To investigate the probability theory of World Series Match and the Japan Series Match is a good topic for high school students. We have found an interesting fact about the World Series. The World Series consists of seven games. If a team wins four out of the seven games, the team becomes the World Champion. If a team has already won four, the rest of the games will be cancelled.

3.1 An analytical solution of the problem when teams A and B are going to play in the World Series and the probability that team A wins each game is x.

```
(* Here we used the argorithm
  by Skiena[4] to make strings. *)
st[l_List, 0] := {{}}
st[l_List, k_Integer?Positive] :=
        Block[
   {oneless = st[l, k - 1], i, n = Length[l]},
                 Apply[Join, Table[
     Map[(Prepend[#, l[[i]]]) &, oneless], {i, n}]]];
Clear[p, q, x]; probability[n_] :=
 Block[{a1, b, c, f}, a = Select[st[{p, q}, n],
    Count[#, p] == 4 && Part[#, n] == p ||
      Count[#, q] == 4 && Part[#, n] == q &];
  f[x_List] := Apply[Times, x]; b = Map[f, a];
  c = Apply[Plus, b]];
t = Table[probability[n], {n, 4, 7}];
p = x; q = 1 - x; average = Simplify[t.{4, 5, 6, 7}];
Print[" The probabilities of
  {4 games, 5 games, 6 games, 7 games} are "];
Print[t]; Print["Average is"]; Print[average]
```

"The probabilities of {4 games,5 games,6 games,7 games} are"

$$\{(1-x)^4+x^4,\ 4(1-x)^4x+4(1-x)x^4,\ 10(1-x)^4x^2+10(1-x)^2x^4,\ 20(1-x)^4x^3+20(1-x)^3x^4\}$$

" Average is "

$$4(1+x+x^2+x^3-13x^4+15x^5-5x^6)$$

3.2 Comparison of the analytical solution with real data for the Japan Series and the World Series.

When we applied the analytical solution to real data for the Japanese Series and the World Series,we found a very courious fact which was not known at the time.The expected number of games is 5.88 in the Japan Series and 5.9 in the World Series compared to 5.8125 for the expected value of analytical solution when x=0.5.These are well known facts. For example Paul Nahin [3] calculates the probability that the World Series will require 4,5,6 or 7 games. He shows that over the years,the the average of the number of games is almost exactly what we would expect for two evenly matched teams. On the contrary, when we calculated the probability distribution,we found the courious fact.The probability that the World Series will require 4,5,6 or 7 games is {0.18,0.16,0.24,0.42} compared to {0.125,0.25,0.3125,0.3125}for the analytical solution when x=0.5. There is too big difference between these. For the Japan Series the actual probabilities are {0.0980392,0.235294,0.352941,0.313725}. For the Japan Series we can conclude that two teams have the almost same chance of winning each game, but to explain the data of the World Series we have to consider the home field advantage and the existence of more than two types of teams. What we got after calculation is the hypothesis that there are two types of teams. One is always very strong, and the other is very strong in its home ground and weak in its opponent's ground.

4 Investigating interesting features of patterns.

Hujimura and Tamura [1] introduced a puzzle in their book A teacher is going to give students a true-false test. The test consists of 5 questions and along with each question there is an answer which may be true or false. The students are expected to write "true" if they think the answer is correct and "false" if they think not. The teacher has two strategies. One is that he never presents three true answers in a row nor three false answers in a row. Another is that he presents more correct ones than false ones. What is the best strategy for students if they do not know anything about the subject? The answer to this puzzle by is quite simple, but when we began to study the case with more than 5 questions, a very interesting feature of the problem suddenly appeared in front of our eyes. We got the following theorem after calculations with *Mathematica*.

4.1 Theorem1. The best strategy for students is {1,1,0,1,1,···,1,1,0,1,1} for any numbers of questions, where "1" stands for "Yes" and "0" stands for "No".

For Proof see Kohno [2].

5 An interesting pattern which appears in the square root of periodic number.

We play ed with numbers, and found a interesting pattern using only natural numbers and $\sqrt{}$. Since it takes too much space, we omitted the most of the output of the following calculation. In the program n runs from 0 to 100, but in the output n runs up to only 8.

$$g[m\_] := \sum_{k=0}^{m} 10^{k}$$

```
TableForm[
 Table[{SqrtBox[g[n]] // DisplayForm, √g[n] },
  {n, 0, 100}]]
```

$\sqrt{1}$	1
$\sqrt{11}$	$\sqrt{11}$
$\sqrt{111}$	$\sqrt{111}$
$\sqrt{1111}$	$\sqrt{1111}$
$\sqrt{11111}$	$\sqrt{11111}$
$\sqrt{111111}$	$\sqrt{111111}$
$\sqrt{1111111}$	$\sqrt{1111111}$
$\sqrt{11111111}$	$\sqrt{11111111}$
$\sqrt{111111111}$	$3\sqrt{12345679}$

You will easily find the patterns in numbers such as 12345679. If you make n big enough, you will find numbers such as 9182736455463728191 and 22675736961451247165532879818594104308 39. Can you find any pattern?

6 Find the last decimal of $n^{n^{n^{n^{\cdot^{\cdot^{n}}}}}}$.

The algorithm to solve this problem is already known, but one of us is working on this and rediscovered Fermat's Little Theorem. We are going to present the result of his study in IMS, because we this is a very good problem for a mini-research for high school students using Mathematica.

7 References

[1] Hujimura, K. and Tamura, S. Introduction to mathematical puzzle, 1997, Kodansha (in Japanese).

[2] Kohno, K. A teacher's habit. Preprint.

[3] Nahin, P. Duelling Idiots and Other Probability Puzzlers, 2002, Princeton University Press.

[4] Skiena, S. Implementing Discrete Mathematics, 1990, Addison-Wesley.

A Tutoring System of Symbolic Calculations Supported byWebMathematica

Hitoshi Nishizawa,Yasuto Kajiwara,Takayoshi Yoshioka
Toyota National College of Technology,Japan

{nisizawa, yoshioka}S@toyota-ct.ac.jp

1 Abstract

This paper describes a WebMathematica-based tutoring system for algebraic calculations, such as simplifications of rational expressions, for high school students. The system has two learning paths to accommodate various student learning styles: interactive exercises and step-by-step instructions. It also has an explanation page to help with calculation errors based on misconceptions, which beginners and slow learners tend to have. The pages are linked to each other and inherit the student's expression from one page to the next. Based on the expression, the system shows a comment, a list of the next steps in the calculation and the explanation for any incorrect calculation step. The system uses WebMathematica to display the dynamically changing contents, *Mathematica* to evaluate the symbolic expressions, based on rules described in custom *Mathematica* functions, and a J-Link program to retrieve the necessary data from a database.

2 Introduction

We think it necessary for our students to learn more algebraic calculations as white boxes, which act as the first stage of the white-box/black-box principle [1], as we have observed an increasing number of students who just try to memorize theorems and formulae as black boxes when they first meet them. This phenomenon is not due to the influence of Computer Algebra Systems (CAS) like Mathematica, which are not commonly used in Japanese primary and secondary schools, but is the result of something that might be called an "efficient" learning style for examinations. It would be disastrous for engineering education to be overlooked. We have to do something to stop them. The phenomenon is well observed in the calculation errors made in manipulating rational expressions. Simplifying rational expressions or rewriting them into partial fractions used to be easy tasks for new students (aged 15-16) of our College of

Technology, but they are not for some students in recent years. Although it is partly because of reduced teaching hours for mathematics in primary and secondary schools in Japan, we are perplexed by those students who can calculate fractions numerically by several methods but fail to rewrite rational expressions correctly. Many of them seem to overestimate their competence in symbolic calculations because of their success in numerical calculations. They believe their errors are careless mistakes caused by a lack of concentration,and pay little attention to rewriting rules. As a result, they repeat the same types of incorrect calculations caused by misconceptions of rewriting rules. Their lack of attention to calculation rules even hinders their further learning in the upper grades. The first thing the teachers must do is make them aware of calculating processes and rewriting rules, but it takes many hours to teach them all to change their approach. We need an adaptive tutoring system, which assesses each step in the students' calculations and gives them some explanations. In this paper, we describe our web-based tutoring system for symbolic calculations, which has recently been improved with the help of WebMathematica. The improvements include easier to read web pages, contents which adapt to each student, a new function to categorize the mistakes made by the students and integration of the tutoring system into our learning environment, where students and teachers receive other services in addition to mathematics teaching and learning. The development takes advantage of Mathematica, WebMathematica and J-Link. The original tutoring system had been built on a Mac-OS operating system using Web-DB (database) software and *Mathematica 3*,connected by a small MathLink program [2]. It provided interactive exercises in factorizing or expanding polynomials, and the system gave the position of any calculation errors and explanations of why the students' expressions differed from the correct answer. The explanations were produced by custom *Mathematica* functions that analysed the students' expressions and categorized them. The system has been migrated to a Linux-based server for better performance and shorter response time [3]. Now, the system is mainly used by our 15-16 year old students in their supplementary lessons for exercises in calculating polynomials and rational expressions and for finding the algebraic expressions of quadratic functions from given graphs. The students like the quick response and find helpful the feedback on each input, but they don't like the input method where they have to input their rational expressions in a single line using layered parentheses. With Mathematica, although it was possible to have two-dimensional graphic images of fractions in our web pages, the quality of our graphic images was low, since our MathLink program couldn't use the *Mathematica* FrontEnd, which is necessary for high quality images, simultaneously with the *Mathematica* kernel. Before we improved the MathLink program, WebMathematica solved the problem.

3 Advantages of Technology in Teaching Algebra

3.1 A Web-based System with Mathematica

A web-based system with a CAS like *Mathematica* has an advantage over direct use of a CAS in its flexibility of customization for educational purposes. Instead of full access to the power of a CAS, a web page can provide each student with information or services which are limited but appropriate to his current need. It is especially convenient for a beginner or an infrequent user of CAS because it doesn't require knowledge of the CAS itself. On the interactive web page, the amount of calculation the system does, instead of the student, is carefully designed. On an instruction page, the system does all the calculations and shows the result. On the other hand, on an exercise page, the students do the calculations and the system analyzes them and returns a comment explaining their appropriateness. The quality of the response information determines the usefulness of the pages. In a web-based system, as well as interactive pages that change their contents in reaction to the students, it is also effective to have static pages that have fixed explanations like textbooks. By linking the static pages with the interactive pages, the system can be integrated to form a complete educational system. Preparing the interactive pages usingMathematica is easier for mathematics teachers than using other programming languages like C or Java. It is more like describing mathematical procedures than programming. Updating the contents or services is easy because the only thing we must do is rewrite the source file on the server. After that,every student is guaranteed access to the latest services.

3.2 WebMathematica

Our main technical reason for using WebMathematica is to have high quality two-dimensional graphic images for symbolic fractions on the web pages,which are easy for students to recognize. For displaying the graphic images, WebMathematica handles the necessary connections to the *Mathematica* kernel and FrontEnd, and also does the garbage collection. The only thing the author of the web pages has to do is add a special tag in his HTML. Then, the functions ofMathematica become accessible from the web pages. The extension of HTML is called MSP (*Mathematica* Server Page) and is based on JSP (Java Server Page). In the case of our tutoring system, the conversion from a CGI-based system to a WebMathematica-based one was not difficult because most of the services are implemented as custom *Mathematica* functions. If the services are accessible from *Mathematica* notebooks,they can be called from either a CGI-program or an MSP page. Some difficulty was met in the rather complicated installation process of WebMathematica.

3.3 *Database Connection*

An adaptive tutoring system needs domain knowledge and user models, which are either written in programs or stored as data. Questions for exercises and the current status of each student are the data regularly referred to or updated. It is desirable that updates are handled by a database management program. It is possible to use a DBMS from *Mathematica* with the help of a MathLink program [4], but we chose J-Link and JDBC to connect our system to a DBMS because the development is easier. The small Java program for the database connection is wrapped up with a custom *Mathematica* function, and can be used in *Mathematica* notebooks or in MSP source codes.

4 The Tutoring System

Our tutoring system for the calculation of rational expressions has two learning paths: on one of them a student has to simplify rational expressions on the exercise pages, with the help of the system's comments on every intermediate step. On the other path he has to follow the step-by-step instructions for various expressions given by the system [5]. The courses differ in the activity of the student but share the same user data and questions,which range from cancellation of a simple fraction to rather complicated sums of fractions with different denominators. The student may change his path from one to the other as he learns. Each page also refers to others as an additional aid to the student on the learning path. For example, a step-by-step instruction page links to a hint for the student in trouble on an exercise page. An exercise page gives an example exercise to confirm what he has just learnt on an instruction page, and the student continues his learning on the instruction page with another expression.

4.1 *Interactive Exercise*

The interactive exercise pages provide a student with the opportunity to develop his manipulation skills through their quick responses and meaningful comments, which tell the student what types of errors he has made, (if any), and how far the intermediate expressions are from the answer. All the comments are produced by custom *Mathematica* functions.

On the exercise pages, the student is asked to input the next symbolic expression after his calculation. He is encouraged to describe steps in as much detail as possible, because it helps him to learn to use mathematical rewriting rules, and makes it easier for the system to assess the current understanding level of each student for its adaptive actions.

After each calculation step, the system displays the two-dimensional image of the input expression on the page,with a comment on the expression like "includes calculation error", "needs more simplification" or "is the final answer". The system separates calculation errors into three categories; errors based on misconceptions, careless mistakes and others. If it finds an error based on a misconception i.e. if an expression is apparently formed from the previous expression by use of an incorrect rule, it automatically displays the explanation page described in section 4.3 on top of the exercise page to alert the student.

4.2 Step-by-Step Instruction

A step-by-step instruction page shows a list of rewriting rules applicable to the current expression. It starts with an expression selected by the student from the menu, an expression closely related to the student's level of understanding, or the latest intermediate expression inherited from the exercise page. The page shows the next step of the calculation according to the rewriting rule selected by the student. The student continues the calculation steps until he gets the answer,by selecting rules from the list. The list includes only appropriate rules so the calculation step is guaranteed to be correct.Inquisitive students are allowed to explore as much as they want and gradually learn the appropriate calculation steps from the experience. Through these pages, the student learns to relate rewriting rules to calculation steps, which most of the slow learners don't do with their own calculations. After learning the procedure on the instruction pages, the interactive exercises become more meaningful to the student. The instruction pages are popular among slow learners. However, they aren't suitable as the main path for all students because the pages don't appeal to students who know the calculations fairly well or to fast learners. On the instruction pages, the role of the student is passive, just observing the changes in the expressions on his screen, so we set the instruction pages as an alternative path to the interactive exercise pages for those students who hesitate to start their own calculations at first. The pages also act as an adaptive hint for the student who can't work out the next step of his calculation on an exercise page. After he has learnt the step,he returns to his exercise page and continues his exercises. The instruction page inherits the current expression from the exercise page so that the student can receive the instruction he needs and learn the next calculation step.

4.3 *Explanation of an Error Caused by a Misconception*

Because a student who makes a calculation error based on an incorrect rewriting rule usually believes his calculation is correct, or doesn't see any reason to look for an error, it is not easy to convince him of the incorrectness of his calculation. Indication of the erroneous step is not enough and more detailed explanation, with evidence, is necessary. The explanation page tries to convince the student by showing a numerical counter example as the evidence which conflicts with the student's calculation step. The counter example is prepared by substituting small integers for the variables in the expressions before and after the calculation step. Both numerical expressions and their evaluations are displayed, along with the incorrect rule suspected to have been used. If the student is convinced, he may feel the need to get instructions for rule-based calculations. To show the explanation page at the right moment, it is necessary to separate the errors caused by incorrect rewriting rules from other errors. We have written a custom function for the purpose.

5 The Effectiveness of the System

In the 2002/2003 school year, six students in the first grade (age 15), group A, used the tutoring system in their supplementary lessons to learn about the calculation of rational expressions. They had been the lowest performers in their class in the pencil-and-paper pre-test. Their performances in the post-test were compared with those of four other slow learners, group B, who had only pencil-and-paper supplementary exercises during the same period. Before the supplementary lessons, both groups had used incorrect rewriting rules from time to time. Group A used the exercise page for five hours in July 2002, and used both the instruction page and the explanation page as well as the exercise page for three hours in November 2002, on average. In the post-test, group A performed as well as group B in the pencil-and-paper calculations, but they did better in the special test, where the students were required to assess a printed answer sheet with four correct calculations, three calculations with careless mistakes and two calculations with errors caused by incorrect calculating rules. They worked out whether the calculation steps were correct or included some errors. When they found an error in any of the calculation steps, they were instructed to mark the position of the error and to describe what kind of error it was. In the special test, all the students succeeded in identifying the correct calculations, but failed to find some of the incorrect calculations. See Table 1 for details.

The better performance of students in group A in the special test suggests the promising potential of the tutoring system, especially for slow learners. The system might have taught them to pay more attention to the rewriting rules applied at each calculation step, which we think is the right attitude and a good step towards mastering the subject.

group	student	pre-test		post-test		special test: correct calculations		special test: with careless mistakes		special test: incorrect calculating rules	
	A_1	0.33		0.50		1.00		1.00		1.00	
	A_2	0.50		0.38		1.00		0.67		0.50	
A	A_3	0.50	0.43	0.75	0.48	1.00	1.00	0.33	0.83	1.00	0.83
	A_4	0.56		0.50		1.00		1.00		1.00	
	A_5	0.17		0.50		1.00		1.00		1.00	
	A_6	0.50		0.25		1.00		1.00		0.50	
	B_1	0.83		0.50		1.00		0.33		0.50	
B	B_2	0.89	0.74	0.50	0.50	1.00	1.00	0.33	0.50	0.50	0.63
	B_3	0.67		1.00		1.00		0.67		1.00	
	B_4	0.56		0.00		1.00		0.67		0.50	

Table 1

6 Conclusion

A WebMathematica-based tutoring system for algebraic calculation, the aim, the implementation and the preliminary effectiveness,are described. The system has two learning paths of interactive exercises and step-by-step instructions. It also has an explanation page for calculation errors based on misconceptions, which beginners and slow learners tend to make. The pages are linked to each other by inheriting the student's expression through the pages. The experience of using the system helped low performers to improve their calculation of rational expressions, mainly because they paid more attention to the rewriting rule for every calculation step.

Acknowledgement

This paper is the result of a research project supported by Grant-in-Aid for Scientific Research of the Ministry of Education, Science, Sports and Culture of Japan, No.(C2) 13, 680, 269.

7 References

[1] Buchberger B., Should Students Learn Integration Rules?, SIGSAM Bulletin 24(1), pp10-17 (1990).

[2] Nishizawa H.,Nakayama K.,Kanaya T., A Remote Tutoring System for Algebra Enhanced by Mathematica, in Innovation in Mathematics (eds. Keranen,V., Mitic,P. and Hietamaki,A.), Proc.of 2nd IMS, Rovaniemi, Finland, pp361-168 Computational Mechanics Publications (1997).

[3] Yoshioka T.,Higuchi Y.,Nishizawa H., A Web-based Exercise System for Learning Mathematical Functions, Proc.of ICCE/ICCAI 2000, pp1574-1576 (2000).

[4] Ohgi Y.,Kimura H.,Miyaji C., Mathgres-A MathLink Connection between PostgreSQL and Mathematica for Experimental Data Analysis and Storage, in Symbolic Computation: New Horizons (eds. Tazawa,Y et al), Proc.of IMS2001, Tokyo, pp393-400, Tokyo Denki University Press (2001).

[5] Nishizawa H.,Kajiwara Y.,Yoshioka T., Partial Evaluation of Mathematical Expressions as a Teaching Tool, in Symbolic Computation: New Horizons (eds. Tazawa,Y et al), Proc.of IMS2001,Tokyo, pp375-382, Tokyo Denki University Press (2001).

Editors' note

Applying a *t*-test for the difference of means, $\mu_A - \mu_B$, none of the cases 'post-test', 'special with careless mistakes' and 'special with incorrect calculating rules', is significant at the 5% level. The only significant difference at the 5% level is in $\mu_A - \mu_B$ for the pre-test. The interpretation is that Group A performed significantly worse in the pre-test but not in the post- or special tests, and that this was due to the tutoring system. However, the question of whether or not any difference in means is due to motivational factors when using the computer is open.

Partial Differential Equations with webMathematica

Ünal Ufuktepe
Izmir Institute of Technology, Department of Mathematics

unalufuktepe@iyte.edu.tr

1 Abstract

The growing popularity of the internet, and the increasing number of computers connected to it, make it an ideal framework for remote education. Many disciplines are rethinking their traditional philosophies and techniques to adapt to the new technologies. Web-based education is an effective framework for such learning, which simplifies theory understanding, encourages learning by discovery and experimentation and undoubtedly makes the learning process more pleasant. There is a need for adequate tools to help in the elaboration of courses that might make it possible to express all the possibilities offered by www teaching. *webMathematica* is a web-based technology developed by Wolfram Research that allows the generation of dynamic web content with *Mathematica*. With this technology, distance education students should be able to explore and experiment with mathematical concepts. In this paper we present a sample lecture for Partial Differential Equations in *webMathematica* for the distance learning environment.

2 Instruction

The purpose of this web-based lesson is to show what partial differential equations (PDEs) are, how they are classified and how they are solved. We will give *Mathematica* commands with the package MSP. To see the details of the our MSP files (PDE.msp, heat.msp, heat2.msp, and step1.msp), please check the following URL address:

http://gauss.iyte.edu.tr:8080/webMathematica/math

Typical material is organized in modular form with the instructional objective clearly spelled out at the beginning of each lesson. The topic of discussion is the concept of PDEs, which students face difficulty understanding. With *webMathematica*, a visual representation can be useful in the form of animation and interactivity. The MSP scripts require some knowledge of HTML, including FORM and INPUT elements, and

Mathematica [7].

2.1 Partial Differential Equations

A PDE is an equation that contains partial derivatives. In contrast to ordinary differential equations (ODEs), where the unknown function depends on just one variable, in PDEs, the unknown function depends on several variables. Some well-known PDEs are: $u_t = k\,u_{xx}$ (the Heat Equation in one dimension), $u_t = k\,(u_{xx} + u_{yy})$ (the Heat Equation in two dimensions), $\nabla^2 u = u_{xx} + u_{yy} = 0$ (Laplace's Equation), and $u_{tt} = k^2\,\nabla^2 u$ (the Wave Equation).

How do we solve a PDE ? There are many techniques. The most important methods are those that change PDEs into ODEs. PDEs are classified according to order of the PDE, number of variables, linearity, homogeneity, kinds of coefficients [2]:

$u_t = t\,u_{xx} + \cos t$ (second order linear)

$u_t = u\,u_{xx} + x\,u_{yy}$ (nonlinear)

$u_{xx} + 2\,u_{xy} - 3u = 0$ (second-order linear equation with constant coefficient)

In this paper we present two lectures. The first is about classifying second-order equations and transforming them into the canonical forms. The second is about the solution of the heat equation by separation of variables[2][8].

3 Classification of Second-Order Equations

The general linear second-order PDE in two independent variables *x,y* has the form

$$A\,u_{xx} + B\,u_{xy} + C\,u_{yy} + D\,u_x + E\,u_y + F\,u = G \tag{1}$$

where the coefficients *A*, *B*, *C*, *D*, *E*, *F*, and *G* are functions of *x* and *y* in some domain of the *xy*-plane. If all these coefficients are constants, with *A*, *B*, *C* not all zero, it is possible by means of a coordinate transformation to reduce the equation to one of three canonical forms, in each which the second-order derivative terms appear in a simple form. In an attempt to simplify the form of the second-order terms of this equation, we introduce new variables ξ and η by means of the linear transformation

$$\xi = \alpha\,x + \beta\,y,\ \ \eta = \gamma\,x + \delta\,y \tag{2}$$

The coefficients α, β, γ, and δ are constants, which will be determined subject to the condition on the Jacobian, $\alpha\,\delta - \beta\,\gamma \neq 0$. Writing $u(x,y) = U(\xi\,,\eta)$, we find u_x, u_y,

u_{xx}, u_{xy} and u_{yy} by means of the chain rule. Substituting these in equation (1) and collecting similar terms, we have

$$a\,U_{\xi\xi} + b\,U_{\xi\eta} + c\,U_{\eta\eta} + d\,U_{\xi} + e\,U_{\eta} + f\,U = g \qquad (3)$$

where a, b, c, d, e, f, and g depend on A, B, C, ..., α, β, γ, and δ . We choose the constants α, β, γ, and δ in such a way that at least one of the second-order terms in (3) drops out. In fact we shall show that according as **(a)** $B^2\text{-}4AC > 0$, **(b)** $B^2\text{-}4AC = 0$, **(c)** $B^2\text{-}4AC < 0$, the constants α, β, γ, and δ can be chosen such that **(a)** $a = 0$, $c = 0$, **(b)** $b = 0$, $c = 0$, **(c)** $a = c$, $b = 0$, thus simplifying the equation to one of the three canonical forms: **a)** $U_{\xi\eta}+ ... = 0$ **b)** $U_{\xi\xi}+ ...= 0$ **c)** $U_{\xi\xi}+ U_{\eta\eta}+ ...= 0$. [8] We say the PDE (1) is of the

1) Hyperbolic type, if $typ = (\frac{B}{2})^2 - AC > 0$, (then $\xi = (-\frac{B}{2} + \sqrt{typ}\,)x + Ay$ and $\eta = (-\frac{B}{2} - \sqrt{typ}\,)x + Ay$)
2) Parabolic type, if $typ = (\frac{B}{2})^2 - AC = 0$, (then $\xi = x$ and $\eta = \frac{B}{2}x - Ay$)
3) Elliptic type, if $typ = (\frac{B}{2})^2 - AC < 0$, (then $\xi = -Bx + Ay$ and $\eta = \sqrt{-typ}\;x$

Example 3.1

Classify the following PDE and reduce it to canonical form:

$u_{xx} - 2\,u_{xy} + 5\,u_{yy} = 0$

Solution: http://gauss.iyte.edu.tr:8080/webMathematica/MSP/Application/PDE

```
Needs["MSP`"]

$$a = 1 ; $$b = -2  ; $$c1 = 5 ;
 $$d = 0 ;  $$e = 0 ; $$f = 0 ; $$g = 0;
typ = ($$b / 2) ^2 - $$a * $$c1;

Equationtype =
 Which[typ > 0, "The given PDE is Hyperbolic",
  typ < 0, "The given PDE is Elliptic",
  typ == 0, "The given PDE is Parabolic"]
```

The characteristic equations are

```
Which[typ > 0, ki[x_, y_] = (-$$b / 2 + (typ) ^ (1 / 2))
     * x + $$a * y, typ < 0,
  ki[x_, y_] = -$$b * x + $$a * y,
  typ == 0, ki[x_, y_] = x]

Which[typ > 0, nu[x_, y_] = (-$$b / 2 - (typ) ^ (1 / 2)) * x
    + $$a * y, typ < 0,
  nu[x_, y_] = (-typ) ^ (1 / 2) * x, typ == 0,
  nu[x_, y_] = ($$b / 2) * x - $$a * y]

ux = uk * D[ki[x, y], x] + un * D[nu[x, y], x];
uy = uk * D[ki[x, y], y] + un * D[nu[x, y], y];

uxx = Expand[(ukk * D[ki[x, y], x] + ukn * D[nu[x, y], x])
      * D[ki[x, y], x] + uk * D[ki[x, y], {x, 2}] +
    (ukn * D[ki[x, y], x] + unn * D[nu[x, y], x]) *
     D[nu[x, y], x] +
    un * D[nu[x, y], {x, 2}]];

uxy =
  Expand[(ukk * D[ki[x, y], y] + ukn * D[nu[x, y], y]) *
     D[ki[x, y], x] + uk * D[ki[x, y], x, y] +
    (ukn * D[ki[x, y], y] +
        unn * D[nu[x, y], y]) * D[nu[x, y], x] +
    un * D[nu[x, y], x, y]];

uyy =
  Expand[(ukk * D[ki[x, y], y] + ukn * D[nu[x, y], y]) *
     D[ki[x, y], y] + uk *
     D[ki[x, y], {y, 2}] +
    (ukn * D[ki[x, y], y] + unn * D[nu[x, y], y]) *
     D[nu[x, y], y] + un * D[nu[x, y], {y, 2}]];
```

The canonical form of the equation is

```
Simplify[$$a * uxx + $$b * uxy +
    $$c1 * uyy + $$d * ux + $$e * uy + $$f * u == $$g]
```

4 Separation of Variables

Separation of variables is one of the oldest techniques for solving initial boundary-value problems (IBVPs) and applies to problems where the PDE is linear and homogeneous. It dates back to the time of Joseph Fourier and is probably the most widely used method of solution. Let us now consider the problem of finding a solution of the homogeneous heat equation. We must specify $u(x,t)$ at some time and describe how the two ends of the rod exchange heat energy with the surrounding medium. Thus, with reference to the case in which there is no heat source, if the temperature of the rod at $t = 0$ is described by the function $f(x)$, $0 \le x \le L$, and the two ends are maintained at zero temperature at the time, then the temperature distribution u in the rod at any later time $t>0$ is found by solving

$$u_t - k\,u_{xx} = 0 \tag{4}$$

subject to the conditions

$$u(x, 0) = f(x) \qquad (0 \le x \le L), \tag{5}$$

$$u(0, t) = 0, \quad u(L, t) = 0 \qquad (0 \le t). \tag{6}$$

This problem is called an initial-boundary value problem for the heat equation. The conditions (5) and (6) are called initial and boundary conditions, respectively. We shall solve this problem by the method of separation of variables. Accordingly, we seek particular solutions of (4) in the form

$$u(x, t) = X(x)\,T(t) \tag{7}$$

which satisfy the boundary conditions (6). Substituting (7) in (4) and separating the variables, we obtain

$$\frac{T'}{k\,T} = \frac{X'}{X} = -\lambda. \tag{8}$$

This leads to the two ordinary differential equations $X'' + \lambda X = 0$ and $T' + \lambda\, kT = 0$. In order for (7) to satisfy the boundary conditions (6), we must have $X(0) = 0$ and

$X(L) = 0$. Thus, the function X is determined by solving the Sturm-Liouville problem. We find the eigenvalues

$$\lambda_n = \frac{n^2 \pi^2}{L^2} \qquad (n = 1, 2, \dots) \tag{9}$$

and the corresponding eigenfunctions

$$X_n(x) = \sin \frac{n \pi x}{L}. \tag{10}$$

Now, for each λ_n, a solution of the equation $T' + \lambda kT = 0$ is given by

$$T_n(x) = e^{-k \lambda_n t}. \tag{11}$$

Thus, the functions

$$u_n(x, t) = e^{-k \lambda_n t} \sin \frac{n \pi x}{L} \tag{12}$$

with λ_n given by (9), are all the particular solutions of (4) satisfying the homogeous boundary conditions (6). To obtain a solution of our problem, we consider a series of the functions (12) in the form

$$u(x, t) = \sum_{n=1}^{\infty} b_n e^{-k \lambda_n t} \sin \frac{n \pi x}{L}, \tag{13}$$

and determine the coefficients b_n so as to satisfy the initial condition (5). Thus, setting t=0 in (13), we see from (5) that the coefficients b_n must satisfy the relation

$$f(x) = \sum_{n=1}^{\infty} b_n \sin \frac{n \pi x}{L} \qquad (0 \le x \le L). \tag{14}$$

This is precisely the Fourier sine series representation of f on the interval *[0, L]*. Therefore, the coefficients b_n are given by the formula

$$b_n = \frac{2}{L} \int_0^L f(x) \sin \frac{n \pi x}{L} \, dx \qquad (n = 1, 2, \dots) \tag{15}$$

The function u represented by the series (13), with b_n given by (15), is the solution of our problem. In order to verify this, we now assume that the function f is continuous and piecewise smooth on $[0, L]$ and that $f(0) = f(L) = 0$.

Example 4.1

Find the solution of the initial-boundary value problem

$u_t - u_{xx} = 0 \quad (0 < x < \pi,\ t > 0)$, with

$u(x, 0) = \sin x \ (0 \le x \le \pi)$, $u(0, t) = u(L, t) = 0 \quad (0 \le t)$

Solution: http://gauss.iyte.edu.tr:8080/webMathematica/MSP/Application/heat

```
$$k = 1 ; $$L = Pi; $$m = 40; f[x_] := Sin[x];

la[n_] = n Pi / $$L;
Unprotect[Sin, Cos]; Sin[n Pi] = 0;
Cos[2 n Pi] = 1; Cos[n Pi] = (-1) ^n;

b[n_] =
 2 / $$L Integrate[f[x] Sin[ la[n] * x], {x, 0, $$L}]
b[1] = 2 / $$L Integrate[f[x] Sin[la[1] * x], {x, 0, $$L}]

u[x_, t_] =
 Sum[b[j] Exp[-$$k * la[j] ^2 * t ] Sin[la[j] * x],
  {j, 1, $$m}]

tplot = Table[u[x, t], {t, 0, 5, 0.5}];
grays = Table[GrayLevel[i], {i, 0, 0.7, 0.7 / 20}];
MSPShow[Plot[Evaluate[tplot],
  {x, 0, 3 Pi}, PlotStyle -> grays]]
MSPShow[DensityPlot[u[x, t], {x, 0, π},
  {t, 0, 5}, PlotPoints -> 30]]
```

In the case where both ends of the rod are insulated, so that the initial-boundary value problem consists of (4),(5) and the boundary conditions

$$u_x(0, t) = 0, \quad u_x(L, 0) = 0 \qquad (t \ge 0), \tag{16}$$

the method of separation of variables leads to the Sturm-Liouville problem: $X'' + \lambda X = 0\ X'(0) = 0,\ X'(L) = 0$. We recall that the eigenvalues of this problem are $\lambda_n = \frac{n^2 \pi^2}{L^2}$ $n = 0,1,..$, with the corresponding eigenfunctions $X_n(x) = \cos \frac{n \pi x}{L}$. Thus, the functions $u_n(x, t) = e^{-(k\lambda_n t)} \cos \frac{n \pi x}{L}$ are all particular solutions of (4) satisfying the boundary conditions (16). We consider a series consisting of the functions in the form

$$u(x, t) = \frac{a_0}{2} + \sum_{n=1}^{\infty} a_n\, e^{-(k \lambda_n t)} \cos \frac{n \pi x}{L} \tag{17}$$

The initial condition (5) then requires that the coefficients a_n are given by the formula

$$a_n = \frac{2}{L} \int_0^L f(x) \cos \frac{n \pi x}{L}\, dx \qquad (n = 0, 1, \ldots). \tag{18}$$

Example 4.2

Find the solution of the following initial-boundary value problem

$u_t - u_{xx} = 0 \quad (0 < x < \pi,\ t > 0)$ with

$u(x, 0) = \sin x \quad (0 \le x \le \pi),\quad u_x(0, t) = 0,\ u_x(L, 0) = 0 \quad (0 \le t)$

Solution: http://gauss.iyte.edu.tr:8080/webMathematica/MSP/Examples/heat2

5 References

[1] M. L .Abell and J.P.Braselton, Differential Equations with Mathematica, 1997, Academic Press.

[2] S.J.Farlow, Partial Differential Equations for Scientists& Engineers, 1982, John Wiley & Sons.

[3] J.W.Gray, Mastering *Mathematica*, 1998, Academic Press.

[4] Tom Wickham-Jones, webMathematica: A User Guide, 2001, Wolfram Research.

[5] P.K.Kythe and P.Puri - M.R.Schäferkotter, Partial Differential Equations and Mathematica, 1997, CRC Press.

[6] Y. Tazawa, Symbolic Computation New Horizons IMS 2001, Tokyo Denki Univ. Press.

[7] U.Ufuktepe, An Application with webMathematica, ACA 2002, Volos, Greece, Univ. of Thessaly

[8] E.C.Young, Partial Differential Equations:An Introduction, 1972, Allyn and Bacon.

New Developments in Web Applications using J/Link

Hidekazu Takahashi
Kawase High School , Japan
hidekazu@pop.biwako.ne.jp
http://www.biwako.ne.jp/~hidekazu/
http://www3.kawase-h.ed.jp/Teachers/~Takahashi/

1. Abstract

In the 4th IMS, held in Japan in 2001, I talked about "Web Applications using J/Link for Math Education" and introduced some web applications that calculate a client's request with the *Mathematica* kernel on the web server and return the result back to the client. In this paper, I describe two main improvements I have made on these applications. The first is the integration of different web applications. Combining several separate applications dealing with 3D figures enables you to carry out more interesting and effective operations for 3D figures. For example, consider two applications: one for drawing quadrics and the other for cutting solids. If they are joined together, the user can freely draw a quadrics and cut through its surface. This can help students to learn mathematical properties of quadrics and conic section.

The second improvement consists of four new web applications added to the site. On these applications, the user can draw semi-transparent polyhedra, plot beautiful chaos images, enjoy "Origami" on the web and so on. They are based on my new packages and the powerful function of "JavaView". I hope that students enjoy these applications and get interested in mathematics.

2. Integration of some separate web applications

The following four applications I introduced in **[1]** are integrated here and user can export the graphics to the "Knife Page" from "Quadrica Page", "Revolver Page" and "Polyhedron Page" for cutting them.

< Knife Page >

This page allows you to cut through the selected solid by an arbitrary plane.

< Quadrica Page >

This page allaws you to draw any quadric simply by giving its equations.

< Revolver Page >

This page allows you to draw a solid of revolution of a selected figure around an arbitrary axis of revolution.

< Polyhedron Page >

This page allows you to cut off all edges and vertices of a regular polyhedron with arbitrary cutting ratios.

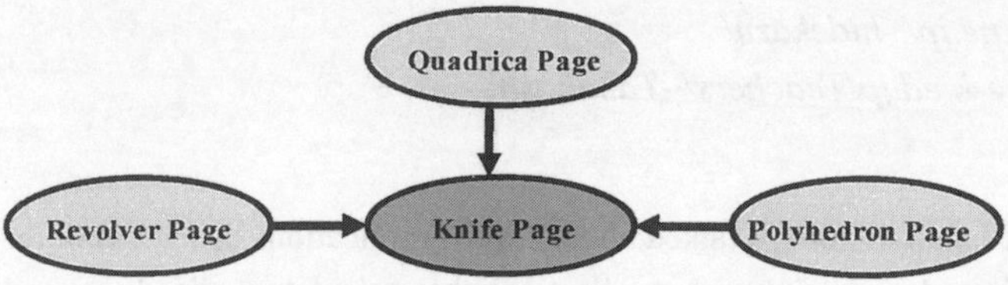

2.1 Mechanism

The mechanism of the web application using J/Link in my original servlet is shown below. The information given by the user is sent to the *Mathematica* kernel through J/Link in the servlet program and the result calculated in *Mathematica* is returned back to the user in the dynamic HTML file generated by the servlet. In the above four web applications, the results returned from the kernel are all text files of Graphics3D objects. They are kept in the temporary graphics pool on the web server and displayed on the user's web browser by the **[3] "LiveGraphics3D"** applet. The graphics files stored in the temporary graphics pool can be held in common by several web applications and make it possible to give and take the data between defferent web applications. Thus, for example, it is possible to cut the solid generated by other web applications in the "Knife Page".

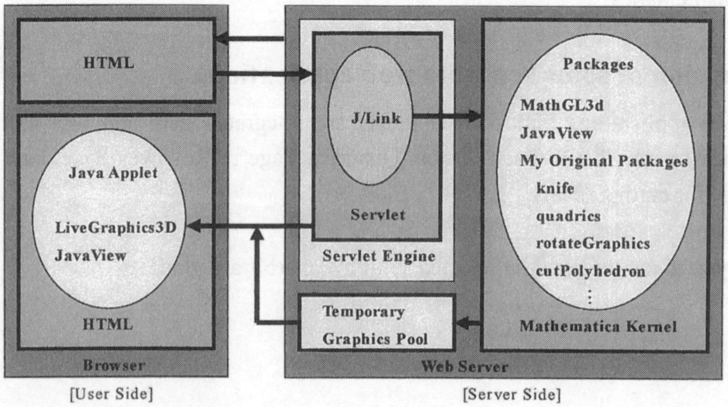

2.2 *Examples*

Quadrica Page → Knife Page

This is an example of cutting a quadric. The quadric $x^2 + y^2 - z^2 - 1 = 0$ drawn in the "Quadrica Page" is exported to the "Knife Page" and cut by the plane $x - z = 0$. Two parallel straight lines appear on the section.

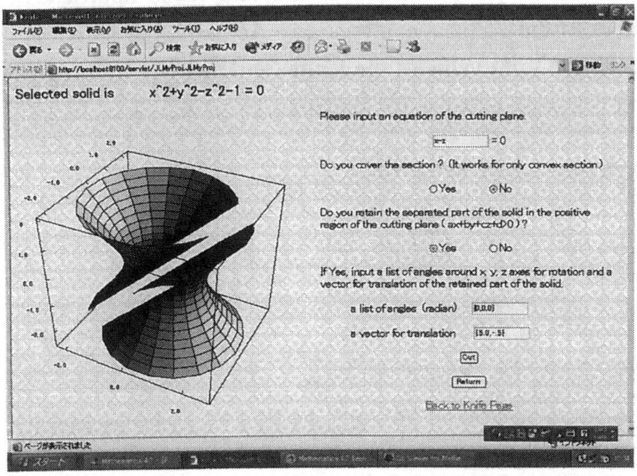

Revolver Page → Knife Page

This is an example of cutting a solid of revolution. This solid is generated in the "Revolver Page" by revolving a square around an axis which crosses the square's diagonal at 45 degrees. The cutting plane is $y = 0$ and the shape of the section gives you a hint to calculate the volume of it.

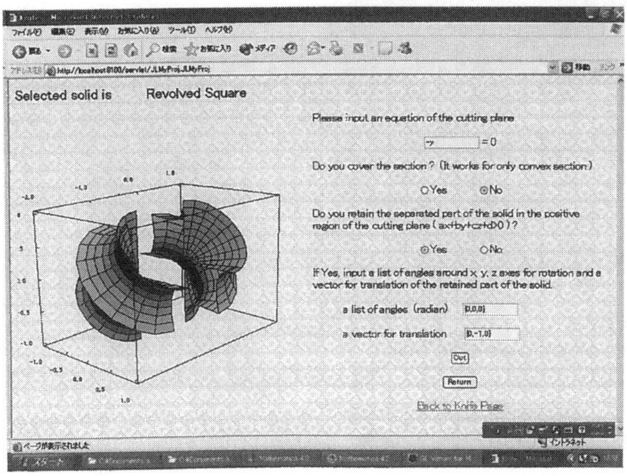

3. New web applications added to the site

3.1 Rolling Page

This page allows you to roll various 2D figures along a straight line and see the trace of the point on them. It is based on my package "**spin.m**" which rolls arbitrary 2D figures along a line and plot a locus of the point on them. The following is an example of rolling a regular polygon. Seeing the animation, students notice that the locus approches a cycloid gradually by as the number of the vertices is increased. Actually, the length and the area of a cycloid are calculated as a limit value of the length and the area of a regular polygon's locus. This calculation is a good exercise for students.

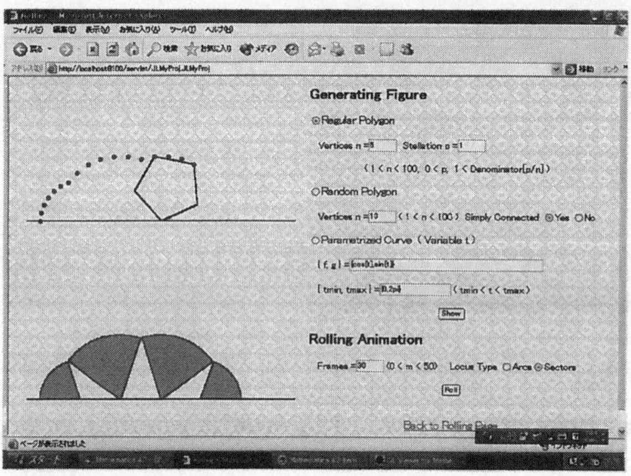

You can roll a parametrised closed curve too. This is an example of rolling an ellipse.

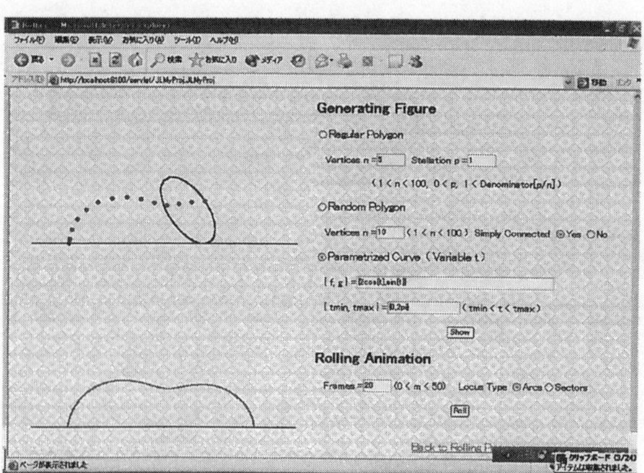

3.2 *Transparent Page*

This page allows you to draw two polyhedra that overlap each other. The outside one is semi-transparent and the inside one can be given a specified size and orientation. Though *Mathematica* can't draw semi-transparent graphics by itself, the powerful package **[4] "JavaView"** makes this possible. It is also used as an applet in this page and produces an interactive graphic which can be rotated with a mouse. Students can find interesting relations among regular polyhedra, stellated polyhedra and rhombic polyhedra in this page. The following is an example showing duality of the Dodecahedron and the Icosahedron.

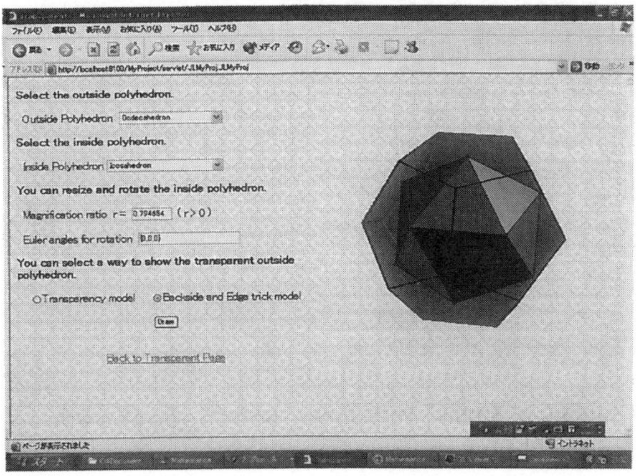

The following is an inscribed SmallStellatedDodecahedron in the RhombicTriacontahedron.

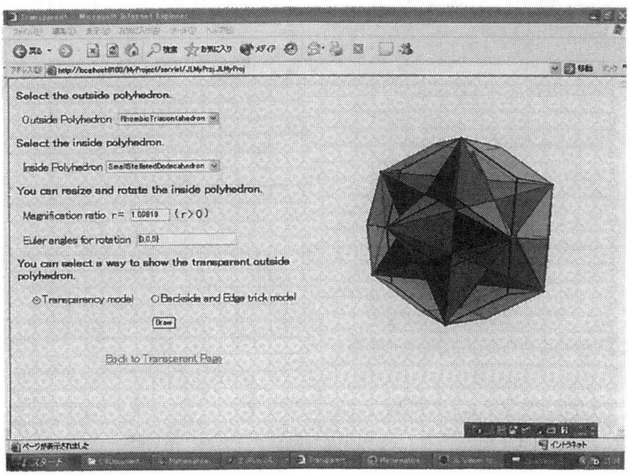

3.3 Origami Page

This page allows you to do paper folding ("**Origami**"). Two different foldings "Valley or Mountain Fold" and "Inside Reverse Fold" are supported, and various paper models can be made by using them repeatedly. This page is based on the Origami System developed in the joint research **[2]** **"Origami Programming Environment"**. The following is an example of folding an "Airplane".

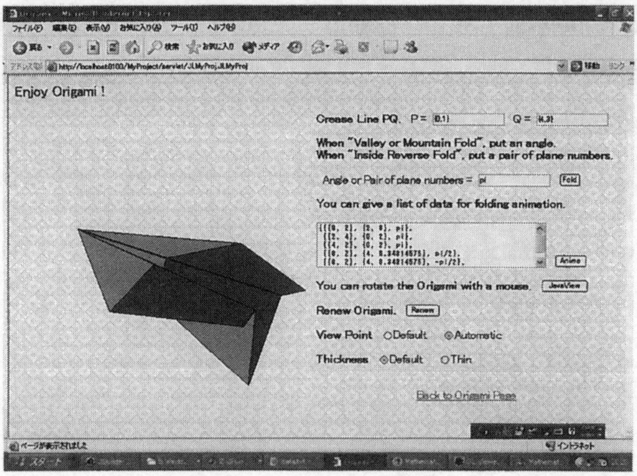

The following is an example of folding a "Crane" with flower texture. JavaView has a lot of interesting effects for 3D images. Texture mapping is one of them.

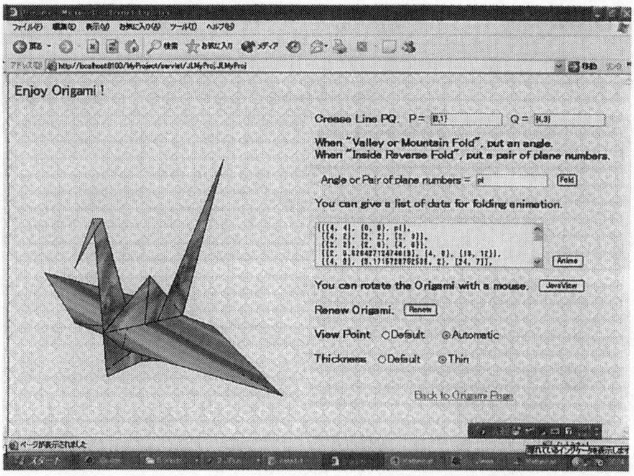

3.4 Chaos Page

This page allows you to plot a beautiful chaos image by applying a specified rule to an initial point repeatedly. You can give the rule, initial point and the number of iteration. This page uses **[5]** **"webMathematica"** technology instead of the original servlet. The following is a chaos image generated by the rule $f(x, y) = \{ax + by + c + (d\,x + e)/(x^2 + 1), -x\}$ where $\{a, b, c, d, e\} = \{-1.97, 1, 0, 5, 0\}$. The iteration depth is 50000.

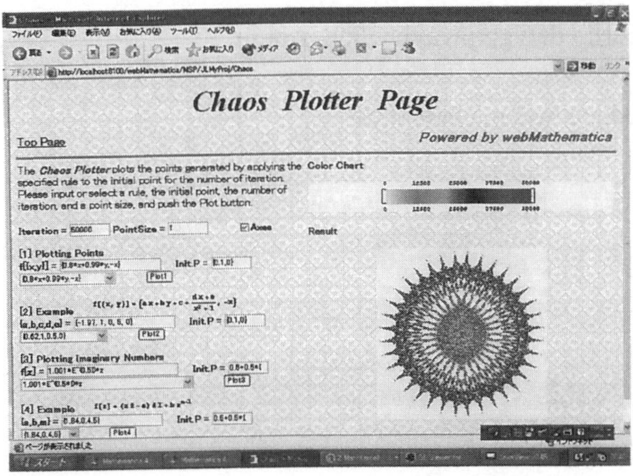

The rule can be given as a complex function. The function $f(z) = (z\,\bar{z} - a)\,\bar{z}\,i + b\,z^{m-1}$ where $\{a, b, m\} = \{2.15, 0.3, 3\}$ creates the beautiful chaos image shown below. The initial point is $0.5 + 0.5\,i$.

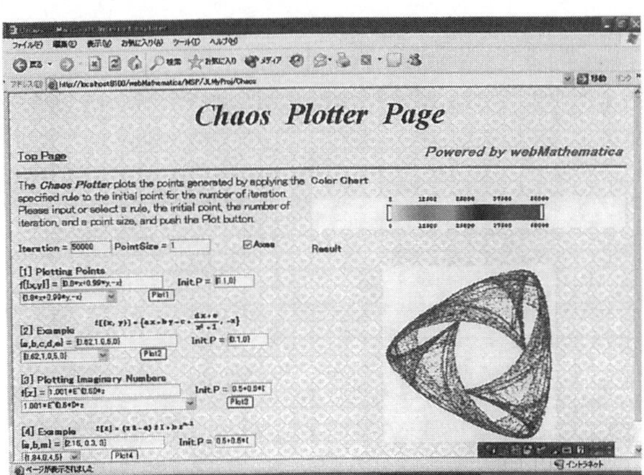

4. References

[1] **Takahashi, H.** ***Web Applications using J/Link for Math Education***, Proceedings of the fourth International Mathematica Symposium, pp.477-484, 2001

[2] **Ida,T and Takahashi, H.** *Origami Programming Environment*, Proceedings of the fifth International Mathematica Symposium, 2003

[3] **LiveGraphics3D**
http://wwwvis.informatik.uni-stuttgart.de/~kraus/LiveGraphics3D/index.html

[4] **JavaView**
http://www-sfb288.math.tu-berlin.de/vgp/javaview/index.html

[5] **webMathematica**
http://www.wolfram.com/products/webmathematica/

Experimental Education System of Mathematics based on Mathematica

Y. Tazawa
Tokyo Denki University
tazawa@cck.dendai.ac.jp

1 Abstract

The School of Information Environment of Tokyo Denki University was founded in 2001, and at this school we adopted a new system of mathematics education based on *Mathematica*. In this talk, I will introduce some of the features of this system through *Mathematica* and video-files. All *Mathematica* commands related to this article are included in the appendix *Mathematica* notebooks. on the CD-ROM Other notebooks are also available at
http://math.kn.dendai.ac.jp/tazawa/.

2 Introduction

The courseware at the new school of Tokyo Denki University consists of
(1) ordinary textbooks including instructions for *Mathematica*,
(2) CD-ROMs of *Mathematica* notebooks,
(3) CD-ROMs of video files of five minute summary of each section,
(4) video files of each lecture downloadable through the network,
(5) *Mathematica* exercises that should be handed in through the network.
It covers standard engineering mathematics, from basic linear algebra and calculus to applied mathematics including information mathematics. We use *Mathematica* here for various purposes, for example, to draw illustration figures, to create simulation tools, and of course as computer algebra and a programming tool for students.

In this talk, I will introduce some of the features of this system through *Mathematica* notebooks and video-files on the following three topics:
(1) Simulation of oscillations by numerical solutions of ODE's,
(2) Visualization of eigenvalues and eigenvectors,
(3) Finite fields and cryptography.

By clicking an underlined part of this notebook in the CD-ROM, the related *Mathematica* notebook will be opened via a hyperlink.

3 Simulation of oscillations

Oscillation is a nice example of the motion expressed by ODEs of the second order. Making use of *Mathematica*'s `NDSolve`, we can construct virtual experiment tools of oscillations as seen in the following.

3.1 Simple oscillation

`SimpleOscilation[m,c,x0]` returns the animation of the simple harmonic oscillation, where `m` is the mass of the weight, `c` is the spring constant, `x0` is the initial deviation. The motion is expressed by the solution of the following ODE.

$$m\,x''(t) + c\,x(t) = 0.$$

3.2 Damped oscillation (1)

`DampedOscilation[m,c,k,x0,n]` returns the animation of the damped oscillation, where `m` is the mass of the weight, `c` is the spring constant, `k` is the dashpot-constant, `x0` is the initial deviation, and `n` is the number of vibration. The motion is expressed by the following ODE.

$$m\,x''(t) + c\,x(t) + k\,x'(t) = 0.$$

The solution is classified into the three types, *B*, *C*, *D* (*A* is a simple oscillation).

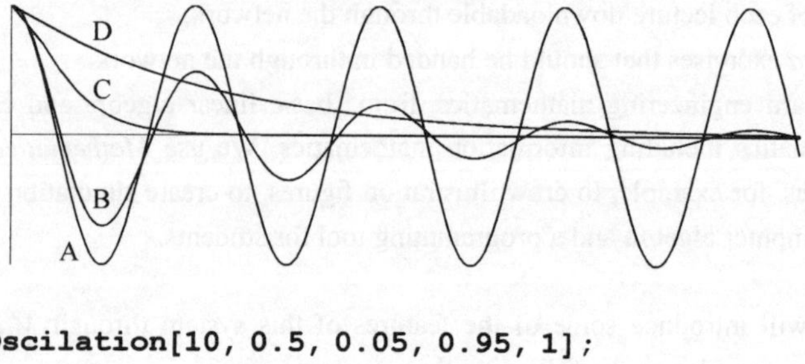

Eg B;

```
DampedOscilation[10, 0.5, 0.05, 0.95, 1];
```

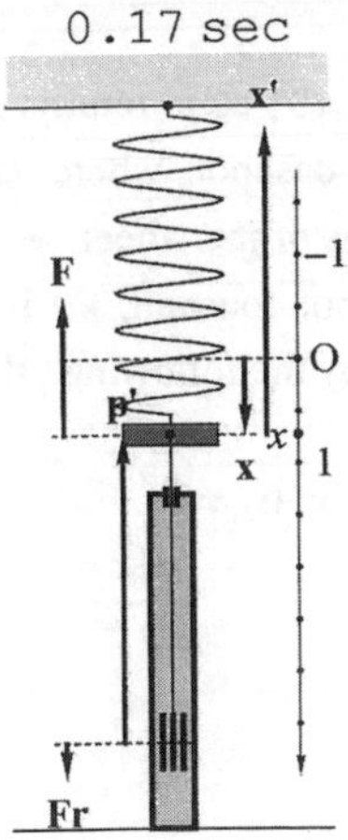

3.3 Damped oscillation (2)

Eg C;

```
DampedOscilation[0.8,18,8,1.0,4];
```

3.4 Damped oscillation (3)

Eg D;

```
DampedOscilation[1,20,4Sqrt[5],1.0,4];
```

3.5 Forced oscillation (1)

`ForcedOscilation[cc,mm,rr,kk]` returns the animation of the oscillation forced by a wheel, where `mm` is the mass of the weight, `cc` is the spring constant, `rr` is the radius of the wheel, `ww` is the rotation speed of the wheel, `w=Sqrt[cc/mm]`, and `kk` is the ratio of `ww/w0`. The motion is expressed by the following ODE.

$$mm\, x''(t) + cc\, x(t) + rr\, \mathrm{Sin}(ww\, t) = 0.$$

3.6 Forced oscillation (2)

ForcedOscilation22[cc,mm,rr,kk,ww,x0,t1] returns the animation of the oscillation forced by a wheel and damped by a dashpot, where mm is the mass of the weight, cc is the spring constant, rr is the radius of the wheel, ww is the rotation speed of the wheel, w=Sqrt[cc/mm], kk is the dashpot constant, x0 is the initial deviation, t1 is the maximal time. The motion is expressed by the following ODE.

$$mm\,x''(t) + cc\,x(t) + rr\,\mathrm{Sin}(ww\,t) + kk\,x'(t) = 0.$$

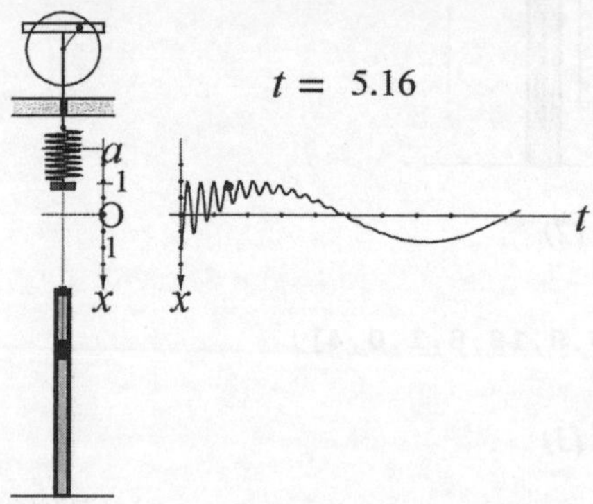

3.7 Coupled oscilation

CoupledOscilation[mm,cc,x0,y0] returns the animation of the coupled oscilation with three springs of the same spring constant and two weights of the same mass, where mm is the mass of the weights, cc is the spring constant, x0 and y0 are the initial deviations. The motion is expressed by the following system of two ODE's.

$$mm\ xx''(t) = -2\,cc\ xx(t) + cc\ yy(t),\ mm\ yy''(t) = cc\ xx(t) - 2\,cc\ yy(t)$$

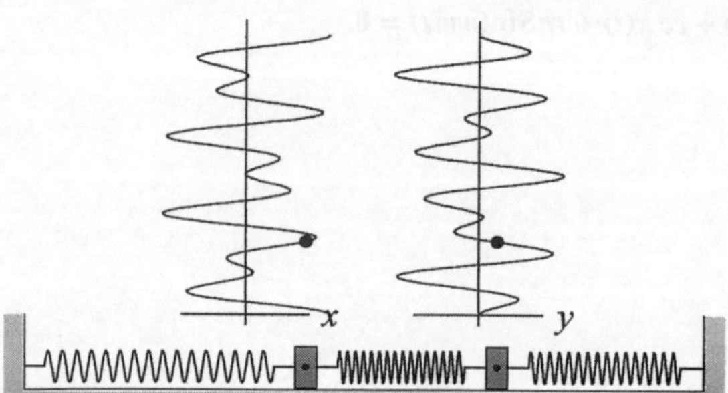

4 Eigenvectors and eigenvalues in animation

The notion of the eigenvalues and eigenvectors of matrices is sometimes hard for students to follow. This is partly because the multiplicity of the eigenvalues and that of eigenvectors are different. In the following, **eigenvectoranimation[mat]** returns an animation which shows the situation of the real eigenvalues and eigenvectors of a real 2×2 matrix **mat**. The unit vector represented by a blue arrow is mapped by **mat** into a vector represented by a red arrow.

Example 1

$$\begin{pmatrix} 2 & 0 \\ 0 & 2 \end{pmatrix}$$

Homothety. One real eigenvalue 2 (multiplicity = 2), all vectors are eigenvectors.

Example 2

$$\begin{pmatrix} -2 & 0 \\ 0 & -2 \end{pmatrix}$$

Homothety. One real eigenvalue -2 (multiplicity=2), all vectors are eigenvectors.

Example 3

$$\begin{pmatrix} 0 & 0 \\ 0 & 0 \end{pmatrix}$$

0-matrix. One real eigenvalue 0 (multiplicity=2), all vectors are eigenvectors.

Example 4

$$\begin{pmatrix} \frac{\sqrt{3}}{2} & -\frac{1}{2} \\ \frac{1}{2} & \frac{\sqrt{3}}{2} \end{pmatrix}$$

Rotation. No real eigenvalue.

Example 5

$$\begin{pmatrix} 1.7 & -1 \\ 0.5 & 1 \end{pmatrix}$$

No real eigenvalue.

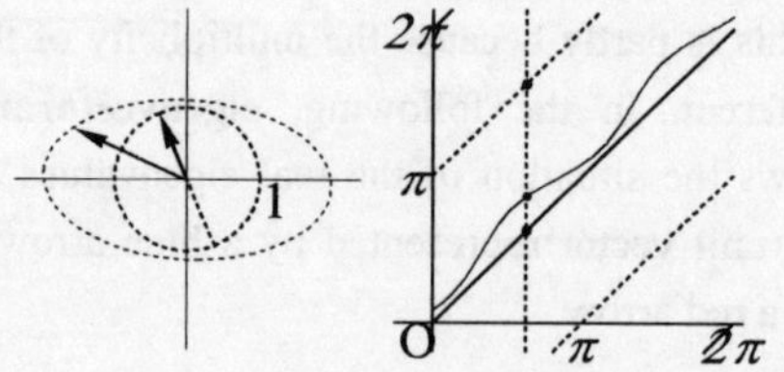

The matrix has
no real eigenvectors.

Example 6

$$\begin{pmatrix} 1.7 & -2 \\ -2 & 0.8 \end{pmatrix}$$

Two distinct real non − zero eigenvalues. Two eigenvectors

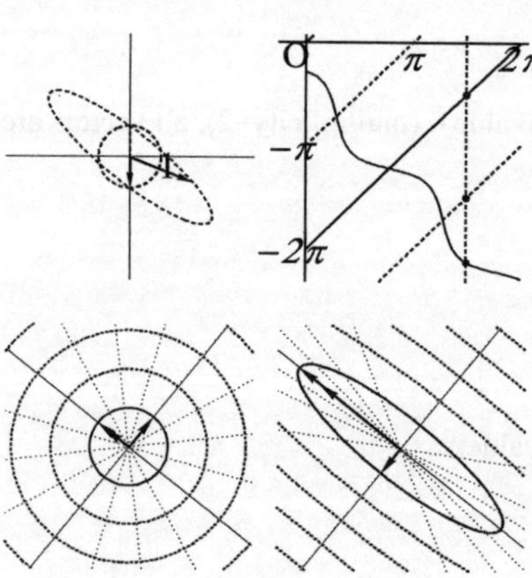

Example 7

$$\begin{pmatrix} 1 & 1.2 \\ 1.2 & 1.44 \end{pmatrix}$$

projection to a straight line. Two distinct real eigenvalues (0, and non-zero), two independent eigenvectors.

Example 8

$$\begin{pmatrix} 1 & 0 \\ 0 & 0 \end{pmatrix}$$

projection to the *x*-axis. Two distinct real eigenvalues (0, and non-zero), two independent eigenvectors

Example 9

$$\begin{pmatrix} 0 & 1 \\ 0 & 0 \end{pmatrix};$$

One real eigenvalue 0 (multiplicity = 2), one real eigenvector.

Example 10

$$\begin{pmatrix} 0 & 0 \\ 2 & 0 \end{pmatrix}$$

One real eigenvalue 0 (multiplicity = 2), one real eigenvector

5 Cryptography

The data manipulation and algebraic computation of *Mathematica* enables us to handle cryptography. I will introduce the classical cryptography by Julius Caesar and the method based on the linear algebra on a Galois field.

5.1 Strings & digits

`stringtodigits[strng]` converts the string of letters into a list of digits. **`digitstostring[lst]`** converts the list of digits into a string of letters.

5.2 Caesar's method (1)

crypt[num][txt] encodes the text **txt** by Caesar's method shifting each letter by the number **num** in the alphabetical list. **uncrypt[num][txt]** decodes the encoded text **txt**.

5.3 Caesar's method (2)

An example of a long text

5.4 Primes

findprime[a,b] returns the prime numbers between two natural numbers **a** and **b**.

5.5 Euclidean algorithm

mygcd[a,b] returns three integers {d, x, y}, where d is GCD of a and b, and x,y are integers satisfying ax+by=d.

5.6 Galois field

The addition and subtraction $a \pm b$ in the finite abelian group $\mathbb{Z}_n=\{0,1,2,\ldots,n-1\}$ are calculated by **Mod[a±b,n]**. If n is a prime number, then $\mathbb{Z}_n$ is a finite field (Galois field). The multiplication is calculated by **Mod[a*b,n]**. For Division, **invmod[i,n]** returns the inverse of **i** in $\mathbb{Z}_n$. **invmatmod[mat,n]** returns the inverse of a 2×2 matrix in $GL(2,\mathbb{Z}_n)$.

5.7 Matrix method

cryptmat[num,mat][txt] encode the text **txt** by 2×2 matrix **mat** in the general linear group $GL(2,\mathbb{Z}_{\mathbf{num}})$. **uncryptmat[num,mat][txt]** decodes it.

6 References

[1] Y. Tazawa et al, Calculus for Information Science with *Mathematica*, to appear from Tokyo Denki University Press, 2003 (in Japanese).

[2] Y. Tazawa et al, Linear Algebra for Information Science with *Mathematica*, to appear from Tokyo Denki University Press, 2003 (in Japanese).

[3] Y. Tazawa, Applied Mathematics for Information Science with *Mathematica*, to appear from Tokyo Denki University Press, 2004 (in Japanese).

Education, Design and Implementation Aspects of a Generic Step-by-Step Solver based on *Mathematica*

Bernhard Zgraggen
Distance University of Applied Sciences of Switzerland
bzgraggen@fernfachhochschule.ch

1 Abstract

The starting point of this paper is a project at the Distance University of Applied Sciences of Switzerland, which deals with the development and deployment of programs interactively generating detailed and dynamic step-by-step solutions to typical problems in higher mathematical education, based on *Mathematica*. The first part of the paper explains organizational and didactic aspects, empirical values, advantages and disadvantages as well as proposals for improvements. The second part puts the emphasis on an approach to a generic step-by-step solver and its interface, enabling tutors to create step-by-step solutions according to the mathematical context by setting appropriate parameters. The functioning of such an algorithm and its parameters is explained in detail, based on an example from Boolean algebra.

2 Educational aspects

2.1 Organizational background

It is an important part of the basic diploma courses, (economics, engineering and computer science), at the Distance University of Applied Sciences of Switzerland to teach basic knowledge of applied mathematics. There is a permanent and serious lack of time for seminars in the classroom and for students' homework because the university puts the main emphasis on distance learning and because the students have to work for their living. Generally, the time for homework is roughly five times the time spent in the classroom. Electronic mail and a professional web-based learning management system ensure support outside the classroom. The teaching material consists mainly of textbooks, accompanied by a precise syllabus.

Because of the acute shortage of time and because the students live in many different places, it is unfortunately not practicable to introduce powerful mathematical software like *Mathematica* or to carry out mathematical laboratory software experiments in the classroom. For this reason the university decided to support its students with web-based mathematical tools. In a first project, about seven web-based training modules for applied mathematical topics were developed, using technologies such as HTML and Java [1]. The second and current project is more demanding and is called *webSolutions* [2,3]. It is powered by Wolfram's *webMathematica* [4]. In this project the main emphasis is put on the development of programs generating detailed and dynamic step-by-step solutions to problems in higher mathematical education, covering more advanced topics like multi-dimensional analysis and calculus, linear algebra, linear and non-linear optimization and mathematical logic.

2.2 Didactic aspects

Many mathematical textbooks have clear disadvantages when used in a distance education environment. We often find that solutions are incompletely explained, and prime examples, exercises and corresponding solutions are static. Furthermore, the electronic reprocessing of parts of mathematical textbooks, (as attachments to an e-mail or as spontaneous presentations), cannot be carried out easily within a reasonable amount of time.

Lecturers often complain about the impossibility of presenting complete solutions to more difficult exercises at home or in the classroom in the available time.

For technical, syntactical and semantic reasons it is normally difficult for students in a distance learning environment to ask specific questions concerning an actual or imagined mathematical problem. On the other hand, tutors and lecturers are often confronted with technical and syntactical difficulties when trying to answer mathematical questions by electronic means.

Successful applications of the current tools of *webSolutions* [3] require that students have prepared the theory corresponding to a problem area well enough to be familiar with the corresponding problems, mathematical terms and formulation. When this condition is fulfilled the tools may help to reach the following goals: **1)** Students learn and practise applying structured methods of solution based on concrete problems; **2)** Students control and correct their own handwritten solutions to exercises; **3)** Tutors and students present, reprocess and communicate complex solutions within an appropriate time.

2.3 Advantages and disadvantages

The educational advantages of application of the current tools can be summarised as follows: **1)** Students and tutors can receive step-by-step solutions to mathematical problems in their browsers. The solutions are probably more structured, detailed and dynamic than solutions in ordinary textbooks. Thus, the tools may help to alleviate the shortage of solutions in textbooks and time in the classroom. **2)** Students and tutors are not restricted to the exercises in their textbooks; the tools are designed to be flexible so they can be combined with a wide range of teaching materials. **3)** Accessing the tools is easy and mainly independent of the users' platforms and browsers. The tools are based on a technology which supplies interfaces to the most important internet technologies. Thus it is not at all difficult to combine the tools, e.g. with electronic platforms. **4)** The tools may support tutors when preparing lessons, presenting mathematical topics and coaching their students.

Of course, the current versions of the tools also have a great many disadvantages: **1)** Creating and testing programs computing dynamic step-by-step solutions within *Mathematica* takes a lot of time. **2)** Sometimes computed step-by-step solutions may be rather complex - although mathematically correct. Such results may confuse or discourage students. **3)** The level of interactivity is rather low. **4)** The user-friendliness is still not sufficient. **5)** Tutors do not have the possibility of adding their own material, modifying the existing material or manipulating the program fundamentally.

2.4 Possible improvements

The most important things to do, from a student's point of view, must be to raise the level of interactivity and to improve the user-friendliness. These goals can be reached by, e.g., **1)** allowing students to propose rules or interim results between the steps of the solution; **2)** enabling students to browse backwards and forwards through the separate steps of the solution; and **3)** interactively providing more information on certain interim results or terms when desired, (e.g. step-by-step explanations on how the zeros of the partial derivatives are found when considering an optimization problem in several variables).

From the tutor's viewpoint - in addition to the suggestions above - the most interesting improvements concern the possibility of adding one's own material, modifying the existing material and manipulating the functioning of the programs. These improvements may be realized by developing a kind of generic algorithm, allowing tutors to add mathematical transformation rules, conditions on rules, formatting rules, output comments and further functionality to the programs, to influence the foreground sequence of the rules and to set parameters which fundamentally control the program run.

3 Design and implementation aspects

Many mathematical problems can be solved using transformation rules. By repeated and careful application of such rules a mathematical expression is transformed until it fulfills a specific condition or matches a specific form. Typical examples of such problems are; solving a system of linear equations, solving a linear optimization problem, finding the limit of a real function, computing the derivative or the integral of a function and simplifying an expression in a specific algebraic context.

Of course, *Mathematica* contains a large number of efficient and ingenious generic algorithms, such as 'Solve' or 'FullSimplify', which are based on an extensive and cleverly devised collection of transformation rules. As far as we know, design and implementation of these algorithms and the underlying collection of transformation rules are not public. In our opinion, a detailed and didactically practical representation of a step-by-step solution using *Mathematica*'s tracing algorithms, for example, is therefore a difficult programming undertaking.

In this section, we would like to show design and implementation possibilities for a parameterizable procedure, computing and representing detailed step-by-step solutions based on existing algorithms of *Mathematica*. The procedure should be usable in different mathematical contexts by providing parameters for transformation rules, additional rule conditions and rules for formatting the output, as well as by providing parameters allowing tutors to influence the program run.

3.1 Example: Simplifying Boolean Expressions

In this subsection we roughly follow the lines developed in [5].

A small sample of rules for transforming Boolean expressions, implemented in *Mathematica*, might look like this:

```
CommutativityRule2 = a_ || b_ :→ b || a;
CommutativityPrintRule2 = bool[a || b] :→ bool[b || a];

DistributionRule1 = a_ && (b_ || c_) :→
a && b || a &&c /; a==!b || a==!c;
DistributionPrintRule1 = bool[a && (b || c)] :→
bool[a] && bool[b] || bool[a] && bool[c];

TautologyRule4 = a_ || b_:→ True /; a==!b || b==!a;
TautologyPrintRule4 = bool[a || !a] :→ True;
```

We see that some rules contain additional conditions, e.g. the condition **a==!b || a==!c** for DistributionRule1.

More complete lists of rule names and rules, respectively, are given in the lists below:

```
BooleanRulesNames =
{"TautologyRule1", "TautologyRule2", "TautologyRule3",
"TautologyRule4", "ContradictionRule1",
"ContradictionRule2", "ContradictionRule3",
"ContradictionRule4","AbsorptionRule1",
"AbsorptionRule2","AbsorptionRule3","AbsorptionRule4",
"IdempotenceRule1","IdempotenceRule2",
"DoubleNegationRule","DeMorganRule1","DeMorganRule2",
"DistributionRule1","DistributionRule2",
"DistributionRule3","DistributionRule4",
"CommutativityRule1", CommutativityRule2"};
```

(A similar list for the print rules is **BooleanPrintRulesNames**; it is not shown here.)

```
ListBooleanRules = ToExpression[BooleanRulesNames]
{a_||b_/;a==True||b==True:→True, a_&&b_:→b/;a==True,
a_&&b_:→a/;b==True, a_||b_:→True/;a==!b||b==!a,
a_||b_:→b/;a==False,a_||b_:→a/;b==False,
a_&&b_:→False/;a==False||b==False,
a_&&b_:→False/;a==!b||b==!a, a_&&(b_||c_):→a/;b==a||c==a,
(b_||c_)&&a_:→a/;b==a||c==a, a_||b_&&c_:→a/;b==a||c==a,
a_&&c_||a_:→a/;b==a||c==a, a_&&a_:→a, a_||a_:→a,
!(a_&&b_):→!a||!b, !(a_||b_):→!a&&!b,
a_&&(b_||c_):→a&&b||a&&c/;a==!b||a==!c,
(b_||c_)&&a_:→b&&a||c&&a/;a==!b||a==!c,
a_||b_&&c_:→(a||b)&&(a||c)/;a==!b||a==!c,
b_&&c_||a_:→(b||a)&&(c||a)/;a==!b||a==!c,
a_&&b_:→b&&a,a_||b_:→b||a}
```

The rules in **ListBooleanRules** are sequentially and repeatedly applied to a Boolean expression using *Mathematica*'s ReplaceAll function, until the expression's structure does not change any longer although each rule has been considered. Whenever

a rule is applied, (changing the structure of the boolean expression), it is appended to the list of output strings using the special output formatting rule:
`{bool[p_]:>StandardForm[p]}/.{a→𝒜, b→ℬ, c→𝒞}`.
(* The wrapping head ‚bool' has the advantage, e.g., that bool[a||b] =!= bool[b||a] *)

Moreover, the transformed boolean expression is appended to the list of output strings using the formatting output rule `{bool[p_]:>StandardForm[p]}`.

This procedure is very simple, but has the obvious disadvantage of creating, in many cases, endless loops. Starting with the expression (a || b), for example, the program will end up in an endless loop, always applying the second commutativity rule. The variable ‚booleanexpr' alternately takes the values bool[a||b] and bool[b||a].
Possible solutions to avoid these problems are:

1) to do without critical rules, e.g. commutative rules, and to add commutative variations of the other rules, (such as `DistributionRule2 = (b_ || c_) && a_:> a && b || a &&c /;a==!b || a==!c;`).
2) to add conditions for the application of the rules, e.g. the condition of non-atomic parts with respect to commutativity rules (`CommutativityRule2 = a_ || b_:> b || a /;!AtomQ[a] || !AtomQ[b];`).
3) to introduce an evaluation function *eval*, (comparable to *Mathematica*'s ComplexityFunction in FullSimplify), playing a crucial role in conditions for rules:
`CommutativityRule2 = a_ || b_:> b || a/;`
`(eval[(b || a) < eval[(b || a));`
A typical example for such a function is *Mathematica*'s LeafCount.
4) to introduce controlling parameters in order to influence the program run, e.g. a parameter bounding the depth of the outer iteration, or the use of an evaluation function as mentioned above, so that rules leading to expressions with a low evaluation value have priority over other rules.

A combination of these suggestions was used for Bernouilli-l'Hospital rules when programming a step-by-step solver for limit problems of real functions [3].

It is a serious consequence of the proposals mentioned above that in many cases optimal solutions may not be found any more. Changing rules and adding conditions to rules or controlling parameters may have unpredictable or unverifiable impacts on the solubility of a mathematical simplification or transformation problem.

3.2 *Approach to a generic algorithm*

A generic approach should take advantage of *Mathematica*'s performance and be based on its functions to the widest possible extent. Firstly, proposal solutions, (maybe pattern-like or conditional), are computed using appropriate *Mathematica* functions. Then, lists S_i, $(i = 0, 1, 2, \ldots)$, consisting of transformed expressions with corresponding history tables, are created by systematic and iterative application of all the rules. The iteration stops when an expression which corresponds to the proposed solution is created, an empty list S_i occurs or a more general stop condition is fulfilled.

The procedure will be illustrated by the following example:

Initial expression: a || !a
S_0**= { { (a || !a), { } };**
S_1**= { { True, {TautoloyRule3} }, { (!a || a), {CommutativeRule2} } } -** S_0**;**
S_2**= { { True, {TautoloyRule3} }, { True, {CommutativeRule2, TautoloyRule3} }, { (a || !a), {CommutativeRule2, CommutativeRule2} } }**
- $(S_1 \cup S_0)$ **= { };**

The lists S_i consist of pairs whose first parts represent the expressions and whose second parts represent the history tables of applied rules. The difference operator is based only on structural coincidences of the first (expressional) parts. In our example, the procedure would stop after computing S_1, provided that the proposed solution is the same as **True**.

Generally, the sequence of lists S_i $(i = 0, 1, 2, \ldots)$ is defined as follows:

S_0**= { {initial-expression}, { } };**
S_n is computed by applying sequentially all the transformation rules to the expressional parts of the elements in S_{n-1}, yielding a list of pairs consisting of the transformed expressions and extended history tables, while obeying the condition that the transformed expressions are not in any of the former lists S_i, $(i = 0, 1, \ldots, n-1)$.

3.3 *Parameters of a generic algorithm*

This section gives an outline of the parameters of a generic approach to a step-by-step solver in *Mathematica*.

General stop conditions: List of conditions, e.g. controlling the maximum depth of iterations or lengths of the lists S_i mentioned above. The default value for the general stop conditions is an empty list.

Lists of pre-functions and post-functions: The purpose of pre-functions or post-functions is to do things before or after the beginning of the problem solving procedure, respectively, such as printing comments or setting attributes for operators. The default value for these lists is { }.

Function for possible solution: The purpose of this function is to compute a proposal for the solution of the underlying problem. The default for this function is *Mathematica*'s FullSimplify, with appropriate transformation rules, (built from the user's parametric list of transformation rules mentioned below), and an appropriate complexity function determined by the user's parametric complexity function mentioned below.

Complexity function: This function allows us to compare the complexity of expressions during the program run. It may serve as an ordering function for storing the expressions in the lists S_i, as a default argument for the ComplexityFunction in *Mathematica*'s FullSimplify or as an ordering function for the application of the rules.

The parameters, list of transformation rules, list of conditions on rules, list of formatting rules for the output of expressions, list of formatting rules for the output of applied transformation rules and list of additional textual comments or explanations were essentially explained in section 3.1.

4 References

[1] Bernhard Zgraggen, Web site of the first project: http://www.ffhs.ch/zgraggen/webmath/index.html.

[2] Bernhard Zgraggen, Interactive and dynamic step-by-step solutions to mathematical problems in the World Wide Web, to appear in Proc. 5th ICNEE, Lucerne, Switzerland, May 2003.

[3] Bernhard Zgraggen, Web site of the project *webSolutions* http://www.ffhs.ch/zgraggen/webmath/webmathematica-ffhs-publicity.html.

[4] Web site of Wolfram's *webMathematica*: http://www.wolfram.com/products/webmathematica.

[5] George Beck, Step-by-Step Differentiation, http://documents.wolfram.com/v4/GettingStarted/Step-by-StepDifferentiation.html

The Development of the *Mathematica* Package '*StandardPhysicalConstants*'

V. Ezhela and V. Larin
Institute for High Energy Physics, Russia

{ezhela, larin}@mx.ihep.su

1 Abstract

We report on further development of the *'StandardPhysicalConstants'* package, presented at IMS2001. We address these issues: package structure; current status of the physical constant database; data sources; current data collection and data structure; the main modules of the data management system; the first version of "error propagator"; usage examples in calculations for high precision tests of physics theories. We also indicate future developments of the basic package.

2 The package structure

The general package structure has not changed much since the first presentation [1]. The current version also consists of two files -- the expanded constants database (*Dbfpc110.m*) and the expanded constant manipulation module (*Manager.m*). We add the module *ErrorPropagator* to *Manager.m*. It is designed to calculate the average values of algebraic functions depending upon fundamental constants, their variances, and their standard absolute and relative uncertainties. To simplify manipulation with constants when designing formulae for the physical observables in *Mathematica* we have created two palettes: the file with package names of the physical constants (*CPalette.nb*) and the other one with the names of the package functions (*FPalette.nb*).

3 Current status of the physical constant database

The structure of the database is slightly changed compared with the previous version. We added one new table with the standard constant names as presented by CODATA [2]. The number of constants stored in the new version of database is increased from 60 to 110, and this is reflected in the name *Dbfpc110.m*.

3.1 The data structure

All data on fundamental physical constants are grouped in the unique list named as *dataBase:*

dataBase = { basicTable, standardNames, ccTable },

where: *basicTable* is the table of the package constant names, constant average values with units, and their standard uncertainties with units; *standardNames* is the table (list) of the "String" type elements representing the standard CODATA constant names; *ccTable* is the table of the pair-wise correlation coefficients stored as the list of upper off-diagonal matrix elements of the correlation matrix which is symmetric and real by definition. The structures of the first and third tables are the same as in the first version of the package where they were named as *"basicl"* and *"cct"*. Only dimensions are increased in each table in accordance with increased number of constants in the database.

3.2 Current data collection, sources, and maintenance

The data set of the current version of the database was collected from the same sources as the first version: source [2] (unchanged) and source [3] (renewed as of 2002 edition). Unfortunately the data collection still were performed "by hand" for a few reasons. The basic problems are the absence of data in a computer readable form in [3] in a , and the "misanthropic manner" of web access to the correlation data on the CODATA and NIST sites. Furthermore, when running the *ErrorPropagator* check on real calculations for precision physics we met serious problems with usage the "CODATA recommended values of the fundamental physical constants:1998", as they presented in both paper and electronic forms.

3.3 Warning

It seems that the final CODATA recommended values were seriously damaged at the release and publication preparation stage, and they should be used with great caution. This is a very preliminary warning that we are clarifying. The report on this clarification and our suggestions to improve the situation will be presented elsewhere soon. In the forthcoming sections we mention the problem once again with an example. Thus, the problem of automated maintenance of the database is postponed to the next iteration of the NIST re-evaluation of the fundamental physical constants.

4 Current status of the data manipulation package

The data manipulation package now consists of 17 modules, 6 of which were presented in [1]. Their functional options remain unchanged but performance is improved. Here we describe new modules and describe the *ErrorPropagator* in detail.

4.1 *Module 'Access'*

This module is the service module for the other modules of the package, but its functions to extract and convert the internal representation of the constants into the standard list might be useful for different applications. Here is an example of usage of this module to extract data on muon-electron mass ratio:

```
In[3]:= Access[basicTable, "muonElectronMassRatio"]
```

```
Out[3]= {206.7682657, 6.3×10^-6, 3.0×10^-8}
```

Here the first element of the list is the average value of the ratio, the second element is the standard uncertainty of the ratio, and the third one is the standard relative uncertainty. In the database these data are held as a string "206.7682657(63)".

4.2 *Module 'Shift'*

This module is to move specified constants to an another place in the *basicTable* with simultaneous corresponding rearrangements in the *standardNames* and *ccTable* structures.

4.3 *Module 'PrintDB'*

This module is to print all or part of the database, as well as the user-defined working sub-base created by the *NewListFPC* function and indicated as a parameter of *PrintDB*, in conventional form.

5 Module *'ErrorPropagator'*

The main purpose of the module is to calculate the average value and variance (standard uncertainty) of any function of any reasonable number of physical constants recognizable by *Mathematica*. In due course, this module will be able to calculate average values, variances, and the covariance (correlation) matrix for any vector function of physical constants recognizable by *Mathematica* .

5.1 *Format*

The *ErrorPropagator* function has the format:

***ErrorPropagator[formula, database, {control_keys}]*,**

where the first parameter is the formula expressing the dependence of an observable (function) on physical constants, the second parameter is the name of the database used, and the third parameter is an optional list of control keys. The database can be the main database, the working database prepared by user on the basis of the main *dataBase*, or a user defined database with an structure identical to the *dataBase*.

5.2 *General algorithm*

The general algorithm comprises these steps:

- Parsing of the formula for an observable Q, expressed in terms of the physical constant package names from the database to extract the names of constants needed for calculations;
- Forming the temporal working sub-base and extracting the corresponding values;
- Calculating the average value $\langle Q\rangle$, variance $u(Q)^2$, standard uncertainty $u(Q)$, and the relative uncertainty $u_r(Q) = u(Q)/\langle Q\rangle$ if $\langle Q\rangle$ is nonzero.

The results of calculations are collected as the standard list $\{ \langle Q\rangle, u(Q), u_r(Q) \}$.

5.3 *Statistical procedure*

Let a physical observable Q depend on N random variables (physical constants in our case) $x_i : Q = Q(x_1, x_2, ..., x_N)$. Then the variance $u(Q)^2$ is defined as

$$u(Q)^2 = \sum_{i=1}^{N} \left(\frac{\partial Q}{\partial x_i}\right)^2 u(c_i)^2 + \sum_{i\neq j}^{N} \frac{\partial Q}{\partial x_i}\frac{\partial Q}{\partial x_j} u(c_i)\, u(c_j)\, r_{ij},$$

where partial derivatives calculated analytically are evaluated at values of $x_i = c_i$, $u(c_i)$ is the standard uncertainty of the constant c_i and r_{ij} is the corresponding correlation matrix element from the current working database. The covariance matrix, if non-degenerate, should be positive definite by definition. In the case when some of the random quantities are fully correlated ($|r_{ij}| = 1$), the correlation matrix will have some zero eigenvalues. Hence, before starting the variance calculation, a test for positive definiteness is done. If matrix is positive definite, the module will produce the standard output. If not, a warning is generated with a printout of the minimal eigenvalue. Calculations are performed in the standard way and then, in addition, all off-diagonal matrix elements are set to zero and calculations are repeated with the identity correlation matrix. In this case the output will be a list with the correlation submatrix used and two lists of calculated values with and "without" off-diagonal matrix elements.

5.4 Control parameters

The third optional parameter in the *ErrorPropagator* is, in general, a two-component list. The first is a control key to automate changes of the values of the temporal correlation matrix elements. The values of the first component are as follows:

- **(0)** - to set all off-diagonal matrix elements to zero;
- **("Null")** - to replace all unknown matrix elements r_{ij} (that are set to zero by default) by new "package" variables cC[i,j], for which one can assign any reasonable values;
- **({{i,j,value}, ...})** - to replace the value of the specified by (i, j) correlation coefficient with a new value without changing in the temporal correlation matrix.

The second component is to control the output. If it is absent then the output will be the standard one. In the current version the values and meanings of the second component of the control key are as follows:

- **("CM")** means that in addition to the standard output the correlation matrix used will be presented;
- **("EM")** means that in addition to the standard output the list of values of the individual terms in the sum defining the variance will be presented, in order to analyze the source of major contributions to the variance;
- **("All")** means that all the above lists will be in the output.

6 Examples of the *ErrorPropagator* applications

We produced a few checkout calculations using this module. For the checkout we decided to reproduce some recent calculations [5] of the hyperfine muonium splitting (HFS) and present here only the result for the lowest order (fine structure constant) term.

6.1 *The muonium hyperfine splitting (HFS)*

The detailed discussion of the re-calculation results will be done elsewhere soon [4]. We have reproduced the numerical results obtained in [5] with values of the relevant fundamental constants without correlations, and failed to reproduce it exactly when using CODATA-recommended values with correlations. For illustration we quote the result for the lowest order term calculated by *ErrorPropagator*

$$\nu_F = \frac{16}{3}\,\alpha^2\, cR_\infty\, \frac{\mu_\mu}{\mu_B}\left(\frac{m_R}{m_e}\right)^3,$$

where: α - the fine structure constant, c - velocity of light in vacuum, R_∞-Rydberg constant, $\mu_\mu/\mu_{§£}$ - muon magnetic moment to Bohr magneton ratio, m_e - electron mass, m_μ - muon mass, and $m_R = m_e m_\mu/(m_e + m_\mu)$.

```
In[4]:= ErrorPropagator[νF, dataBase] /. Second → 10^3 kHz^-1

Out[4]= {4.4590321×10^6 kHz, 0.131638 kHz, 2.95217×10^-8}
```

6.2 *Rydberg constant*

As a rule the physical observable might be expressed in terms of physical constants in different algebraically equivalent forms. For self-consistent values of the constants with proper estimated correlations the numerical results should be statistically consistent. This means that the results can deviate from each other no more than one standard deviation. We have performed a few calculations to check "the re-parameterization consistency" on the CODATA recommended values of the fundamental physical constants (1998 evaluation). One example is the Rydberg constant :

$$R_\infty = \alpha^2\,\frac{m_e\, c_0}{2\,h}$$

```
In[6]:= ErrorPropagator[R∞, dataBase] /. Joule →
            Kilogram Meter^2 Second^-2 /. Meter -> 1
```

```
(* We have removed units here for brevity *)

        *** WARNING:  correlation
         matrix is non-positive definite!

                          Minimum eigenvalue
          of this matrix is  -0.000425096

Out[6]= {{{1., -0.092, 0.996},
     {-0.092, 1., 0.002}, {0.996, 0.002, 1.}},
    {{1.09737316×10^7, 0.0251319 i, 2.29018×10^-9 i},
     {1.09737316×10^7, 1.22497, 1.11628×10^-7}}}
```

From the results presented it is seen that when calculating correlations with the CODATA values we obtain nonsense - the variance is negative (!). If we ignore correlations, we obtain a standard uncertainty which is five order of magnitude larger than the value quoted by CODATA.

7. The future of the package

The considered examples showed that correlations are very important in precision calculations. We will continue to develop the package to solve the remaining problems:

- The automatic control of the accuracy and precision of the calculations at all stages, and especially in the presentation of the results when working with vector functions;
- The organization of the automatic database refreshment after new releases of the CODATA recommended values. The simplest way is to ask NIST or CODATA to create a compact computer-readable file without unjustified rounding of numerical values, accessible by FTP or by Web browsers, for convertion into the *Mathematica* fundamental constants data base;
- The creation of the database containing units and conversion factors. Organization of an automated handling of units during calculations.

In conclusion, we summarize the preliminary results obtained so far in our activity with fundamental constants in *Mathematica*:

- The current standard *Mathematica* package *PhysicalConstants* is not enough and should be extended to include handling the standard uncertainties and correlations. Our package is aimed to fill this gap;

- The "CODATA recommended values of the fundamental physical constants:1998" are badly conditioned, i.e. some non-degenerate sub-matrices of the CODATA correlation matrix published on the Web are not positive definite, presumably because of the unjustified independent rounding of the numerical values of the correlation coefficients;
- It will help if the notion of "Hermitian matrix" will appear in *Mathematica,* with the assurance of conservation of the main features of Hermitian matrices in all numerical transformations, especially for Hermitian matrices close to degenerate.

8 References

[1] Ezhela, V. and Larin, V. The physical constants for *Mathematica*, in Proc. of the fourth International *Mathematica* Symposium, Tokyo, Japan, June, 2001. (eds. Tazawa, Y.), June 2001, Tokyo Denki University Press.

[2] Mohr, P. and Taylor B. CODATA recommended values of the fundamental physical constants: 1998, in Rev. Mod. Phys., 72, 351, (April 2000), The American Physical Society. http://physics.nist.gov/cuu/Constants/

[3] Hagiwara, K. et al. Review of particle physics. Particle Data Group, in Phys. Rev., D66, 010001-1, (July 2002) . The American Physical Society.

[4] Ezhela, V., Larin, V., and Siver, A. *Mathematica* System and Precision Tests: Muonium ground-state hyperfine splitting, (to be published).

[5] Czarnecki, A., Eidelman, S. and Karshenboim, S. Muoniun hyperfine structure and hadronic effects, in Phys. Rev., D65, 053004-1, (January 2002) , The American Physical Society.

Boole-Bell-type inequalities in Mathematica

Stefan Filipp
Atominstitut der Österreichischen Universitäten, Stadionallee 2, 1020 Wien, Austria
sfilipp@ati.ac.at

Karl Svozil
Institut für Theoretische Physik, Technische Universität Wien, Wiedner Hauptstrase 8-10/136, 1040 Wien, Austria
svozil@tuwien.ac.at

1 Abstract

The violation of Bell-type inequalities by quantum probabilities represents an important aspect of quantum mechanics which is linked to the mind-boggling quantum features of nonlocality, complementarity and contextuality. Formally, such bounds on the classical probabilities from consistency arguments have already been investigated by Boole in the middle of the nineteenth century. Boole called them "conditions of possible experience." We introduce CddIF, a *Mathematica* package to compute all the Boole-Bell-type inequalities associated with an arbitrary physical setup. We have also computed with *Mathematica* the tractable special cases of two particles with up to three possible detection angles per particle, and the three-particle/two directions (Greenberger-Horne-Zeilinger) case.

2 Boole-Bell Type Inequalities

Would you believe someone telling you that the chances of rain in Vienna and Budapest are 0.1 in each one of the cities alone, and the joint probability of rainfall in both cities is 0.99? I believe you would not. It just does not make any intuitive sense to claim that it rains almost never in one of the cities but almost always in both of them. The nagging question appears, though: which numbers would you believe in? Of course one could argue that the joint probability should not exceed any single probability. This certainly appears to be a necessary condition, but is it a sufficient one? In the middle of the 19th century George Boole, in response to such queries, formulated a theory of "conditions of possible experience" [1,2,3,4,5]. These conditions on the (joint) probabilities of logically connected events are expressed by certain equations or inequalities relating those probabilities. More recently, similar inequalities for a particular setup which are relevant in the quantum mechanical context have been discussed by Clauser and Horne and others [6,7,8]. Pitowsky has given a geometrical interpretation in terms of correlation polytopes [9, 4, 10, 5].

2.1 *Geometrical interpretation [4,5,9,10]*

Consider the truth table (Table 1) of the above example, where a_1 and b_1 represent the statements that *"it rains in Vienna"* and *"it rains in Budapest"*, and $a_1 b_1$ represents the proposition *"it rains in Vienna and in Budapest ,"* respectively; i.e.,

```
In[1]:= TruthTable[2, 1]

Out[1]//MatrixForm=
```

$$\begin{pmatrix} a_1 & b_1 & a_1 b_1 \\ 0 & 0 & 0 \\ 1 & 0 & 0 \\ 0 & 1 & 0 \\ 1 & 1 & 1 \end{pmatrix}$$

Table 1

In the weather example, there are two places with one observable each; hence the arguments 2,1. The third "component bit" of the vector is a function of the first components. Since we are dealing with the classical logical "and" operation here, this function is a multiplication. This amounts to interpreting the rows of the truth table as extremal points; the entries of the rows being the components. The vectors have the following meaning: {0,0,0} stands for the case of neither rain in Vienna nor in Budapest, {1,0,0} stands for the case of rain in Vienna but not in Budapest, and so on. Any probability distribution characterized by the three numbers {a1, b1, a1b1} should be a weighted linear average over all these (weather) conditions.

Interpret the set of all numbers {a1, b1, a1b1} as a set of points in a three-dimensional real space. As pointed out by Pitowsky [4], this set is the polytope formed by a weighted linear average over all extremal points of Table 1 representing the possible (weather) conditions. More precisely, the surface of the convex polytope with vertices {0,0,0}, {1,0,0}, {0,1,0} and {1,1,1} (cf. Figure 1) represents all conceivable classical probability distributions. The faces of the polytope represent the conditions by which classical probabilities are bound; and these "inside-outside" conditions are expressed as inequalities. These inequalities are called Boole-Bell inequalities.

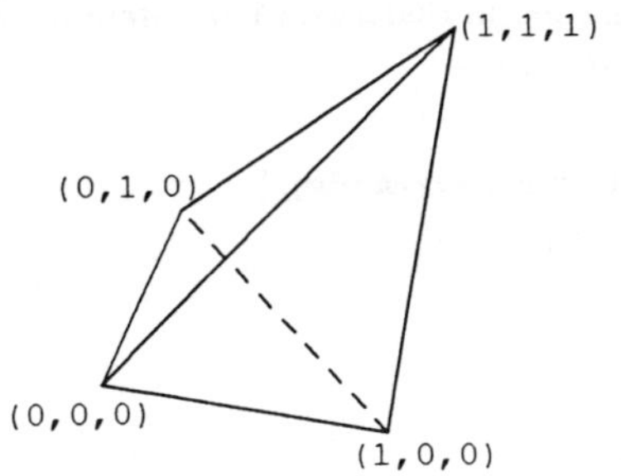

Figure 1

2.2 *Minkowski-Weyl representation theorem*

Generally, according to the Minkoswki-Weyl representation theorem (eg.,[10,p.29]), every convex polytope in an Euclidean space has a dual description: either as the convex hull of its extreme points (vertices), or as the intersection of a finite number of half-spaces, each one given by a linear inequality. The problem to obtain all inequalities from the vertices of a convex polytope is known as the hull problem. For the Boole-Bell type problems at hand, the Minkowski-Weyl representation theorem represents a direct, provable method to obtain a complete set of optimal inequalities for classical probability distributions characterizing any physical setup or arrangement.

3 Calculating Inequalities using CddIF

This package acts as an interface to the cdd-program by Komei Fukuda [11] for a transformation from a vertex to a half-space representation and vice versa, making use of the Minkowski-Weyl representation theorem by means of the Double Description Method [12], which we shall use but not review here. A complete description can be found in [13] and the package can be downloaded from *http://tph.tuwien.ac.at/~svozil/cdd/*. Available *Mathematica* functions are for example **TruthTable[x,y]** to build the corresponding truth table for the configuration *x-particles/y-measurements*, **ConvToHRep** to construct the half-spaces representation or **PlotInequalities** to plot the calculated inequalities.

3.1 *Computation of inequalities corresponding to {{0,0,0}, {1,0,0}, {0,1,0} and {1,1,1}}*

The appropriate inequalities for the probabilities of rain in Vienna (a1), rain in Budapest (b1) and rain in Vienna and in Budapest (a1b1) from the example above can easily be calculated with the use of *CddIF:*

```
In[3]:= InequToRead[GetInequFromHRep[
           ConvToHRep[2, 1]]]
```

```
Out[3]//MatrixForm=
```

$$\begin{pmatrix} a1 - a1b1 + b1 \leq 1 \\ -a1 + a1b1 \leq 0 \\ a1b1 - b1 \leq 0 \\ -a1b1 \leq 0 \end{pmatrix}$$

Here **ConvToHRep[2,1]** calculates the half-space representation for this particular configuration. Again, as in the weather example discussed above, there are two places with one observable each; hence the arguments 2, 1. The functions **InequToRead[GetInequFromHRep[...]]** display the inequalities in a human-readable form.

3.2 *Quantum mechanical context*

In the quantum mechanical case elementary events and their probabilities refer to clicks in particle detectors. Quantum probabilities have to be calculated using the formalism of quantum mechanics based on Hilbert space entities (i.e., Gleason's theorem) rather than on Boolean algebras. There exist complementary observables which cannot be measured simultaneously. And unlike classical Boolean algebras, Quantum Hilbert lattices are non-boolean (non-distributive). It is by no means trivial that quantum probabilities should satisfy the consistency conditions for classical probabilities; and indeed Bell has discovered physical configurations where Boole's "conditions of possible experience" are violated by quantum probabilities.

As an example, consider a source that produces pairs of spin-1/2 particles in a singlet-state, meaning that, no matter in what direction the spins of the two particles are measured, they will always be opposed to each other. The particles fly apart along the *z* axis, and after the particles have separated, measurements on spin components along one out of two directions are made. If, for simplicity, the measurements are made in the *x-y* plane perpendicular to the trajectory of the particles, the direction of the measurement can be given by angles measured from the vertical x axis (α_1 and α_2 on one side, β_1 and β_2 on the other side). On each side the measurement angle is chosen randomly for each pair of incoming particles and each measurement can yield two results - in $\hbar/2$ units: "+1" for spin up and "-1" for spin down (cf. Figure 2).

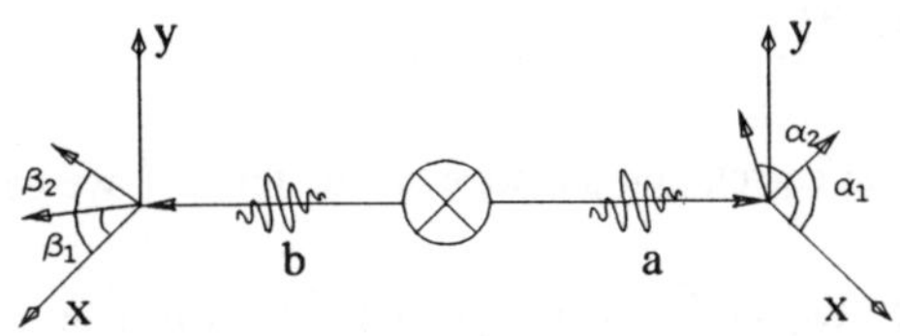

Figure 2

Deploying this configuration we get probabilities to find a particle measured along the axis specified by the angles α_1, α_2, β_1 and β_2 either in spin up or in spin down state denoted as a1, a2, b1, b2 - and we also take the joint event of finding a particle on one side at the angle α_1 (α_2) in a specific spin state and the other particle on the other side along the vector β_1 (β_2) in a specific spin state, denoted by a1b1, a2b1, a1b2 and a2b2.

```
In[4]:= GetInequFromHRep[ConvToHRep[2, 2]] // InequToRead

Out[4]//MatrixForm=
                    [...]
 a1 - a1b1 - a1b2 + a2b1 - a2b2 + b2 ≤ 1
 a1b1 - a1b2 + a2 - a2b1 - a2b2 + b2 ≤ 1
 -a1b1 + a1b2 + a2 - a2b1 - a2b2 + b1 ≤ 1
 a1 - a1b1 - a1b2 - a2b1 + a2b2 + b1 ≤ 1
                    [...]
 -a1 + a1b1 + a1b2 - a2b1 + a2b2 - b2 ≤ 0
 -a1b1 + a1b2 - a2 + a2b1 + a2b2 - b2 ≤ 0
                    [...]
                 -a1b1 ≤ 0
               -a2 + a2b2 ≤ 0
                    [...]
               a1 - a1b1 + b1 ≤ 1
```

The first inequalities correspond to the well known CHSH inequalities, which are thereby obtained in a very natural way.

3.3 *Two particles - three measurement directions*

In the case of two particles (*a* and *b*) with three properties (whereas the properties are three different angles of the detectors for each particle) denoted by $a_1, a_2, a_3, b_1, b_2, b_3$ 15 different events can be found: $\{a_1, a_2, \ldots, b_3, a_1 \wedge b_1, a_1 \wedge b_2, \ldots, a_3 \wedge b_3\}$. Using

```
In[5]:= TruthTable[2, 3]
```

all vertices of the corresponding correlation polytope can be found - we get a dimension of 15 and 64 vertices as result.

A representation in terms of inequalities of the polytope described by the truth table above can be created by

```
In[6]:= h = ConvToHRep[2, 3]
```

This results in 684 hyper-planes or 684 inequalities, respectively, from the 64 vertices limiting the polytope. All inequalities can be displayed by

```
In[7]:= GetInequFromHRep[h] // InequToRead
```

```
Out[7]//MatrixForm=
```

$$\begin{pmatrix} [\ldots] \\ a1b1 - a1b3 + a2 - a2b1 - a2b3 + b3 \le 1 \\ [\ldots] \\ -a2b3 \le 0 \\ a2 - a2b1 + b1 \le 1 \\ a1 - a1b1 + b1 \le 1 \end{pmatrix}$$

To show all inequalities that are violated at the specific measurement directions $a_1 = b_1 = 0$, $a_2 = b_2 = 2\pi/3$, $a_3 = b_3 = 4\pi/3$ with the functions **P[{x_}]:=1/2** and **P[{x_,y_}]:=1/2 Sin² [(x-y)/2]** yielding the quantum probabilities to find two spin-1/2 particles (in a singlet state) both either in spin up or both in spin down, we execute

```
In[8]:= P[{x_}] := 1/2;
        P[{x_, y_}] := Sin[(x - y)/2]^2/2;
        GetViolInequalities[h,
          {{0, 2π/3, 4π/3}, {0, 2π/3, 4π/3}},
          P, All] // TableForm
```

```
Out[10]//TableForm=
                        [...]
    -a1 + a1b2 + a1b3 + a3b2 - a3b3 - b2 ≤ 0      0.125
    a2b1 - a2b2 - a3 + a3b1 + a3b2 - b1 ≤ 0       0.125
    -a2 + a2b1 + a2b3 + a3b1 - a3b3 - b1 ≤ 0      0.125
```

A two-dimensional graphical representation of all inequalities (cf. Figure 3) can be generated for example by

```
In[11]:= PlotInequalities[h, {x, 0, π},
           {{0, x, 4π/3}, {0, x, 4π/3}}, P];
```

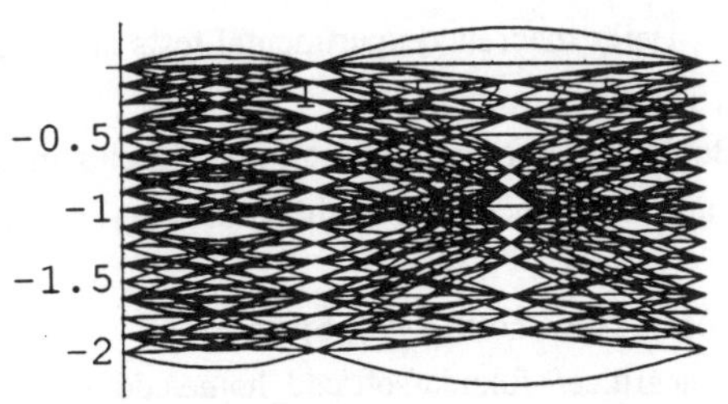

Figure 3

A very similar procedure involving **TruthTable[3,2]** yields all Boole-Bell inequalities for the three-particle and two measurement directions (Greenberger-Horne-Zeilinger) case.

4 Summary

In summary we have implemented a systematic method to compute all optimal Boole-Bell type inequalities satisfied by classical probabilities, given a physical setup. By optimal we mean that there is no condition such as equality or inequality which is satisfied by classical probabilities and which is a less stringent bound thereof. We have used this method to compute well known as well as new types of bounds on probabilities in configurations relevant to quantum physics. We believe that this represents a powerful new tool for the investigation of the nonclassical features of quantum mechanics.

5 References

[1] Boole, G. An investigation of the laws of thought, New York, 1958, Dover Edition.

[2] Boole, G. On the theory of probabilities, in Phil. Trans. R. Soc. London, 152:225-252, 1862.

[3] Hailperin, T. Boole's logic and probability (Studies in logic and the foundations of mathematics; 85), Amsterdam, 1976, North-Holland.

[4] Pitowsky, I. Quantum Probability - Quantum Logic, Berlin, 1989, Springer.

[5] Pitowsky, I. George Boole's 'conditions od possible experience' and the quantum puzzle, in Brit. J. Phil. Sci, 45:95-125, 1994.

[6] Clauser, J. F. and Horne, M. A. Experimental consequences of objective local theories, in Phys. Rev D, 10: 526-535, 1974.

[7] Clauser, J. F. *et al.* Proposed experiment to test local hidden-variables theories, in Phys. Rev. Lett., 23:880-884, 1969.

[8] Clauser, J. F. and Shimony, A. Bell's theorem: experimental tests and implications, in Rep. Prog. Phys., 41:1881-1926, 1978.

[9] Pitowsky, I. From George Boole to John Bell: The origin of Bell's inequality, in Bell's Theorem, Quantum Theory and the Conception of the Universe, Dordrecht, 1989 (ed. Kafatos, M.), Kluwer.

[10] Ziegler, G. Lectures on Polytopes, New York, 1994, Springer.

[11] Komei Fukuda, http://www.cs.mcgill.ca/~fukuda/soft/cdd_home/cdd.html.

[12] Motzkin, T. S. *et al.* The Double Description Method, in Contributions to theory of games, Vol. 2, New Jersey, Princeton, 1953, Princeton University Press.

[13] Filipp, S. and Svozil, K. Boole-Bell-type inequalities in *Mathematica*, e-print: quant-ph/0105083.

Reversible Computation and a Toolkit for Quantum Turing Machine Simulation

Joachim Hertel
jhertel@h-star.com

1 Abstract

This talk begins with some background information on reversible computation from a Thermodynamics and Quantum Theory point of view, and then introduces a Mathematica® Package *Quantum Turing Machine Simulator* (QTS) that can serve as an educational tool to explore the principles of a reversible computational process within the context of Quantum Mechanics.

2 Reversible Computation

The computational process has long been an interesting and challenging subject of basic research in Mathematics, Physics and Computer Sciences, see [2]. One of the fundamental insights is that Information is indeed *physical* [1,4] and therefore subject to the basic laws of Physics. Well known milestones of that research are Landauer's [3] entropy counting argument that erasure of information decreases the entropy of a system by $k \log 2$ per bit and Bennett's [5] proof, that every computational irreversible process can be re-designed as a reversible one. Turning to Quantum Mechanics, we recognize that the physical laws are reversible in time [7]. Following Deutsch's [6] architecture of a Quantum Turing Machine (QTM) , the machine state advances in time according to $\psi(t) = U(t)\,\psi(0)$, U(t) being the unitary operator acting on the Hilbert space of the machine. Typically, U(t) is given as $U(t) = e^{-iHt}$ with H being the Hamiltonian of the system. For our purposes we are interested in the properties of H as they relate to the computational process. A computational process can be described as a sequence q0,q1,... of steps , and to each such process a linear, bounded operator T ("step operator") can be associated. In general, the step operator T is not associated with a finite time interval; instead T is linked to an infinitesimal time step. According to Feynman [8] one can construct a time-independent Hamiltonian H(T) from the step operator T by $H(T) := K(2-T-T^*)$. (Here, T^* is the adjoint of T.) H(T) is a generator of a unitary time evolution $U(t) = e^{-iH(T)t}$. Benoff [9] proved, that iff T is *distinct path generating* (dpg) [9], then T can be used to model the time evolution of a quantum system whose elementary steps are modelled by T and T* respectively. More precisely, Benioff's structure theorem [9] says, U(t) = $e^{-iH(T)t}$ is a *quantum ballistic* [9] time evolution, iff T is dpg. We use the Benioff step operator approach [9] to implement a *Quantum Turing Machine Simulator* (QTS), [11] as a tool to study aspects of the computational process such as reversibility, locality of the Feynman Hamiltonian, quantum parallelism and entanglement. In QTS, a Turing head exists in any one of a final number of states. The Turing head moves along a two way infinite chain of qubits. The head scans one qubit at a time. The action taken depends on the state of the Turing head and the state of the qubit scanned. Possible elementary actions include changing the internal state of the head, state of the qubit or moving the head one lattice point to the right or to the left.

3 QTS Package

3.1 Notations and Definitions

The state of a Quantum Turing Machine is described as a state in a large, but separable Hilbert space $\mathcal{H}$. The computational basis $\mathcal{B}$, which spans the Hilbert space $\mathcal{H}$ is given by the set of states $\{|l,j,\mathcal{S}\rangle\}$, where

j denotes the lattice site
l denotes the internal state of the head
$\mathcal{S}$ denotes the tensor product over all qubit states $|s>_j$, $-\infty \le j \le +\infty$

To keep $\mathcal{H}$ separable, choose $|s>_j \neq 0$ for a **finite** number of lattice sites only. According to Benioff [9], we specify the dynamics of a specific QTM by a step operator T and its adjoint T*. Both operators are defined as the sum of elementary step procedures $T_{l,s}$:

$$T = \sum_{l,s} T_{l,s} \text{ and } T_{l,s} = \gamma_{l,s} \sum_{j=-\infty}^{+\infty} w_{l,s}\, Q_l\, v_{l,s}\, P_{s,j}\, u_{l,s}\, P_j \tag{1}$$

and

Q_l projector for the head in state l
P_j projector for the head being at site j
$P_{s,j}$ projector for the qubit at j being in state s

The operators w, v and u must be **unitary** and have the following meaning:

$w_{l,s}$ describes head state changes and operates in the Hilbert space of the QTM head

$v_{l,s}$ describes qubit state changes and operates in the Hilbert space of a qubit
$u_{l,s}$ describes the motion of the head depending on head state l and qubit state s.

To ensure **reversibility** of the computation, the step operator T and its adjoint T* must be **Distinct Path Generating (dpg)** [9]. That means, iterations of T and T* on any state$|l,j,\mathcal{S}\rangle$ either annihilate that state or generate a finite or infinite set of states which do **not** join,branch or intersect. Each such set of states is called a **computational path** $\mathcal{P}$ [9].

3.2 Functions and Templates

The QTS package provides all basic functions to deal with QTM related functional analysis in a Hilbert space or in tensor product Hilbert spaces. The creation of new QTM models is easy by using templates for step operators and predefined projectors, shift operators such as **`ProjHeadAtSite`**, **`ProjHeadInState`**, **`ProjQubitAtSiteInState`**, **`ModShiftHeadPos`**, **`UpdHeadState`**, **`UpdSiteState`**, **`FmtState`**. To initialize the ground state of a QTM use **`InitQTM[s-dim,s-label,ex_interval,exstate_list,h-dim,h-label,h-pos]`**, where **`s-dim`** specifies the Hilbert space dimension of the lattice spins, **`s-label`** is the label used to enumerate the ON basis for that Hilbert space, **`ex_interval`** specifies the lattice positions of excited spins, **`exstate_list`** specifies the states of the excited spins, **`h-dim`** and **`h-label`** define the dimension and labels for the Hilbert space of the Turing header and **`h-pos`** is the initial header lattice position.

3.3 An Example

This example is due to Benioff [10]. The step operator is dpg and generates linear superpositions of input states. The Turing head lives in a 4-dimensional Hilbert space $\mathcal{H}_4$ with basis $\mathcal{B}_{\text{Head}}$= {h[0],h[1],h[2],h[3]}.The lattice spins are living in a 3-dimensional Hilbert space $\mathcal{H}_3$ with basis $\mathcal{B}_{\text{Spin}}$={b[0],b[1],b[2]}. The basis state **b[2]** serves as a marker state to trigger certain actions while theTuring head is scanning the lattice sites.

The step operator T is defined as:

$$T := \sum_{j=-\infty}^{+\infty} [Q_0 P_{0,j} u + w Q_0 P_{2,j} u + Q_1 \, vP_{0,j} u + w Q_1 P_{2,j} u^* + Q_2 \sigma_x P_{1,j} u^* + wQ_2 \sigma_x P_{0,j} u + wQ_2 P_{2,j} u + Q_3 P_{0,j} u + wQ_3 P_{2,j} u] P_j \quad (2)$$

The unitary operators u , w and v are defined as follows:

u: shift operator, shifts the head position one lattice site to the right, i.e. $u P_j = P_{j+1} u$:

u^*: shift operator, shifts the head position one lattice site to the left, i.e. $u^* P_j = P_{j-1} u^*$

w: shift operator, shifts the head state mod4 and is implemented in MatrixForm as:

$$w = \begin{pmatrix} 0 & 0 & 1 & 0 \\ 0 & 0 & 0 & 1 \\ 1 & 0 & 0 & 0 \end{pmatrix}$$

v: is implemented as the **Hadamard** transformation on the {b[0],b[1]} subspace and leaves the marker state b[2] invariant. v is given as

$$v = \begin{pmatrix} 1 & 0 & 0 \\ 0 & \frac{1}{\sqrt{2}} & \frac{1}{\sqrt{2}} \\ 0 & \frac{1}{\sqrt{2}} & -\frac{1}{\sqrt{2}} \end{pmatrix}$$

The **initial** QTM state is defined as:

head in state h[0] at lattice site position $j = -3$

lattice state $\mathcal{S} \doteq \otimes_{j=-\infty}^{-1} b[0]_j \otimes b[2]_0 \otimes b[0]_1 \otimes b[0]_2 \otimes b[0]_3 \otimes b[2]_4 \otimes \otimes_{j=5}^{+\infty} b[0]_j$

hence $u[0] \doteq | \, h[0], -3, \mathcal{S} \rangle$

After 5 computational steps the state of the QTM is:

$$u[5] = (h(1))_2 \otimes \otimes_{j=-\infty}^{-1} (b[0])_j \otimes (b(2))_0 \otimes \left(\frac{b(1)}{\sqrt{2}} - \frac{b(0)}{\sqrt{2}}\right)_1 \otimes (b(0))_2 \otimes (b(0))_3 \otimes (b(2))_4 \otimes \otimes_{j=5}^{+\infty} (b[0])_j \quad (3)$$

4 Analysis of Computation

4.1 Is Computation Reversible?

Tthe degree to which a given step operator produces a **reversible** computation can be studied numerically. In order to produce a reversible computation, a step operator T (and its adjoint T*) must be **dpg**, a property closely related to the theory of orthogonality preserving, stable partial isometries on a Hilbert space [9]. It is crucial to show [TT*,T*T] = 0. Use the QTS adjoint operator template to define the adjoint T* of the step operator T:

Let q[n] = T^n q[0] and set

c = NLimit[|| T[T*[q[n]]] -T*[T[q[n]]] ||, n -> Infinity] (4)

|| || is the norm on the tensor product Hilbert space. c = 0 provides numerical evidence that T is dpg.

4.2 Reconstructing the Unitary Dynamics

The unitary dynamics U can be reconstructed from the step operator T. Take the Feynman Hamiltonian [8] H(T) ≈ T + T*. The unitary dynamics for a time step t is U(t) := $e^{-iH(T)\,t}$. Let $U_N(t)$ be the Nth order approximation of U(t),

U_N(t) : = (5)

U is completely de-localized, as infinitely many powers of H are contributing at any time.

4.3 Locality of the Feynman Hamiltonian

Let T be a **dpg** step operator. T is local by construction, i.e. a single computational step changes at most the qubit at the lattice position of the head. Following Feynman [8] the associated Hamiltonian H(T) is proportional to T + T*. Define a quantitative measure for locality of powers of H by using the **overlap function**:

$$l_i[n] := \frac{\text{Abs[QInner}[q[0],\, H^n[q[i]]]]}{\sqrt{\text{Abs[QInner}[H^n[q[i]],\, H^n[q[i]]]]}} \tag{6}$$

$l_i[n]$ measures the overlap of q[0] and $H^n[q[i]]$ for given i,n. As a consequence of the locality of H we expect, that the overlap for fixed n decreases fast with increasing *i*, because the head positions of q[0] and q[i] become spatially more separated and we need more and more iterations of H to produce any significant overlap between q[0] and q[i]. As an example we plot $l_i[5]$:

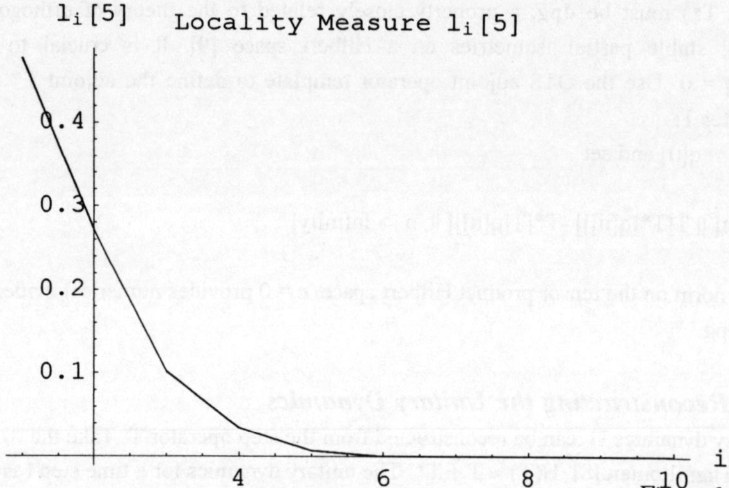

Figure 1

We expect to have a significant overlap when n ≥ i ,if we take $l_i[n]$ for fixed i and increase n. As an example we plot $l_5[n]$:

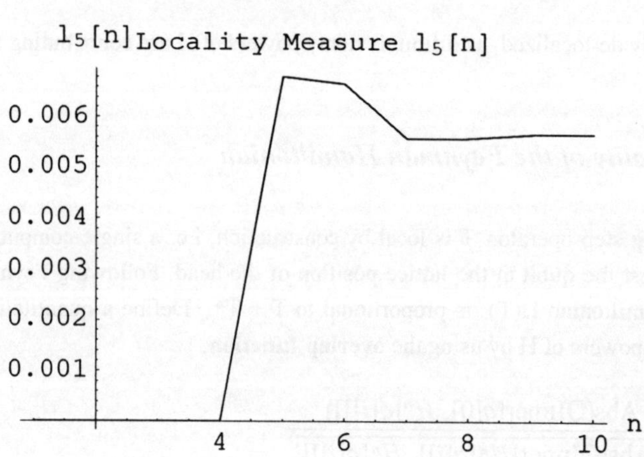

Figure 2

4.4 *Create Entanglement*

QTS can be used to demonstrate quantum entanglement. We use the Mahler&Kim [12] model for a QTM, which they used to study pattern formation in Quantum Turing Machines [12]. To limit complexity and CPU resources we use a simplified version here. The Turing head periodically scans 2 qubits and the step operator T is given by an iterative Mahler & Kim map [12] of the form $T_i = T_2^{2\cdot i} \circ T_1^{2\cdot i-1}$, with T_1 and T_2 partial step operators similar to the first terms of eqn (2) but with shift and update operators w,v

$$w = -I \, \mathrm{Sin}\left[\frac{\pi}{2\sqrt{3}}\right] \mathrm{IdentityMatrix}[2]$$
$$v = \mathrm{Cos}\left[\frac{\pi}{2\sqrt{3}}\right] \mathrm{IdentityMatrix}[2] \qquad (7)$$
$$u = \sigma_x$$

Running the QTM for 10000 steps and plotting the trajectory of the Bloch vector $\vec{\lambda} = (\lambda_x, \lambda_y, \lambda_z) = (0, \lambda_y, \lambda_z)$ for the Turing head reveals this pattern:

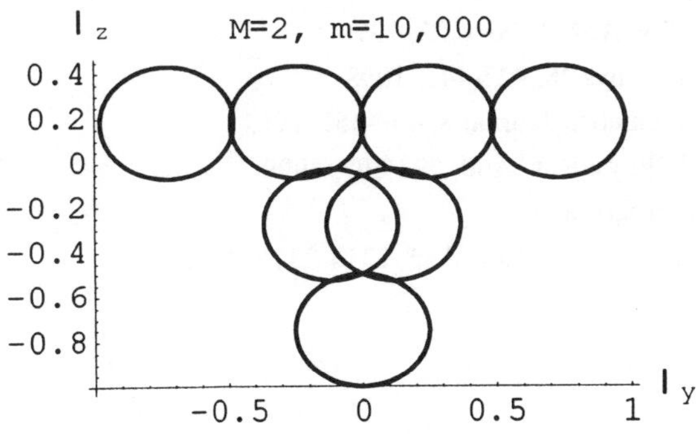

Figure 3

The 2^3-1 = 7 circles can be interpreted (see also Mahler & Kim [12]) as parallel trajectories of pure Turing head states all occurring simultaneously in superpositions ("quantum parallelism"). The overlap indicate entanglement in the mixed state between Turing head and tape. Work is currently in progress [13] to explicitly calculate the measure of entanglement for these mixed states by using an effectively computable measure recently introduced by Vidal & Werner [14].

5 Notes

The QTS package is available for download at:

http://www.mathematica-journal.com/issue/v8i3/features/hertel/

The QTS package includes a User Manual and a Primer on Reversible Computation.

6 References

[01] Bennett, C.H. : Int. J. Theor.Phys. **21**, 905-940,1982

[02] Feynman Lectures on Computation by R.Feynman, AW, 1996

[03] Landauer, R. : IBM J. Res. Dev. **5**, 183-191, 1961

[04] Keyes, R.W. , Landauer, R. : IBM J. Res. Dev. **14**, 152-157, 1970

[05] Bennett, C.H.: IBM J. Res. Dev. **17**, 525-532, 1973

[06] Deutsch, D.: Proc.Roy. Soc. London Ser **A400** , 96 - 117, 1985

[07] Lectures on Phy ,Vol 3 by R.Feynman, R.B.Leighton,M.Sands, AW, 1977

[08] Feynman, R.P.: Opt. News **11**, 11-20, 1985

[09] Benioff, P.: Phys. Rev. **A54**, 1106-1123, 1996

[10] Benioff, P.: Fortschr. Phys.**46**, 423-441, 1998

[11] Hertel, J.: The Mathematica Journal 8, 440-457, 2002

[12] Mahler, G., Kim, I.: Phys. Rev. **A60**, 692-695, 1999

[13] Hertel, J.: Work in Progress

[14] Vidal, G.,Werner, R.F.: Phys. Rev. **A65**, 2314-2315, 2002

The action of the Steenrod algebra in the cohomology of BF_4

Andrzej Kozlowski

Toyama International University

andrzej@mimuw.edu.pl

1 Abstract

The aim of this article is to demonstrate the use of *Mathematica* in computations in algebraic topology. While *Mathematica,* which aims to be a general purpose numeric-symbolic program, can never be as efficient a tool for research level symbolic computation algebra as more specialized programs, the dramatic increase in the speed of personal computers has made such use entirely feasible. We choose as an example a rather arcane subject: the computation of the action of the Steenrod operations in cohomology of the classifying space of the exceptional Lie group F_4.

2 Background

The computation of the cohomology of the classifying spaces of exceptional Lie groups was began by Toda ([5],[6]) and remains an active area of research. The computation we perform has been done by hand by Vavpetic [7] and required a certain amount of mathematical cleverness. Here we shall use nothing but the "brute strength" of *Mathematica*. In order to perform our computation we have to implement the Steenrod squaring operations in mod 2 cohomology. The simplicity and elegance of *Mathematica* make such an implementation almost trivial. The computation we perform here would be impossible to perform "by hand" directly.

Lie groups, which generalize groups of invertible matrices and operators, are important in both mathematics and physics. A Lie group is simultaneously a topological object (a smooth manifold) and an algebraic one (group) and these two structures are compatible in the sense that the operations of multiplication and inversion are given by smooth maps. The natural generalization of finite Lie groups is compact Lie groups. These have been completely classified into several infinite families and several so called "exceptional" Lie groups, denoted by G_2, F_4, E_6, E_7, E_8. Associated with each compact Lie group G there is a topological space BG. The characterization of such spaces is an important problem in algebraic topology. An step in that direction is the computation of their cohomology rings.

A famous theorem of Armand Borel makes it possible to compute cohomology groups of compact Lie groups with rational coefficients, and with coefficients in finite fields of characteristic relatively prime to the Weyl group. However, the case of field coefficients with characteristic dividing the order of the Weyl group is an active area of research, in which computer aided algebraic computations are very useful. Most of these involve computing the cohomology of the manifold G and passing to the cohomology of BG using the fibration $G \longrightarrow EG \longrightarrow BG$ and the associated Moore-Eilenberg spectral sequence. However, there is another approach which makes use of invariant theory. In his doctoral thesis Vavpetic [7] computed the cohomology of the exceptional Lie group F_4 with coefficients in $\mathbb{F}_2$ using this approach. The basic idea of the proof is to consider a maximal elementary abelian 2-subgroup of F_4. One can show that up to conjugation there is just one such group $E_{32} = (\mathbb{Z}/2)^5$. Vavpetic has shown that the subalgebra of invariants (fixed points) $\mathbb{F}_2[t1, \ldots t5]^{W_{F_4}(E_{32})}$ is isomorphic to the cohomology of BF_4. He also explicitly constructed the invariants that generate the cohomology ring. In addition to having the structure of a graded ring the cohomology groups $H^*(X, \mathbb{F}_2)$ of a topological space X admit the action of certain natural operations, known as the Steenrod squares [4]. This action is a powerful invariant (meaning that topologically equivalent spaces not only have isomorphic cohomology rings but also the Steenrod operations act in the same way). Multiplicative properties of the Steenrod operations (the so called "Cartan formula") imply that one only needs to compute their action on the generators of the cohomology ring. This is what we propose to do. First however, we need to describe these generators and implement the Steenrod operations in *Mathematica*.

3 Notation

We would like to use the usual superscript notation Sq^i for the i-th Steenrod square operation. We avoid a clash with powers by means of the Notation package. In addition we format the first Steenrod operation Sq^1 as β, which is the usual thing to do.

```
<< "Utilities`Notation`"
Notation[Sqi_ ⟺ Sq[i_]];
Format[Sq[1]] := β;
```

4 Dickson Invariants

The algebra of invariants of $GL_n(\mathbb{F}_2)$ was calculated by Dickson, who showed that

$$\mathbb{F}_2(t_1, \ldots, t_n)^{GL_n(\mathbb{F}_2)} = \mathbb{F}_2(c_{n,0}, \ldots, c_{n,n-1})$$

The generators $c_{n,i}$, which are known as Dickson invariants, are given by:

$$\prod_{v=\sum_{i=1}^{n} k_i t_i}^{f_n(X)} (X - v) = X^{q^n} + \sum_{i=0}^{n} (-1)^{n-i} c_{n,i} X^{q^i} = 0,$$

where k_i are either 1 or 0. We implement Dickson invariants as follows.

```
c[n_, i_, t_ : t, k_ : 0] :=
 Block[{ls = Distribute[Table[{0, 1}, {n}], List], x},
  PolynomialMod[Coefficient[Times @@
     (x - (Array[Subscript[t, k + #1] &, {n}].#1 &) /@
       ls), x^(2^i)], 2]]
```

The third argument is the letter we use to denote the variable in our invariant, and the fourth determines the initial value of the index of this variable (t and 0 by default). For example

```
c[3, 2, t, 1]
```

$$t_2^4 + t_2^2 t_3^2 + t_3^4 + t_2^2 t_3 t_4 + t_2 t_3^2 t_4 + t_2^2 t_4^2 + t_2 t_3 t_4^2 + t_3^2 t_4^2 + t_4^4$$

Dickson invariants tend to be very complicated:

```
In[4]:= Length[c[4, 2]]

Out[4]= 46
```

We shall be working in graded rings, so we assign all the indeterminates the degree 1.

```
dim[Subscript[t, i_]] = 1;
```

5 Working mod 2

We shall need to use the Steenrod operations in cohomology. These exist whenever the coefficient are taken in a field of finite characterisitc, but we shall only consider the case of coefficients in F_2. We would like to work mod 2 but also to be able to switch to characteristic 0 when convenient. First, we make a list of functions which accept the option Modulus:

```
functions = ToExpression /@
  Select[DeleteCases[Names["System`*"],
    _?(StringTake[#1, 1] == "$" &)],
   ! FreeQ[Options[Symbol[#1]], Modulus] &]
```

```
{Apart, ApartSquareFree, Cancel, Coefficient,
 CoefficientList, Collect, Decompose, Denominator,
 Det, Expand, ExpandAll, ExpandDenominator,
 ExpandNumerator, Exponent, Factor, FactorList,
 FactorSquareFree, FactorSquareFreeList,
 FactorTerms, FactorTermsList, GroebnerBasis,
 Inverse, LinearSolve, LUBackSubstitution,
 LUDecomposition, NullSpace, Numerator,
 PolynomialGCD, PolynomialLCM, PolynomialMod,
 PolynomialQ, PolynomialReduce, Resultant, Roots,
 RowReduce, Together, TrigReduce, Variables}
```

The next function will automatically set the modulus to p for all these functions.

```
FunctionsMod[p_] :=
 Scan[SetOptions[#1, Modulus → p] &, functions]
```

We can now switch all the functions in the above list to working, say, mod 2:

```
FunctionsMod[2]
Expand[(x + 1)^4]
```

6 The Steenrod Squares

Next we implement the Steenrod operations . This implementation contains several superfluous rules, whose presence however seems to improve performance. The implementation is obtained basically just by copying the properties of the Steenrod operations from any text on algebraic topology and writing them down using *Mathematica* syntax. The naturalness and ease of this construction seem to me unmatched by other symbolic algebra programs. The last two rules are the famous Cartan and Adem formulas. (The latter one is not needed for the computations performed below, but we include it for completeness).

```
c[n_, m_] := Mod[n!/((n - m)! m!), 2];
Sq[l_List][x_] := Fold[Sq[#2][#1] &, x, l];
dim[Sq[i_][x_]] := i + dim[x];
Sq[0] = Identity;
Sq[1][Sq[1][x_]] := 0;
Sq[1][Sq[i_?EvenQ][x_]] := Sq[i + 1][x];
Sq[1][Sq[i_?OddQ][x_]] := 0;
Sq[i_][0] = 0;
Sq[i_][k_Integer x_] := Mod[k, 2] Sq[i][x];
Sq /: Sq[i_][x_ + y_] :=
  PolynomialMod[Sq[i][x] + Sq[i][y], 2];
dim[k_?IntegerQ] := 0;
dim[x_ y_] := dim[x] + dim[y];
dim[x_^n_] := n dim[x];
dim[Sq[i_][x_]] := i + dim[x];
Sq[i_][0] := 0;
Sq[i_][a_] /; dim[a] < i := 0;
Sq[i_][a_] /; dim[a] == i := a^2;
Sq[i_][x_^k_Integer?Positive] /; dim[x] == 1
   := [k, i] x^(i+k);
```

```
Sq[k_][x_ y_] := PolynomialMod[
   Sum[Sq[i][x] Sq[k - i][y], {i, 0, k}], 2];
Sq[a_][Sq[b_][x_]] /; 0 < a < 2 b := PolynomialMod[
   Sum[c[b - 1 - j, a - 2 j] Sq[a + b - j][Sq[j][x]], {j, 0, Floor[a/2]}], 2];
```

7 The action of Steenrod Squares on the invariants.

We now apply this to compute the action of the Steenrod operations on the cohomology of the exceptional Lie group F_4. However, it is first of all necessary to construct the invariants in $\mathbb{F}_2[t1, \ldots t_5]^{W_{F_4}(E_{32})}$ that generate the cohomology ring. Vavpetic constructs such invariants out of Dickson invariants in $\mathbb{F}_2[t_3, t_4, t_5]^{GL(3,\mathbb{F}_3)}$ and $\mathbb{F}_2[t_1, t_2]^{GL(2,\mathbb{F}_3)}$. The general nature of this construction is explained in [2]. (In fact the invariance of the elements constructed by Vavpetic can be verified by a direct *Mathematica* computation, which we here omit). First, we define the following three generators of degree 4, 6 and 7 , in which are simply the Dickson invariants.

```
y4 = c[3, 2, t, 2]
y6 = c[3, 1, t, 2]
y7 = c[3, 0, t, 2]
```

Out[8]= $t_3^4 + t_3^2 t_4^2 + t_4^4 + t_3^2 t_4 t_5 + t_3 t_4^2 t_5 + t_3^2 t_5^2 + t_3 t_4 t_5^2 + t_4^2 t_5^2 + t_5^4$

Out[9]= $t_3^4 t_4^2 + t_3^2 t_4^4 + t_3^4 t_4 t_5 + t_3 t_4^4 t_5 + t_3^4 t_5^2 + t_3^2 t_4^2 t_5^2 + t_4^4 t_5^2 + t_3^2 t_5^4 + t_3 t_4 t_5^4 + t_4^2 t_5^4$

Out[10]= $t_3^4 t_4^2 t_5 + t_3^2 t_4^4 t_5 + t_3^4 t_4 t_5^2 + t_3 t_4^4 t_5^2 + t_3^2 t_4 t_5^4 + t_3 t_4^2 t_5^4$

Two more invariants, of degree 16 and 24 and involving both t_1 and t_2 are defined as follows:

```
a[i_] := t[i]^8 + y4 t[i]^4 + y6 t[i]^2 + y7 t[i]
y16 = a[1]^2 + a[2] a[1] + a[2]^2;
y24 = a[1] a[2] (a[1] + a[2]);
```

We do not print out these as they are represented by rather long expressions, even with mod 2 coefficients. We can now simply apply the Steenrod operations to our invariants directly, and even get some useful results, such as

$$\mathrm{sq}^2[y_7]$$

$$0$$

However

$$\mathrm{sq}^4[y_7]$$

$$t_3^8\, t_4^2\, t_5 + t_3^2\, t_4^8\, t_5 + t_3^8\, t_4\, t_5^2 + t_3\, t_4^8\, t_5^2 + t_3^2\, t_4\, t_5^8 + t_3\, t_4^2\, t_5^8$$

is not what we want, since we would like to express each invariant generator in terms of the other ones. Of course this is just the job for GroebnerBasis and PolynomialReduce [1]. The problem is that *Mathematica* will automatically express the output in terms of the t_i rather than the y_i. To cope with this we introduce new variables, Y_4, Y_6, Y_7, Y_{16}, Y_{24} which we think of as new set of names for y_4, y_6, y_7, y_{16} and y_{24}.

```
ideal1 = {Y4 - c[3, 2, t, 2],
   Y6 - c[3, 1, t, 2], Y7 - c[3, 0, t, 2]};
ideal2 = Join[ideal1, {Y16 - y16, Y24 - y24}];
vars1 = Join[Table[t[i], {i, 3, 5}], {Y4, Y6, Y7}];
vars2 =
  Join[Table[t[i], {i, 1, 5}], {Y4, Y6, Y7, Y16, Y24}];
groebner1 = GroebnerBasis[ideal1, vars1];
groebner2 = GroebnerBasis[ideal2, vars2];
```

Now we can compute the actions of the Steenrod operations on the invariants (we only need to do this in dimensions less than the dimension of the invariant; the remaining squares act trivially by definition).

Table[PolynomialReduce[Sq$^{2^i}$[y_4], groebner1, vars1][[2]], {i, 0, 2}]
Table[PolynomialReduce[Sq$^{2^i}$[y_6], groebner1, vars1][[2]], {i, 0, 2}]
Table[PolynomialReduce[Sq$^{2^i}$[y_7], groebner1, vars1][[2]], {i, 0, 2}]
Table[PolynomialReduce[Sq$^{2^i}$[y_{16}], groebner2, vars2][[2]], {i, 0, 4}]
Table[PolynomialReduce[Sq$^{2^i}$[y_{24}], groebner2, vars2][[2]], {i, 0, 4}]

$\{0, Y_6, Y_4{}^2\}$

$\{Y_7, 0, Y_4\, Y_6\}$

$\{0, 0, Y_4\, Y_7\}$

$\{0, 0, 0, Y_{16}\, Y_4^2 + Y_{24}, Y_{16}^2\}$

$\{0, 0, Y_4\, Y_{24}, Y_4^2\, Y_{24}, Y_4\, Y_{24}\, Y_6^2 + Y_{16}\, Y_{24}\}$

These results agree with those obtained by Vavpetic.

8 References

[1] D. Cox, J. Little, D. O'Shea, Using Algebraic Geometry, Springer-Verlag, 1998.

[2] M. Kameko, Rings of Invariants derived from Dickson Invariants (pre-print)

[3] L. Smith, Polynomial Invariants of Finite groups, A K Peters Ltd, 1995.

[4] N. E. Steenrod, Cohomology Operations, written and revised by D.B.A. Epstein, Princeton University Press, 1962

[5] H. Toda, Cohomology of classifying spaces of exceptional Lie groups, Manifolds, pp. 265-271, Tokyo 1973

[6] H. Toda, Cohomology mod 3 of the classifying space BF_4 of the exceptional Lie group F_4, J. Math. Kyoto Univ. 13 (1972), 97-115

[7] A. Vavpetic. Homotopy Characterization of classifying spaces of compact Lie groups. University of Lubljana doctoral thesis, 2000.

[8] C. Wilkerson, A primer on the Dickson Invariants, a corrected version of the paper published in Contemp. Math. 19 (1983), 421-432

Mode-locking in Nonlinear Oscillators

Phil Ramsden
Imperial College London
p.ramsden@imperial.ac.uk

1 Abstract

Motivated, at first, by an interest in nonchaotic strange attractors, I have attempted a complete characterisation of the phenomenon of *mode-locking* in periodically and quasiperiodically driven oscillators, numerically and via perturbation calculations. In both the periodically driven and the quasiperiodically driven cases, it appeared as though certain regions in parameter space associated with mode-locking had points at which their width was zero. Now, this is slightly surprising if true, for two reasons. First, there is no analogue to this pinching effect in skew product circle maps. Secondly, at pinching points, if they exist, the flow is extremely regular, and more so than one might naively expect for this highly nonlinear family of flows. I have not been able to prove analytically that mode-locked regions pinch to zero width. However, I have used *Mathematica*'s arbitrary precision to assemble extremely strong numerical evidence that they do.

2 Why is this problem interesting?

There is no universally accepted definition of "chaos" in dynamical systems, and in fact the chaotic "family" has at least two typical traits. On one hand, there is *sensitive dependence on initial conditions*: orbits that were initially close together diverge in a locally exponential way. On the other, there is *complicated limiting behaviour*: orbits are attracted to limiting sets that are geometrically complicated. It was, for a time, felt that these two properties could be associated with each other fairly reliably: that the occurrence of one without the other was atypical. However, a few years ago, several groups [1, 3, 4, 5] amassed strong numerical evidence that this is not so, and, in particular, that for certain families of dynamical systems *nonchaotic strange attractors* could be found in regions of parameter space with positive measure. Much of this work concerned nonlinear oscillators with quasiperiodic driving and their discrete-time analogues, (skew product circle maps). These attractors occur for parameter values at or close to the boundary of regions in parameter space associated with *mode-locking,* (the tendency for the oscillator to rotate at a rate that is, in the limit, a sum of rational multiples of the driving frequencies). Motivated, at first, by an interest in these attractors, I therefore attempted a complete characterisation of such regions for quasiperiodically driven oscillators and also for the simpler case in which the driving is periodic. On the way, I was able to use *Mathematica* to gather strong numerical evidence for the phenomenon of "pinching" in parameter space which, if real, has some somewhat counterintuitive implications.

3 Nonlinear oscillators with periodic or quasiperiodic driving

The flow equation:

$$\epsilon p \frac{d^2\theta}{dt^2} + \frac{d\theta}{dt} + \epsilon \sin\theta = F(t), \tag{1}$$

represents a nonlinear oscillator such as a damped simple pendulum or a Quantum Interference Device, driven by an external time-dependent forcing function $F(t)$. We here consider two types of driving function: *simple periodic*, in which $F(t) = A + B\sin t$, and *two-frequency quasiperiodic*, in which $F(t) = A + B\sin t + \omega C\sin\omega t$, ($\omega$ being an irrational). In each case, A, B and (where appropriate) C are time-independent parameters. The parameter ϵ will usually be assumed to be close to zero, and for some purposes the parameter p will be assumed to be *equal* to zero. (This latter condition is physically appropriate for quantum interference devices such as Josephson junctions but not, in general, for simple pendulums under typical conditions).

4 Rotation numbers and mode-locking

The state variables for the flow consist, in general, of the triple $(t, \theta, \dot{\theta})$, (unless $p = 0$, in which case $\dot{\theta}$ is not needed). Note that, in the flow equation, the variable θ appears only as the argument of a sine. For this reason, the point $(t, \theta, \dot{\theta})$ may be identified, for all purposes, with the point $(t, \theta + 2\pi, \dot{\theta})$. At least for periodic driving, the time variable t may also be considered to be cyclical in this way, and in fact, through a slight recasting of the problem involving the introduction of a fourth state variable, this can be done for quasiperiodic driving too. We can therefore regard (t, θ)-space as *toroidal*.

The *rotation number*, ρ, of an orbit $\theta(t)$ is the answer to the question: in the limit, how many cycles of θ are executed for each cycle of t? Hence the rotation number is given by:

$$\rho = \lim_{t \to \infty} \frac{\theta(t) - \theta(0)}{t}$$

(where here, θ and t are *not* considered to be cyclical). It can be shown that, at least for $p = 0$, ρ is the same for almost all initial conditions, and depends only on the parameters ϵ, A, B and C; I believe that this is also true, for all practical purposes, for non-zero p.

To illustrate this idea, Figure 1 shows three orbits in toroidal (t, θ)-space, with rotation numbers 0, 1 and 2 respectively. (Each of the orbits happens to be periodic modulo 2π.)

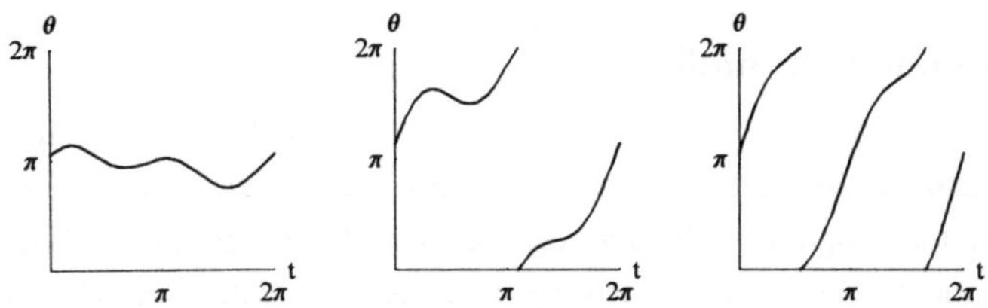

Figure 1

If we make the simplifying assumption $p = 0$, our flow equation (1) may be written

$$\frac{d\theta}{dt} = F(t) - \epsilon \sin\theta. \qquad (2)$$

To zeroth order in ϵ, this is simply $\dot{\theta} = F(t)$. This equation is, of course, trivial to integrate, and it is easy to show that for both types of driving described above, the rotation number is simply equal to the parameter A. Naively, then, we would expect ρ and A to be closely related for general ϵ, and indeed this the case. However, for non-zero ϵ, it turns out [2] that the relationship is not a straighforward one. The relationship is easiest to state for simple periodic driving, $F(t) = A + B\sin t$. As A varies for fixed B, ρ has a strong tendency to take on *rational* values. (More precisely, for each $q \in \mathbb{Q}$, the set $\{A : \rho(A) = q\}$ has positive measure.) This is known as *mode-locking*. In the quasiperiodically driven case, where $F(t) = A + B\sin t + \omega C \sin \omega t$ with ω irrational, mode-locking refers to a similar tendency for the rotation number to take on values that are equal to $q_1 + q_2\,\omega$, where $q_1, q_2 \in \mathbb{Q}$.

The case where the rotation number is zero is attractive for two reasons. The first is that it turns out to be relatively easy to determine numerically whether a flow has rotation number zero. The second is that one can use perturbation calculations to characterise, to first or second order in ϵ, the set in parameter space associated with this value of ρ. Fortunately, it can easily be shown that the parameter values associated with rotation number $\rho = \rho_0$ for the flow (1) are precisely those associated with $\rho = 0$ for the flow:

$$\epsilon\, p \frac{d^2\theta}{dt^2} + \frac{d\theta}{dt} + \epsilon \sin(\theta + \rho_0 t) = F(t) - \rho_0. \tag{3}$$

We can therefore investigate all rotation numbers for the original flow numerically. Moreover, I have developed a fairly general approach for second order perturbation calculations that enables us to characterise the parameter set associated with $\rho = 0$ for all flows of this type (and others).

5 Summary of findings

Using the perturbation technique, I have achieved a complete approximate characterisation of the sets in parameter space associated with all possible rotation numbers, in the simple periodic and quasiperiodic driven cases described above. These characterisations can be confirmed by numerical investigation. However, although *Mathematica* has been indispensible in shaping calculations and in visualising the results, it has not been central to the calculations themselves. I will therefore content myself with Figure 2, which shows, as an example, these sets for rotation numbers 0, 1/6, 1/3, 1/2, 2/3, 5/6 and 1, in the case of periodic driving with $\epsilon = 0.1$. The unbroken lines show the boundaries of the sets as calculated using the perturbation technique, and the dots are the results of numerical calculations.

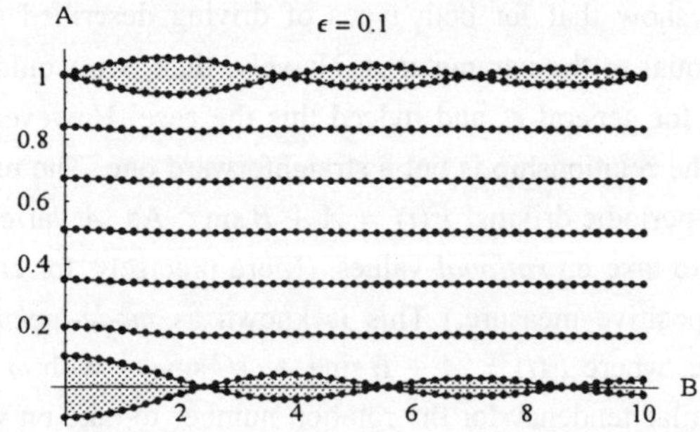

Figure 2

6 "Pinching"

From Figure 2, and indeed from the perturbation calculations that underlie it, it strongly appears that each set in (B, A)-space associated with an *integer* rotation number has points-indeed, an infinite number of points-at which it "pinches" to zero width. The same is true for the quasiperiodically driven oscillator. For the case $\rho = 0$, in the Quantum Interference Device application, this has long been known to physicists on the basis of experiment and first-order perturbation calculations [e.g. 6]. Nonetheless, it is slightly surprising, for two reasons. One is that the standard discrete-time analogue to the quasiperiodically driven oscillator, the *skew product circle map*, exhibits no such structure in parameter space [2]. The second is that I have been able to prove that, for the periodically driven oscillator at least, if "pinching points" exist then the flow associated with the corresponding parameter values is extremely regular. More precisely, for such flows, all orbits, for all sets of initial conditions, can be shown to be periodic with period 2π. For a nonlinear system such as this, one would not expect this to occur.

The problem is best approached through the *stroboscopic map* for the periodically driven flow. The stroboscopic map, (which is the natural *Poincaré section* for the flow), is obtained by, in effect, taking a snapshot of all orbits every 2π time units. Since the driving is periodic, the value of $\theta(2[n+1]\pi)$ depends only on the value of $\theta(2n\pi)$; in particular, it does not depend on n. There is, therefore, a natural map f from the circle $[0, 2\pi)$ to itself, given by $f(\theta_0) = \theta_1$, where θ_1 is the value of $\theta(2\pi)$, reduced modulo 2π, given that the value of $\theta(0)$ is θ_0. The statement that all orbits are periodic with period 2π is equivalent to the statement that the stroboscopic map is the *identity*. Studying the stroboscopic map does not involve investigating any asymptotic behaviour, and it can therefore be done quickly and easily using *Mathematica*'s numerics, which offer the substantial advantage of arbitrary precision, (as well as ease of coding, of course). Some code for generating data on the stroboscopic map appears below.

```
flowEquation[ρ_][ϵ_][A_, B_] =
 θ'[t] == A + B Sin[t] - ϵ Sin[θ[t] + ρ t] - ρ

orbit[ρ_][ϵ_][A_, B_][θ0_] :=
 NDSolve[{flowEquation[ρ][ϵ][A, B], θ[0] == θ0}, θ,
    {t, 0, 2 π}, AccuracyGoal → 14, PrecisionGoal → 14,
    WorkingPrecision → 20][[1]][[1]]

stroboscopicMap[ρ_][ϵ_][A_, B_][θ0_] :=
 stroboscopicMap[ρ][ϵ][A, B][θ0] =
  N[θ[2 π] /. orbit[ρ][ϵ][A, B][θ0], 20]
```

```
stroboscopicMapData[ρ_][ϵ_][A_, B_][n_] := Table[
    {N[θ0], Mod[stroboscopicMap[ρ][ϵ][A, B][θ0], 2 π]},
    {θ0, 0, 2 π - π / n, π / n}]
```

7 Locating pinching points using *Mathematica*

A stroboscopic map for a flow of this type can always be written in the form:

$$f(\theta) = \theta + \alpha + \beta\,\sigma(\theta), \tag{4}$$

where α has been chosen to lie between -1/2 and 1/2, and σ is periodic, with the property that:

$$\sup_{0 \le \theta < 2\pi} \sigma(\theta) = - \inf_{0 \le \theta < 2\pi} \sigma(\theta) = 1. \tag{5}$$

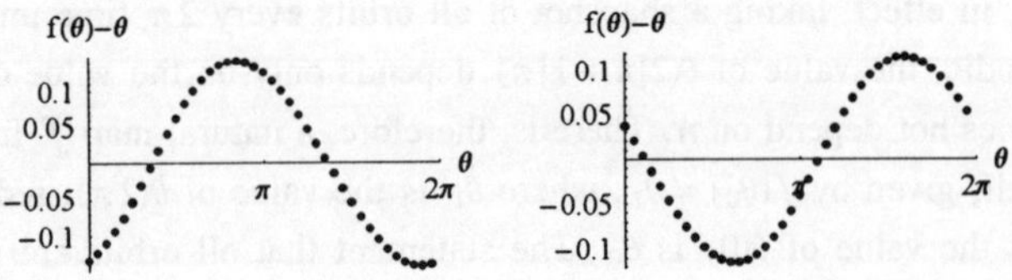

Figure 3

We can think of α and $\beta\,\sigma(\theta)$ as, respectively, the rotational and oscillatory components of the map. If pinching is real, then there ought to exist points at which the map is the identity: that is to say, at which $\alpha = \beta = 0$. Both α and β can easily be estimated from numerical data sets of the form $\{f(\theta_1) - \theta_1, f(\theta_2) - \theta_2, \ldots, f(\theta_N) - \theta_N\}$. If a point in parameter space can be found where both these estimates are extremely small, this would be strong evidence that the stroboscopic map at that point was the identity, and therefore that we really do have a "pinching point". Figure 3 shows the graphs of $f(\theta) - \theta$ against θ for $(A = 0, B = 2.0)$ and $(A = 0,\ B = 2.8)$. We may conjecture that there has been an *amplitude inversion* at the pinching point, (where, of course, $f(\theta) - \theta = 0$); if so, an interval bisection algorithm based on some sort of measure of the *phase* of the curve ought to locate the point as precisely as we like, especially with *Mathematica*'s arbitrary precision. The following is some code for doing that.

```
oscillatoryTermPhase[ρ_][ε_][A_, B_][n_] :=
 oscillatoryTermPhase[ρ][ε][A, B][n] =
  With[{data = nonIdentityComponentData[ρ][ε][A, B][
        n]}, data[[Position[Transpose[data][[2]],
          Apply[Max, Transpose[
             data][[2]]]][[1]][[1]]]][[1]] - 3π/2

nextIterate[ρ_][ε_][A_][n_][
  {{B1_, B2_}, {phase1_, phase2_}}] :=
 N[With[{midPhase = oscillatoryTermPhase[ρ][ε][A,
        (B1 + B2) / 2][n]}, If[midPhase phase1 < 0,
    {{B1, (B1 + B2) / 2}, {phase1, midPhase}},
    {{(B1 + B2) / 2, B2}, {midPhase, phase2}}]], 20]

seedValue = {{2.0, 2.8},
  {oscillatoryTermPhase[0][0.1][0, 2.0][20],
   oscillatoryTermPhase[0][0.1][0, 2.8][20]}}

FixedPoint[nextIterate[0][0.1][0][20], seedValue,
 SameTest → (Abs[#[[1]][[1]] - #[[1]][[2]]] < 10^-15 &)]

{{2.4077321254242148, 2.407732125424215},
{-0.7853981633974483, 1.413716694115407}}
```

At the point ($A = 0$, $B = 2.4077321254242149$) in parameter space, our estimates for α and β are, respectively, 1.55041×10^{-13} and 9.39005×10^{-13}. This is extremely strong evidence that the stroboscopic map here is the identity, that "pinching" is real and that all orbits of the flow for these parameter values are periodic with period 2π. *Mathematica* was clearly the system of choice for these calculations. The top left panel of Figure 4 illustrates several such orbits, on the (t, θ)-torus, for these parameter values; their periodic character is apparent. The right panel shows orbits at a pinching point of the $\rho = 1$ region, found using a similar method; again, each orbit is periodic on the torus. The bottom panels show an embedding of the orbits in three dimensions, making the toroidal structure explicit. Code for generating these figures is available from the author on request.

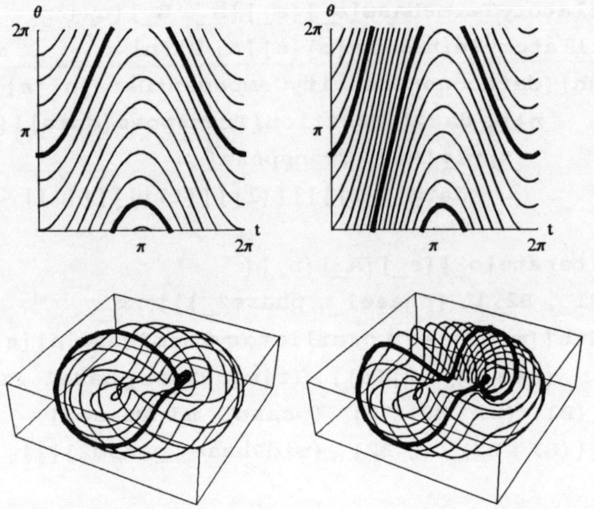

Figure 4

8 References

[1] Feudel, U., Kurths, J, Pikovsky, A. S., "Strange nonchaotic attractor in a quasiperiodically forced circle map", Preprint

[2] Glendinning, P., Feudel, U., Pikovsky, A. S., Stark, J., "The structure of mode-locked regions in quasiperiodically forced circle maps", Preprint

[3] Glendinning, P., "Intermittency and strange nonchaotic attractors in quasiperiodically forced circle maps", in Phys. Lett., A 244, 1996, 545-550

[4] Pikovsky, A. S., Feudel, U., "Characterising strange nonchaotic attractors", in CHAOS 5 (1), 1995, 253-260

[5] Romeiras, F. J., Ott, E., "Strange nonchaotic attractors of the damped pendulum with quasiperiodic forcing", in Phys. Rev. A 35 (10), 1987, 4404-4413

[6] van Vechten, R., Soulen, R. J., Peterson, R. L., "Behavior of the DC impedance of an RF-biased resistive SQUID", in SQUID '80, Walter de Gruyter, NY, 1980, 569-584

Neighbourhoods of Randomness and the Information Geometry of the Mckay Bivariate Gamma 3-manifold

Khadiga Arwini and C.T.J.Dodson
Department of Mathematics, UMIST, Manchester, UK
k.Alrawini@postgrad.umist.ac.uk
dodson@umist.ac.uk

1 Abstract

We showed [3] that gamma distributions provide models for departures from randomness, since every neighbourhood of an exponential distribution contains a neighbourhood of gamma distributions, using an information theoretic metric topology. Here we use *Mathematica* to illustrate these neighbourhoods. Using *Mathematica*, without which the computations would have been prohibitively tedious, we derived also the information geometry of the 3-manifold of Mckay bivariate gamma distributions, which can provide a metrization of departures from randomness and independence for bivariate processes. As in the case of bivariate normal manifolds, we have negative scalar curvature, but here it is not constant; details are provided in our paper [3].

2 Gamma distributions and randomness

The family of gamma probability density functions is given by

$$\left\{p(x;\beta,\alpha)=\left(\frac{\alpha}{\beta}\right)^{\alpha}\frac{x^{\alpha-1}}{\Gamma(\alpha)}\,e^{-\frac{\alpha}{\beta}x}\,\middle|\,\alpha,\beta\in\mathbf{R}^{+}\right\},\quad x\in\mathbf{R}^{+} \tag{1}$$

so the space of parameters is topologically $\mathbf{R}^{+}\times\mathbf{R}^{+}$. It is an exponential family and it includes, as a special case (α=1), the exponential distribution itself, which complements the Poisson process on a line. The property of having sample standard deviation independent of the mean actually characterizes gamma distributions, as shown recently by Hwang and Hu [10]. Of course, the exponential distribution has unit coefficient of variation.The information geometry is known for the gamma family.

Proposition 1.1 [3] *Every neighbourhood of an exponential distribution contains a neighbourhood of gamma distributions, using the subspace topology of* $\boldsymbol{R}^3$ *and information theoretic immersions.*

Outline Proof: Dodson and Matsuzoe [8] have provided an affine immersion in Euclidean $\mathbb{R}^3$ for $\mathcal{G}$, the manifold of gamma distributions with Fisher information metric. The coordinates $(\theta^1, \theta^2) = \left(\mu = \frac{\alpha}{\beta}, \alpha\right)$ form a natural coordinate system (cf Amari and Nagaoka [2]) for the gamma manifold $\mathcal{G}$, so $\mathcal{G}$ can be realized in Euclidean $\mathbb{R}^3$ by the graph of the affine immersion $\{h, \xi\}$ where ξ is the transversal vector field along h. [2, 8]

$$h : \mathcal{G} \to \mathbb{R}^3 : \begin{pmatrix} \mu \\ \alpha \end{pmatrix} \mapsto \begin{pmatrix} \mu \\ \alpha \\ \log\Gamma(\alpha) - \alpha \log \mu \end{pmatrix}, \quad \xi = \begin{pmatrix} 0 \\ 0 \\ 1 \end{pmatrix}. \tag{2}$$

The submanifold of exponential distributions is represented by the curve

$$(0, \infty) \to \mathbb{R}^3 : \mu \mapsto \left\{\mu,\ 1,\ \log \frac{1}{\mu}\right\} \tag{3}$$

and a tubular neighbourhood of this curve will contain all immersions for small enough perturbations of exponential distributions. This is illustrated in the graphics below, which show S, an affine embedding of the gamma manifold as a surface in $\boldsymbol{R}^3$, and R, an $\boldsymbol{R}^3$ −tubular neighbourhood of the curve $\alpha = 1$ in the surface. This curve represents all exponential distributions.

```
S = ParametricPlot3D[{β, ν, Log[Gamma[ν]] - ν Log[β]},
    {β, 0.00001, 4}, {ν, 0.0001, 3},
    ViewPoint → {3, 2, 1}, BoxRatios → {1, 1, 1}];
R = ParametricPlot3D[
    {β - 0.6 (β^2 + 1)^(-1/2) Cos[theta], 1 - 0.6 Sin[theta],
      -0.6 β (β^2 + 1)^(-1/2) Cos[theta] - Log[β]},
    {β, 0.0001, 4}, {theta, 0, 2 Pi},
    PlotPoints → {20, 20}, ViewPoint → {3, 2, 1},
    BoxRatios → {1, 1, 1}];
```

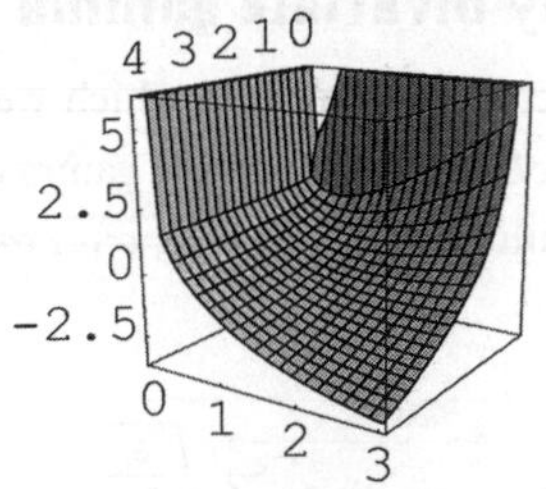

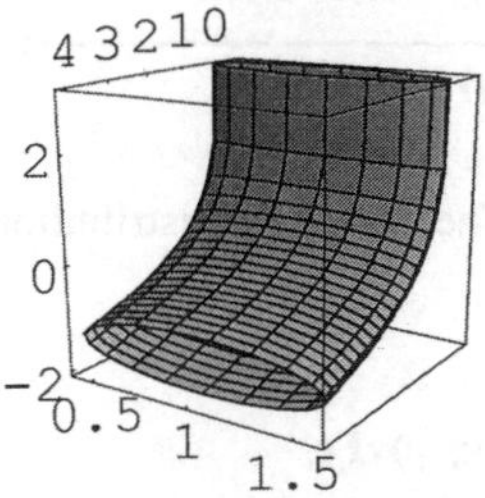


```
Show[S, R];
```

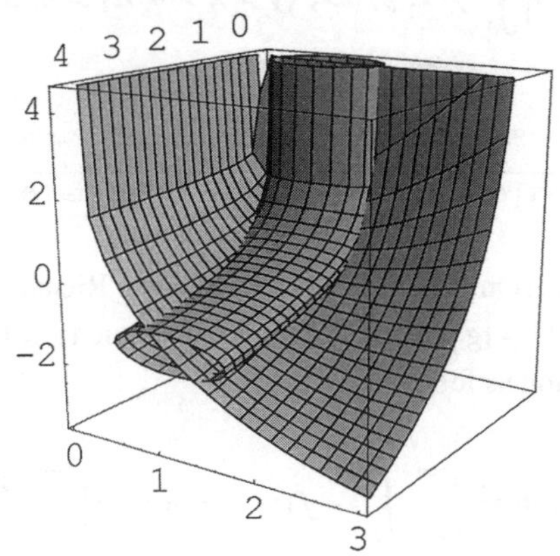

The value of such topological results lies in the fact that they have qualitative consequences that are stable under small perturbations of a process, something that would be important in real applications. It gives confidence in the use of gamma distributions to model near random processes. For comparison purposes, we recall that the information geometry of the univariate and multivariate normal distributions are also known, see for example Sato et al [11] for the bivariate case. We note in particular that the univariate and bivariate normal distributions have constant negative scalar curvature, so geometrically they constitute parts of pseudospheres.

3 Geometry of the Mckay bivariate gamma 3-manifold

In this section we show how to derive a selection of results which were used in our proof of proposition 1.1. The classical family of Mckay bivariate gamma probability density functions defined on $0 < x < y < \infty$ with parameters $\alpha_1, \sigma_{12}, \alpha_2 > 0$ and σ_{12} the covariance of X and Y, is given by:

$$f(\mathrm{x\_}, \mathrm{y\_}) = \frac{\left(\frac{\alpha_1}{\sigma_{12}}\right)^{\frac{(\alpha_1+\alpha_2)}{2}} x^{\alpha_1-1} (y-x)^{\alpha_2-1} e^{-y\sqrt{\frac{\alpha_1}{\sigma_{12}}}}}{\Gamma(\alpha_1)\,\Gamma(\alpha_2)};$$

We denote the set of such pdfs by M. The marginal distributions of X and Y are as follows.

$$\Big\{f_X(x) = \mathbf{FullSimplify}\Big[\int_x^\infty f(x, y)\, dy, \{y > x > 0, \alpha_1 > 0, \sigma_{12} > 0, \alpha_2 > 0\}\Big],\ f_Y(y) = \mathbf{FullSimplify}\Big[\int_0^y f(x, y)\, dx, \{y > x > 0, \alpha_1 > 0, \sigma_{12} > 0, \alpha_2 > 0\}\Big]\ \Big\}$$

$$\left\{\frac{e^{-x\sqrt{\frac{\alpha_1}{\sigma_{12}}}}\, x^{-1+\alpha_1} \left(\frac{\alpha_1}{\sigma_{12}}\right)^{\frac{\alpha_1}{2}}}{\mathrm{Gamma}[\alpha_1]},\ \frac{e^{-y\sqrt{\frac{\alpha_1}{\sigma_{12}}}}\, y^{-1+\alpha_1+\alpha_2} \left(\frac{\alpha_1}{\sigma_{12}}\right)^{\frac{1}{2}(\alpha_1+\alpha_2)}}{\mathrm{Gamma}[\alpha_1+\alpha_2]}\right\}$$

Using standard procedures from [1] and [2], M becomes a Riemannian 3-manifold, with Fisher information metric $G = [g_{ij}]$ obtained from expectations of the second derivatives of the log-likelihood function as follows:

$$\mathbf{Outer}\Big[\mathbf{FullSimplify}\Big[\int_0^\infty \int_x^\infty -f(x, y)\, \mathbf{Simplify}\Big[\frac{\partial^2 \log(f(x, y))}{\partial \#1\, \partial \#2}\Big]\, dy\, dx, \{y > x > 0,\ \alpha_1 > 0, \sigma_{12} > 0, \alpha_2 > 0\}\Big] \&, \{\alpha_1, \sigma_{12}, \alpha_2\}, \{\alpha_1, \sigma_{12}, \alpha_2\}\Big] \mathbin{//} \mathbf{MatrixForm}$$

$$\begin{pmatrix} \mathrm{PolyGamma}[1, \alpha_1] + \frac{-3\alpha_1+\alpha_2}{4\alpha_1^2} & \frac{\alpha_1-\alpha_2}{4\alpha_1\sigma_{12}} & -\frac{1}{2\alpha_1} \\ \frac{\alpha_1-\alpha_2}{4\alpha_1\sigma_{12}} & \frac{\alpha_1+\alpha_2}{4\sigma_{12}^2} & \frac{1}{2\sigma_{12}} \\ -\frac{1}{2\alpha_1} & \frac{1}{2\sigma_{12}} & \mathrm{PolyGamma}[1, \alpha_2] \end{pmatrix}$$

The inverse matrix is $[g^{ij}] = [g_{ij}]^{-1}$ and then the Christoffel symbols are defined by $\overset{i}{\Gamma}_{jk} = \sum_{h=1}^{3} \frac{1}{2} \overset{ih}{g} \left(\partial_{x_k} g_{jh} + \partial_{x_j} g_{kh} - \partial_{x_h} g_{jk}\right)$; $(i, j, k = 1, 2, 3)$. The input is illustrated by:

$$\overset{1}{\Gamma}_{11} = \frac{1}{2}\left(\overset{11}{g}\left(\partial_{x_1} g_{11} + \partial_{x_1} g_{11} - \partial_{x_1} g_{11}\right) + \overset{12}{g}\left(\partial_{x_1} g_{21} + \partial_{x_1} g_{21} - \partial_{x_2} g_{11}\right) + \overset{13}{g}\left(\partial_{x_1} g_{31} + \partial_{x_1} g_{31} - \partial_{x_3} g_{11}\right)\right) \text{ // PowerExpand // FullSimplify}$$

$$-(-4\,\text{PolyGamma}[2, \alpha_1]\,\alpha_1^2 + \text{PolyGamma}[1, \alpha_2]\,(3\alpha_1 - \alpha_2) + 4\,\text{PolyGamma}[1, \alpha_2]\,\text{PolyGamma}[2, \alpha_1]\,\alpha_1^2\,(\alpha_1 + \alpha_2)) / (8\,\alpha_1^2\,(\text{PolyGamma}[1, \alpha_1] + \text{PolyGamma}[1, \alpha_2] - \text{PolyGamma}[1, \alpha_1]\,\text{PolyGamma}[1, \alpha_2]\,(\alpha_1 + \alpha_2)))$$

The components of the curvature tensor $R_{ijkm} = \sum_{h=1}^{3} g_{hm}\left(\partial_{x_i} \overset{h}{\Gamma}_{jk} - \partial_{x_j} \overset{h}{\Gamma}_{ik} + \overset{h}{\Gamma}_{im} \overset{m}{\Gamma}_{jk} - \overset{h}{\Gamma}_{jm} \overset{m}{\Gamma}_{ik}\right)$; $(i, j, k = 1, 2, 3)$ are given by, eg:

$$R_{1212} = g_{12}\left(\partial_{x_1} \overset{1}{\Gamma}_{21} - \partial_{x_2} \overset{1}{\Gamma}_{11} + \overset{1}{\Gamma}_{11} \overset{1}{\Gamma}_{21} - \overset{1}{\Gamma}_{21} \overset{1}{\Gamma}_{11} + \overset{1}{\Gamma}_{12} \overset{2}{\Gamma}_{21} - \overset{1}{\Gamma}_{22} \overset{2}{\Gamma}_{11} + \overset{1}{\Gamma}_{13} \overset{3}{\Gamma}_{21} - \overset{1}{\Gamma}_{23} \overset{3}{\Gamma}_{11}\right) + g_{22}\left(\partial_{x_1} \overset{2}{\Gamma}_{21} - \partial_{x_2} \overset{2}{\Gamma}_{11} + \overset{2}{\Gamma}_{11} \overset{1}{\Gamma}_{21} - \overset{2}{\Gamma}_{21} \overset{1}{\Gamma}_{11} + \overset{2}{\Gamma}_{12} \overset{2}{\Gamma}_{21} - \overset{2}{\Gamma}_{22} \overset{2}{\Gamma}_{11} + \overset{2}{\Gamma}_{13} \overset{3}{\Gamma}_{21} - \overset{2}{\Gamma}_{23} \overset{3}{\Gamma}_{11}\right) + g_{32}\left(\partial_{x_1} \overset{3}{\Gamma}_{21} - \partial_{x_2} \overset{3}{\Gamma}_{11} + \overset{3}{\Gamma}_{11} \overset{1}{\Gamma}_{21} - \overset{3}{\Gamma}_{21} \overset{1}{\Gamma}_{11} + \overset{3}{\Gamma}_{12} \overset{2}{\Gamma}_{21} - \overset{3}{\Gamma}_{22} \overset{2}{\Gamma}_{11} + \overset{3}{\Gamma}_{13} \overset{3}{\Gamma}_{21} - \overset{3}{\Gamma}_{23} \overset{3}{\Gamma}_{11}\right) \text{ // PowerExpand // FullSimplify}$$

$$(\text{PolyGamma}[1, \alpha_2]\,(\text{PolyGamma}[1, \alpha_1] + \text{PolyGamma}[2, \alpha_1]\,(\alpha_1 + \alpha_2))) / (16\,(\text{PolyGamma}[1, \alpha_1] + \text{PolyGamma}[1, \alpha_2] - \text{PolyGamma}[1, \alpha_1]\,\text{PolyGamma}[1, \alpha_2]\,(\alpha_1 + \alpha_2))\,\sigma_{12}^2)$$

The components of the Ricci curvature are given by the symmetric matrix $R = [R_{jk}]$, where $R_{jk} = R_{ijkm} \overset{im}{g}$, eg:

$$R_{11} = R_{1111}\overset{11}{g} + R_{1112}\overset{12}{g} + R_{1113}\overset{13}{g} + R_{2111}\overset{21}{g} + R_{2112}\overset{22}{g} + R_{2113}\overset{23}{g} + R_{3111}\overset{31}{g} + R_{3112}\overset{32}{g} + R_{3113}\overset{33}{g} \text{ // PowerExpand // FullSimplify}$$

```
(-2 PolyGamma[1, α2] (-1 + 2 PolyGamma[1, α1] α1)
    (PolyGamma[1, α1] + 2 PolyGamma[2, α1] α1) -
   (-PolyGamma[1, α1] PolyGamma[1, α2] +
      PolyGamma[2, α2] (4 PolyGamma[2, α1] α1^2 +
         PolyGamma[1, α1] (3 α1 - α2)))
    (-1 + PolyGamma[1, α1] (α1 + α2)) + PolyGamma[1, α2]
    (PolyGamma[1, α1] + PolyGamma[2, α1] (α1 + α2))
    (-1 + PolyGamma[1, α2]
         (α1 (-3 + 4 PolyGamma[1, α1] α1) + α2))) /
 (16 α1^2 (PolyGamma[1, α1] + PolyGamma[1, α2] -
      PolyGamma[1, α1] PolyGamma[1, α2] (α1 + α2))^2)
```

The scalar curvature $R = R_{jk}\overset{jk}{g}$ is given by:

$$R = R_{11}\overset{11}{g} + R_{12}\overset{12}{g} + R_{13}\overset{13}{g} + R_{21}\overset{21}{g} + R_{22}\overset{22}{g} + R_{23}\overset{23}{g} + R_{31}\overset{31}{g} + R_{32}\overset{32}{g} + R_{33}\overset{33}{g} \text{ // PowerExpand // FullSimplify}$$

```
(PolyGamma[2, α2] (-2 PolyGamma[1, α1] +
       (PolyGamma[1, α1]^2 - PolyGamma[2, α1]) (α1 + α2)) +
   PolyGamma[1, α2] (PolyGamma[1, α1]
        (PolyGamma[1, α1] + PolyGamma[1, α2]) -
       2 PolyGamma[2, α1] + PolyGamma[1, α2]
        PolyGamma[2, α1] (α1 + α2))) /
 (2 (PolyGamma[1, α1] + PolyGamma[1, α2] -
       PolyGamma[1, α1] PolyGamma[1, α2] (α1 + α2))^2)
```

4 Applications

The univariate gamma information geometry is known and has been applied recently to represent and metrize departures from randomness of, for example, the processes that allocate gaps between occurrences of each amino acid along a protein chain within the Saccharomyces cerevisiae genome, (see Cai et al [4]), cosmological void distribution and clustering of galaxies and communications, Dodson [5,7,6]. Two new case studies are being developed at present. Both involve bivariate data from measurements on stochastic porous media. Near-planar stochastic fibrous network structures,such as paper and nonwoven textiles, are deposited on a continuous filter bed. Dodson and Sampson [9] used the product of two gamma distributions to obtain the void size distribution, recovering the known random model as a special case. Some commercial processes involve preferential alignment of fibres in the direction of manufacture, resulting in anisotropic void distributions; then the McKay bivariate model is applicable. The 3-dimensional tomographic images of soil samples in hydrology surveys yield 3-dimensional reconstructions of the stochastic porous structure and, interestingly, these generate a bivariate positively correlated pair of random variables $0 < x < y < \infty$. Here, the larger variable represents the size of a void and the smaller variable represents the sizes of the throats that connect it to neighbouring voids. This work is being pursued in collaboration with Professor J. Scharkanski of the Federal University of Rio Grande do Sul, Brazil and will be reported elsewhere.

Acknowledgment: The authors are grateful to Peter Mitic for suggesting improvements in the *Mathematica* formatting and to Paul Abbott for improvements to the *Mathematica* code.

5 References

[1] S-I. Amari, O.E. Barndorff-Neilsen, R.E. Kass, S.L. Lauritzen and C.R. Rao. Differential Geometrical Methods in Statistics, Springer Lecture Notes in Statistics 28, Springer-Verlag, Berlin 1985.

[2] S-I. Amari and H. Nagaoka. Methods of Information Geometry, American Mathematical Society, Oxford University Press, 2000.

[3] Khadiga Arwini and C.T.J. Dodson. Information geometric neighbourhoods of randomness and geometry of the McKay bivariate gamma 3-manifold. Preprint 2002. http://www.ma.umist.ac.uk/kd/PREPRINTS/gamran.pdf

[4] Y. Cai, C.T.J. Dodson, O. Wolkenhauer and A.J. Doig. Gamma Distribution Analysis of Protein Sequences shows that Amino Acids Self Cluster. J. Theoretical Biology 218, 4 (2002) 409-418.

[5] C.T.J. Dodson. Spatial statistics and information geometry for parametric statistical models of galaxy clustering. Int. J. Theor. Phys., 38, 10, (1999) 2585-2597.

[6] C.T.J. Dodson. Information geodesics for communication clustering. J. Statistical Computation and Simulation 65, (2000) 133-146.

[7] C.T.J. Dodson. Geometry for stochastically inhomogeneous spacetimes. Nonlinear Analysis, 47 (2001) 2951-2958.

[8] C.T.J. Dodson and Hiroshi Matsuzoe. An affine embedding of the gamma manifold. Applied Sciences, 5, 1 (2003) 1-6.

http : // www.ma.umist.ac.uk / kd / PREPRINTS / affimm.pdf .

[9] C.T.J. Dodson and W.W. Sampson. Modeling a class of stochastic porous media. App. Math. Lett. 10, 2 (1997) 87-89.

[10] T-Y. Hwang and C-Y. Hu. On a characterization of the gamma distribution: Independence of the sample mean and the sample coefficient of variation. Ann Inst. Statist Math. 51, 4 (1999) 749-753.

[11] Y. Sato, K. Sugawa and M. Kawaguchi. The geometrical structure of the parameter space of the two-dimensional normal distribution. Reports on Mathematical Physics 16, 1 (1979) 111-119.

Parabolic and Non-Parabolic Loci of the Center of Gravity of Variable Solids

Tilak de Alwis
Department of Mathematics
Southeastern Louisiana University
Hammond, LA 70402
USA
talwis@selu.edu

1 Abstract

Consider certain types of solids S in three-dimensions with a variable boundary . It is assumed that this boundary changes with two real parameters s and t. Examples of such include solids bounded by a fixed elliptic cylinder and a variable plane, solids bounded by a fixed elliptic paraboloid and a variable plane, and also solids bounded by a fixed astroidal cylinder and a variable plane. Let $G(\bar{x}, \bar{y}, \bar{z})$ denote the center of gravity of the solid S. As the parameters s and t change, this center of gravity G changes. According to some of the previous studies, for several well-known types of solids S, the locus of G is a paraboloid in three-dimensions. In this paper, we will further investigate such locus problems. In particular, it is of interest to discover familiar types of variable solids, for which the locus of the center of gravity G is not a paraboloid. In order to calculate the triple integrals associated with these center of gravity problems, and also to create visualizations of the loci, one can effectively use *Mathematica*.

2 Introduction

Consider a solid S in three-dimensions. Then its center of gravity $G(\bar{x}, \bar{y}, \bar{z})$ is given by the following equations (see [4], [5], and [6])

$$\bar{x} = I_1 / I; \quad \bar{y} = I_2 / I; \quad \bar{z} = I_3 / I \tag{1}$$

where the integrals I_1, I_2 and I_3 are defined by

$$I_1 = \iiint x\,dV; \quad I_2 = \iiint y\,dV; \quad I_3 = \iiint z\,dV; \quad I = \iiint dV$$

The above integrals (2) are all performed over the three-dimensional region bounded by the solid S. The equations (1)-(2) can be used the calculate the center of gravity of many types of solid objects. For example, in [1] and [2], we investigated the center of gravity of some specific variable solids. In order to make this paper as much self-contained as possible, we will summarize some of these earlier results in the next section.

3 A Summary of Previous Results

3.1 *The center of gravity of a solid bounded by a fixed elliptic cylinder and a variable plane*

In [1], we were able to prove the following theorem:

Theorem 3.1 Consider the solid S_1 bounded by the XY-plane, the fixed elliptic cylinder $\mathcal{E}_1$ with equation $x^2/a^2 + y^2/b^2 = 1$ where a and b are positive constants, and the plane $\mathcal{P}$ through the fixed point (0, 0, c) with variable normal vector <s, t, 1> where s and t are real parameters and $c > 0$. We will assume that s and t change so that the plane $\mathcal{P}$ intersects the cylinder $\mathcal{E}_1$ in the upper-half space $z > 0$. Then the center of gravity of the solid S_1 is given by $G = (-a^2 s/(4c), -b^2 t/(4c), (4c^2 + a^2 s^2 + b^2 t^2)/(8c))$. Furthermore, for changing s and t, the locus of G is an elliptic paraboloid, given by the equation $z = (c/2) + 2c(x^2/a^2 + y^2/b^2)$.

Proof. The theorem can be proved with or without the use of *Mathematica* (see [1]).

In [2], we designed a *Mathematica* program to create an animation of the center of gravity of the solid S_1 for changing parameters s and t. In the following, we include three frames of this animation: As s and t change, the roof of the solid, given by the plane $\mathcal{P}$, tilts around the fixed point (0, 0, c). The center of gravity G of the solid S_1 traverses on an elliptic paraboloid, shown inside the cylinder $\mathcal{E}_1$.

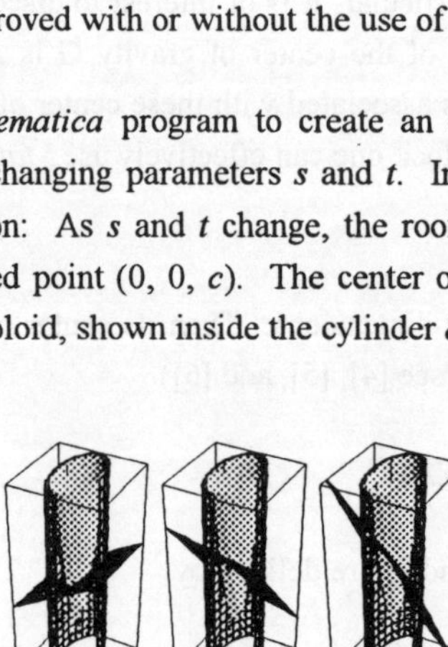

Figure 1 An animation of the center of gravity of the solid S_1

3.2 *The center of gravity of a solid bounded by a fixed elliptic paraboloid and a variable plane*

In [1], we were able to state without proof, a more general version of the following theorem:

Theorem 3.2 Consider the solid S_2 bounded by the fixed elliptic paraboloid $\mathcal{E}_2$ with equation $z = x^2/a^2 + y^2/b^2$ where a and b are positive real constants, and the plane $\mathcal{P}$ through the fixed point (0, 0, c) with variable normal vector $\langle s, t, 1\rangle$ where s and t are real parameters and $c > 0$. We will assume that s and t change so that the plane $\mathcal{P}$ intersects the paraboloid $\mathcal{E}_2$ in the upper-half space $z > 0$. Then the center of gravity of the solid S_2 is given by $G = (-a^2 s/2,\ -b^2 t/2,\ (8c + 5a^2 s^2 + 5b^2 t^2)/12)$. Furthermore, for changing s and t, the locus of G is an elliptic paraboloid, given by the equation $z = (2c/3) + (5/3)(x^2/a^2 + y^2/b^2)$.

In [1], we designed a *Mathematica* program to create an animation of the center of gravity of the solid S_2 described in Theorem 3.2. The following are a few frames of the animation:

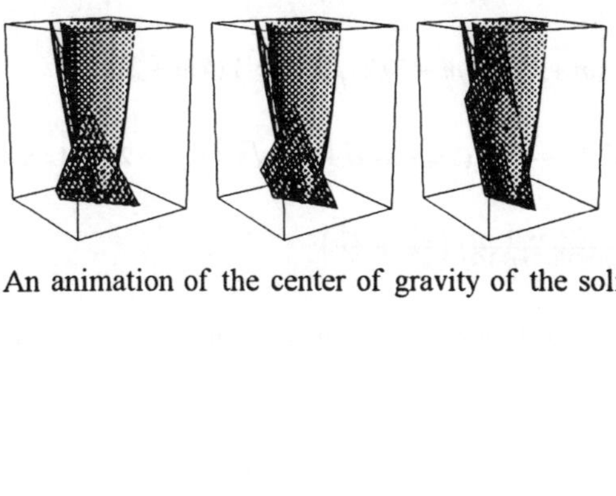

Figure 2 An animation of the center of gravity of the solid S_2

3.3 *The center of gravity of a solid bounded by a fixed astroidal cylinder and a variable plane*

The following is a slight generalization of the astroidal cylinder discussed in [2]. The motivation for the section came from the star-shaped classical Greek curve given by the equation $x^{2/3} + y^{2/3} = a^{2/3}$ where $a > 0$ is a constant. One can now consider the more general equation $(x/a)^{2/n} + (y/b)^{2/n} = 1$, where n is a positive odd integer, and a, b are positive constants. The region $\mathcal{R}$ bounded by this two-dimensional curve will form the base of our solid. Let $\mathcal{E}_3$ be the right-astroidal cylinder with $\mathcal{R}$ as the base. As indicated in Theorems 3.1 and 3.2, let $\mathcal{P}$ be the plane through the fixed point (0, 0, c) with variable normal vector $\langle s, t, 1\rangle$ where s and t are real parameters and $c > 0$. We will assume that s and t change so that the plane $\mathcal{P}$ intersects the astroidal cylinder $\mathcal{E}_3$ in the upper-half space $z > 0$. Let $\mathcal{S}_3$ be the solid bounded by the astroidal cylinder $\mathcal{E}_3$ and the plane $\mathcal{P}$. The following theorem concerns the center of gravity of this solid $\mathcal{S}_3$.

Theorem 3.3 The center of gravity $G(\overline{x}, \overline{y}, \overline{z})$ of the solid $\mathcal{S}_3$ described above is given by the equations (3)-(5) below. Furthermore, for changing s and t, the locus of G is an elliptic paraboloid given by the equation (6).

$$\overline{x} = -2^{2m-1} a^2 s\, \Gamma(m+1)\, \Gamma(3m+3/2) \big/ \left[c\sqrt{\pi}\, \Gamma(4m+2)\right] \tag{3}$$

$$\overline{y} = -2^{2m-1} b^2 t\, \Gamma(m+1)\, \Gamma(3m+3/2) \big/ \left[c\sqrt{\pi}\, \Gamma(4m+2)\right] \tag{4}$$

$$\overline{z} = \left[4^m (a^2 s^2 + b^2 t^2)\, \Gamma(m+1)\, \Gamma(3m+3/2) + 2c^2 \sqrt{\pi}\, \Gamma(4m+2)\right] \big/ \left[4c\sqrt{\pi}\, \Gamma(4m+2)\right] \tag{5}$$

$$z = \frac{c}{2} + \frac{c\sqrt{\pi}}{2^{2m}} \frac{\Gamma(4m+2)}{\Gamma(m+1)\,\Gamma(3m+3/2)} \left(\frac{x^2}{a^2} + \frac{y^2}{b^2}\right) \tag{6}$$

In the above equations, Γ denotes the Euler's Gamma function, and $m = (n-1)/2$ (recall that n was a positive odd integer).

Proof. The proof is similar to that of Proposition 3.1 in [2].

For some details on the function Γ, the reader can refer to [2], [4], and [6]. Note that the equation (6) can be written using the factorial form, using the connections between the gamma function and the factorials (see [2]). For each of the solids $\mathcal{S}_1$, $\mathcal{S}_2$, and $\mathcal{S}_3$ described in this section under (a), (b), and (c), the locus of the center of gravity G was an elliptic paraboloid. So, it is of interest to discover well-known types of solids for which this phenomenon is not true. The next section achieves this task.

4 The Center of Gravity of a Solid Bounded by a Fixed Cone and a Variable Plane

Consider the cone C given by the equation $z = \sqrt{x^2 + y^2}$. As given in the previous section, let $\mathcal{P}$ be the plane through the fixed point (0, 0, c) with variable normal vector $\langle s, t, 1\rangle$ where s and t are real parameters and $c > 0$. We will assume that $s^2 + t^2 \leqslant 1$, so that the trace of the intersection of the cone C and the plane $\mathcal{P}$ is an ellipse or a circle in the XY-plane. Let S_4 be the solid bounded by the cone C and the plane $\mathcal{P}$. We are interested in studying the behaviour of the center of gravity $G(\bar{x}, \bar{y}, \bar{z})$ of the solid S_4 for changing parameters s and t. One can use the equations (1) and (2) to calculate $\bar{x}$, $\bar{y}$ and $\bar{z}$. We can conveniently use the variable transformation $x = r\,\text{Cos}\theta$, $y = r\,\text{Sin}\theta$, and $z = z$ with its Jacobian $J = r^2$ to calculate the integrals I_1, I_2, I_3, and I (see [1], [2], and [5]). For example, with this transformation, the integral I_1 is given by the following equation, where $r_1 = c/(1 + s\,\text{Cos}\,\theta + t\,\text{Sin}\,\theta)$:

$$I_1 = \int_{\theta=0}^{2\pi} \int_{r=0}^{r_1} \int_{z=r}^{c-s\,r\,\text{Cos}\theta - t\,r\,\text{Sin}\theta} r^3\,\text{Cos}\theta\, dz\, dr\, d\theta \tag{7}$$

One can use the "**Integrate**" and "**Simplify**" commands of *Mathematica* to evaluate the above integral (7) (see [7]).

```
In[1]:= Simplify[Integrate[r^3 * Cos[theta], {theta, 0, 2 Pi},
          {r, 0, c/ (1 + s * Cos[theta] + t * Sin[theta])},
          {z, r, c - s * r * Cos[theta] - t * r * Sin[theta]}]]
```

$$\text{Out[62]=} \quad \frac{c^5\,\pi\,(-1+s)\,s\,\sqrt{-\frac{-1+s^2+t^2}{(-1+s)^2}}\,(4+s^2+t^2)}{20\,(-1+s^2+t^2)^4}$$

The above output can be further simplified to obtain the following:

$$I_1 = -\pi\, c^5\, s\,(4 + s^2 + t^2)/\left[20\,(1 - s^2 - t^2)^{7/2}\right] \tag{8}$$

One can similarly show that

$$I_2 = -\pi\, c^5\, t\,(4 + s^2 + t^2)/\left[20\,(1 - s^2 - t^2)^{7/2}\right] \tag{9}$$

$$I_3 = -\pi c^5(s^4 + t^4 - 8 s^2 - 8 t^2 + 2 s^2 t^2 - 8)\big/\left[60 (1 - s^2 - t^2)^{7/2}\right] \tag{10}$$

$$I = \pi c^4(2 + s^2 + t^2)\big/\left[12 (1 - s^2 - t^2)^{5/2}\right] \tag{11}$$

One can now use equations (2) to calculate $\overline{x}$, $\overline{y}$ and $\overline{z}$, the coordinates of the center of gravity G:

$$\overline{x} = 3 c s(4 + s^2 + t^2)/[5 (-1 + s^2 + t^2) (2 + s^2 + t^2)] \tag{12}$$

$$\overline{y} = 3 c t(4 + s^2 + t^2)/[5 (-1 + s^2 + t^2) (2 + s^2 + t^2)] \tag{13}$$

$$\overline{z} = c (s^4 + t^4 + 2 s^2 t^2 - 8 s^2 - 8 t^2 - 8)/[5 (-1 + s^2 + t^2) (2 + s^2 + t^2)] \tag{14}$$

Use the **"Eliminate"** command of *Mathematica* to eliminate the two parameters s ant t between the three equations (12)-(14). Thus we obtain the following equation of the locus of the center of gravity G of the solid S_4:

$$\begin{aligned}125 (2 x^4 + 2 y^4 - z^4) + 125 (4 x^2 y^2 - y^2 z^2 - x^2 z^2) - 25 c z(4 x^2 + 4 y^2 - 13 z^2) \\ + 20 c^2(8 x^2 + 8 y^2 - 15 z^2) + 112 c^3 z - 12 c^4 = 0\end{aligned} \tag{15}$$

By looking at its form, or its graph, one can see that the equation (15) does not represent a paraboloid in three-dimensions. Therefore, the locus of the center of gravity of our conical solid S_4 is not a paraboloid. We can summarize the findings of this section in the following theorem:

Theorem 4.1 The coordinates of the center of gravity G of the solid S_4 defined in this section is given by equations (12)-(14). Furthermore, for changing parameters s and t, the locus of G is a fourth-degree polynomial equation given by equation (15).

Proof. Refer to the discussion preceding the statement of the theorem.

The following *Mathematica* program creates an animation of the center of gravity of the solid S_4:

Program 4.1

```
In[1]:= Clear[i1, i2, i3, i]
        r1 = c / (1 + s * Cos[theta] + t * Sin[theta]);
        x = r * Cos[theta];
        y = r * Sin[theta];
        i1 = Simplify[Integrate[x * r ^ 2,
            {theta, 0, 2 Pi}, {r, 0, r1}, {z, r, c - x * s - y * t}]];
        i2 = Simplify[Integrate[y * r ^ 2, {theta, 0, 2 Pi},
            {r, 0, r1}, {z, r, c - x * s - y * t}]];
        i31 = Integrate[z * r ^ 2, {z, r, c - x * s - y * t}];
        i32 = Simplify[Integrate[i31, {r, 0, r1}]];
        i3 = Simplify[Integrate[i32, {theta, 0, 2 Pi}]];
        i = Simplify[Integrate[r ^ 2,
            {theta, 0, 2 Pi}, {r, 0, r1}, {z, r, c - x * s - y * t}]];
        {x0, y0, z0} = Simplify[{i1 / i, i2 / i, i3 / i}]
        Clear[x, y, z];
        Eliminate[{x, y, z} == {x0, y0, z0}, {s, t}]
        p1 = ParametricPlot3D[Evaluate[
            {x0, y0, z0} /. {c → 2, s → k * Cos[u], t → k * Sin[u]}],
          {k, 0, 1}, {u, 0, 2 Pi}, PlotPoints → 30,
          DisplayFunction → Identity]
        p2 = ParametricPlot3D[{r * Cos[theta], r * Sin[theta], r}, {r, 0,
            10}, {theta, 0, 5 Pi / 4}, DisplayFunction → Identity];
        p12 = Show[{p1, p2}, DisplayFunction → $DisplayFunction,
          ViewPoint -> {0.755, -3.011, 1.125}]
        p3 := Plot3D[c - x * s - y * t /.
            {c → 2, s → 0.8 Cos[u], t → 0.5 Sin[u]},
                {x, -3.5, 3.5}, {y, -3.5, 3.5},
          PlotRange -> {0, 4}, DisplayFunction -> Identity]
        Do[Show[
            Graphics3D[{PointSize[1 / 60], RGBColor[1, 0, 0],
                    Point[ {x0, y0, z0}]} /.
             {c → 2, s → 0.8 Cos[u], t → 0.5 Sin[u]}],
          p12, p3, ViewPoint -> {0.755, -3.011, 1.125},
```

```
PlotRange → {{-3.5, 3.5}, {-3.5, 3.5}, {0, 4}},
DisplayFunction -> $DisplayFunction],
{u, 0, 2 Pi, 2 Pi / 50}]
```

The above program also calculates the coordinates of the center of gravity G as well as the equation of the locus of G, as given by the equations (12)-(14) and (15). The following are a few frames of the animation:

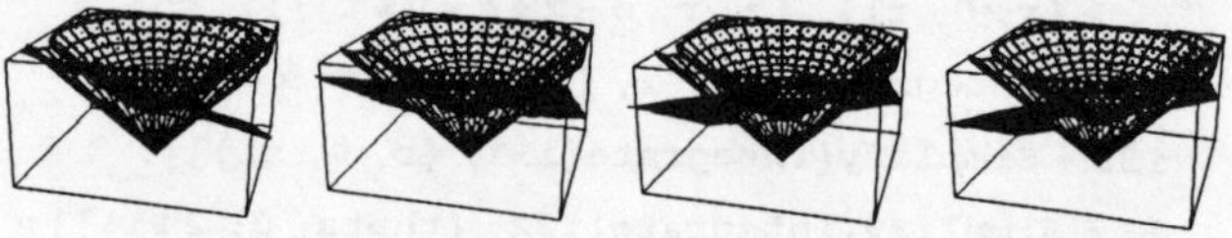

Figure 3 A non-parabolic loci: The center of gravity of the conical solid S_4

References

[1] de Alwis, T. (2001). The Locus of the Center of Gravity of Variable Solids. *Proceedings of the Third International Mathematica Symposium*, Tokyo Denki University, Chiba New Town Campus, Japan.

[2] de Alwis, T. (2001). The Center of Gravity of Classes of Cylindrical Solids via a Computer Algebra System. *Proceedings of the Sixth Asian Technology Conference in Mathematics*, University of Melbourne, Australia.

[3] Gray, T. and Glynn, J. (2000). *The Beginner's Guide to Mathematica*, Version 4. Cambridge, UK: Cambridge University Press.

[4] Kreyszig, E. (1993). *Advanced Engineering Mathematics*. New York, NY: John Wiley & Sons.

[5] Larson, H., Hostetler, R., and Edwards, B. (1994). *Calculus*. Lexington, MA: D.C. Heath.

[6] O'Neil, P. (1991). *Advanced Engineering Mathematics*. Belmont, CA: Wadsworth Publishing Company.

[7] Wolfram, S. (1996). *Mathematica Book*, 3rd ed. Cambridge, UK: Cambridge University Press.

Hero's method: An introduction to *Mathematica* programming

Murray Eisenberg
University of Massachusetts/Amherst
murray@math.umass.edu

1 Abstract

One rich and memorable example can be a more effective tool for learning, or teaching, a body of concepts and techniques than a multitude of separate little examples. Hero's method—a simple iterative algorithm for estimating square roots—serves as an exemplary topic for introducing principles of *Mathematica* in general, and of programming in *Mathematica* in particular. Among the ideas introduced are: *Mathematica*'s symbolic, numeric, and graphical aspects; function definition; list manipulation; functional, procedural, and recursive programming paradigms; and graphical programming.

2 Introduction

Newcomers to *Mathematica* need to learn "the basics": navigating through *Mathematica* notebooks; entering and evaluating input; and consulting the Help Browser. Then what? Working one's way through a multiplicity of unrelated examples can be inefficient and confusing. Wolfram [5] advises that the newcomer soon should "...go through what is needed to solve a complete problem with *Mathematica*." Approximating square roots is such a problem. The solution here, by Hero's method, involves successive approximation. The author has used this example, or its Newton's method generalization, to introduce *Mathematica* and *Mathematica* programming to beginning and advanced undergraduates as well as to college and university faculty. It has proved to be both instructive and eye-opening.The point of this paper is to indicate how this example can serve the stated purpose. For direct use by a novice, this paper is insufficient. Instead, the supplementary *Mathematica* notebook—which includes extended explanations and additional *Mathematica* features—should be adapted as required.

3 Hero's method for approximating square roots

Mathematica can directly approximate the square root of a positive number c is by a built-in method, invoked with the function `N`. Take $c = 2$; then evaluating input `N[√2 ]` yields output `1.41421`. That begs the question of how the numerical approximation is actually obtained. Internally *Mathematica* uses essentially the method about to be considered: *Hero's method*. Described in particular cases by Hero (or Heron) of Alexandria (see Heath [1]), this is an iterative method for approximating $\sqrt{c}$. It rests upon two simple principles: (i) If $0 < x < \sqrt{c}$, then $c/x > \sqrt{c}$ whereas if $x > \sqrt{c}$, then $0 < c/x < \sqrt{c}$; and (ii) the average $g(x) = (x + c/x)/2$ is nearer $\sqrt{c}$ than is either x or c/x. The function g, which will be iterated, is defined as follows:

```
In[1]:= g[x_] := (x + c / x) / 2
```

The value of `g` at 1 may be obtained by evaluating input `g[1]`; the result is the algebraic expression `(1+c)/2`, which is not numeric because so far `c` is just a *symbol*. The input in *In[1]* is not actually a function definition at all, although it can be regarded as such. Rather, it is a *rewrite rule*; evaluating the expression adds this rule to *Mathematica*'s global rule base. According to the rule, anything of the form `g[expr]` encountered during an evaluation is to be replaced by `(expr + c/expr)/2`. The form `x_` on the left hand side of the symbol `:=` is a *pattern* that any expression will match.

Evaluating any expression returns a result, which may be the symbol `Null`. Ordinarily a `Null` result is not displayed. This is the case, for example, with evaluating a `SetDelayed` (`:=`) expression as in *In[1]* or an expression that causes an error; establishing the rule or displaying the error message is a side-effect.

The following definition makes c an additional argument to the function to be iterated:

```
In[2]:= gg[cc_, x_] := (x + cc / x) / 2
```

Then `gg[2, 1]` has the exact, rational value `3/2`. The exact value may be converted to an approximate, real form: `N[gg[2, 3/2]]` has value `1.41667`.

Fixing one argument of a function of two arguments so as to obtain a function of a single argument may be accomplished by forming a *pure function*—an anonymous function, terminated with `&` and having argument `#`. Thus `gg[2, #]&[1]` has value `3/2` again.

To assign a value to a symbol on-the-fly, use the `Rule` operator → along with the `ReplaceAll` operator `/.`. For example, both `g[1.] /. c → 2` and

```
gg[cc, x0] /. {x0 → 1., cc → 2}
```

have value `1.5`. But now `c` will be assigned constant value 2 (the terminal semicolon suppresses the result that would otherwise be displayed):

```
In[3]:= c = 2;
```

Observe that = rather than := was just used. When leftexpr = rightexpr is evaluated, rightexpr is evaluated immediately, before the rewrite rule for leftexpr is added to the rule base. But when leftexpr := rightexpr is evaluated, rightexpr is not evaluated then; instead, each time leftexpr is subsequently evaluated, rightexpr is reevaluated first.

4 Successive approximations through iteration

The function g may be iterated manually. The second iterate `g[g[1]]` has value `17/12`. To avoid distracting nested brackets, use the *prefix form* `g@g[1]`. Then the decimal value `1.41667` for the second iterate is given by `N[g@g[1]]`, which may be written instead in the *postfix form* `g@g[1] // N`. To display machine-precision 16 digits instead of the default 6 digits, use input

```
NumberForm[N[g@g[1], 16], 16]
```

`to` obtain output:

```
1.416666666666667
```

Separate expressions may be combined into a *list*: `{N[g[1]], N[g@g[1]]}` has value `{1.5,1.41667}`. The same value results from

```
N[{g[1], g@g[1]}]
```

because `N` has the *attribute* `Listable`. To avoid applying `N` at each evaluation, make one of the numbers in the calculation an approximate real number. Thus `c` could be changed from 2 to `2.` or `2.0`. Instead, the initial estimate will be changed from the exact integer `1` to the approximate real number `1.` and named `x0`:

```
In[4]:= x0 = 1. ;
```

Numerical values `{1.,1.5,1.41667,1.41422,1.41421}` of the first few iterates of `g` starting at `x0` result from `{x0, g[x0], g@g[x0], g@g@g[x0], g@g@g@g[x0]}`. By contrast a parentheses-enclosed expression such as `(x0; g[x0]; g@g[x0]; g@g@g[x]; g@g@g@g[x0])` is a *compound expression*; each component expression in turn is evaluated, but only the value of the last one is returned.

To reuse the preceding result (`1.`, the value of `x0`), invoke `%` . Thus evaluating `g[%]` and then `g[%]` again now would yield `1.5` and `1.41667`. That way, successive iterates are readily obtained. But such iteration deserves to be automated through one or another of the programming approaches below.

4.1 Iterating functionally: `Nest` and its variants

The "functional" expression `Nest[g, x0, 4]` returns the fourth iterate `g@g@g@g[x0]`, namely, `1.41421`. And `NestList[g, x0, 4]` returns the list of all iterates from 0th to 4th, namely, `{1.,1.5,1.41667,1.41422,1.41421}`. The operation of `Nest` as well as other functions employed here can be grasped visually by using the animations devised by Song [3].

Thanks to listability, several initial estimates may be tried at once, so that `Nest[g, {1., 9.}, 4]` has value `{1.41421,1.43209}`. But to use several different numbers of iterates, another approach must be employed:

```
Map[Nest[g, x0, #]&, {2, 4}]
```

The primitive `Map` causes its first argument—here a pure function—to be evaluated at each entry in its second argument. A special input form with `/@` is frequently used for `Map`, as in `Nest[g, x0, #]& /@ {2, 4}`.

With `FixedPoint` or `FixedPointList`, the iteration can be continued until there is no change. For example,

```
FixedPointList[g, x0] // NumberForm[N[#], 16]&
```

returns a list of seven iterates beginning with `1.` and `1.5` and ending with two instances of `1.414213562373095`.

Because x0 is a *machine-precision* approximate real number, every number calculated from it will have at most machine precision. To obtain higher precision in subsequent results, explicitly specify a higher precision at the start. For example, ``FixedPoint[g, 1.`50]`` has as value the fixed point $\sqrt{2}$ of `g` correct to 50 digits.

4.2 *Iterating procedurally:* `Module` *and* `Do`

Iteration can be programmed explicitly instead of functionally:

```
hero[c_, n_, x0_] := Module[{x = x0}, Do[x = g[x], {n}]; x]
```

For example, `hero[2, 4, 1.]` has value the second iterate `1.41421`. The `Do` expression inside the module is one of several functions for looping (others are `For` and `While`.) In terms of `Do`, a procedural alternative to the `NestList` expression for Hero's method looks like this:

```
herolist[c_, n_, x0_] := Module[{x = {x0}},
    Do[x = Append[x, g[Last[x]]], {n}]; x]
```

Type-checking arguments can be accomplished without having to resort to `If`. Here are two different type-checking mechanisms:

```
herolist[c_,      n_Integer,      x0_?Positive]      :=
Module[{x={x0}},
    Do[x = Append[x, g[Last[x]]], {n}]; x]
```

Both `herolist[2,4.1,1.]` and `herolist[2,4,-1.]` will return unevaluated, whereas `herolist[2,4,1.]` will return the expected list of approximations.

To allow a default initial estimate, say `1.`, for `herolist`, all it takes is to add a single rule to the global rule base for (the original) `herolist`:

```
herolist[c_, n_] := herolist[c, n, 1.]
```

Then `herolist[2, 4]` will return the same result as `herolist[2, 4, 1.]`. Another way to allow default values is as follows:

```
herolist[c_, n_, x0_:1.] := Module[{x = {x0}},
    Do[x = Append[x, g[Last[x]]], {n}]; x]
```

5 Successive approximations through recursion

A recursive version of `hero` (or `herolist`) could be written in a strictly procedural way, with an `If` test for the base case:

```
rhero[n_,x0_]      :=      Module[{},If[n==0,      x0,
g@rhero[n-1,x0]]]
```

A simpler recursive version `hero` exploits pattern-matching and rewrite rules:

```
rh[0, x_] = x;
rh[n_, x_] := g@rh[n-1, x]
```

Which of these two rules applies in a given instance? Always the more specific (the base 0 case here) before the general, recursive rule. If desired, the general rule can be provided with an explicit condition on when it is to be applied (the symbol /; may be read as "provided that"):

```
rhnew[0, x_] = x;
rhnew[n_, x_] := g@rh[n-1, x] /; n > 0
```

To make a recursive function such as rh or rhnew more efficient, it may be modified so as to *cache* values into the rule base as soon as they are calculated. For example:

```
drh[0, x_] = x; drh[n_, x_] := drh[n, x] = g@drh[n - 1, x]
```

6 Visualizing Hero's method with graphics programming

The iterates of g converge to a fixed point of g. In fact, both fixed points can be found exactly by evaluating Solve[g[x] == x, x] and visualized with Plot:

```
In[5]:= Plot[{g[x], x}, {x, -2, 2}];
```

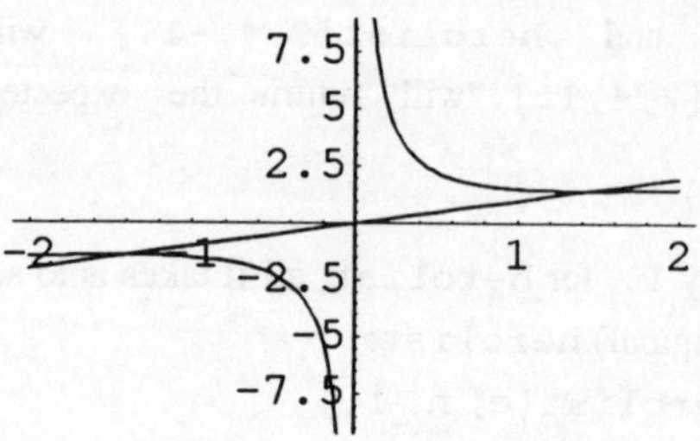

The display is a side-effect of the evaluation. The result that was suppressed by the semicolon is a Graphics object, which may be manipulated like other objects. For example, the following will display two copies of the plot stacked one above the other:

```
both = Plot[{g[x], x}, {x, -2, 2}];
Show[GraphicsArray[Partition[{both, both}, 1]]];
```

Iterating a function *f* twice starting at *x* may be viewed geometrically: In the *xy*-plane, locate the point $(x, f(x))$ on the graph of $y = f(x)$; bounce the point horizontally against the graph of $y = x$ as if it were a billiard ball, and stop the ball when it hits the graph of $y = f(x)$ again, at $(f(x), f(f(x)))$. Repeating the bouncing results in a *cobweb diagram*. The cobweb diagram program below (which enhances one by Wattenberg [4]) is built bottom-up.

```
dash[{{p_, q_}, {r_, s_}}] :=
  {Dashing[{0.005, 0.02}], Line[{{p, q}, {r, s}}]};
solid = Dashing[{0, 0}];
pt[{x_, y_}] := {PointSize[0.02], Point[{x, y}]}
bounce[f_, x_] :=
  Module[{fx, ffx}, fx = f[x]; ffx = f[fx];
     {Line[{{x, fx}, {fx, fx}, {fx, ffx}}],
    dash[{{fx, ffx}, {fx, 0}}], pt[{fx, 0}],
           pt[{f[x], f[f[x]]}]}]
web[f_, n_, x0_] := Module[{x = x0, steps = {}},
   Do[ steps = Append[steps, bounce[f, x]];
    x = f[x], {n}]; steps]
cobweb[f_, lo_, hi_, n_, x0_] :=
  Plot[{f[x], x}, {x, lo, hi},
    PlotRange → {0, hi}, AspectRatio → Automatic,
    Epilog → {pt[{x0, 0}],
       dash[{{x0, 0}, {x0, f[x0]}}], pt[{x0, f[x0]}],
     solid, web[f, n, x0]}];
```

```
In[11]:= g[x_] := 0.5 (x + 2 / x); cobweb[g, 0, 10, 4, 9];
```

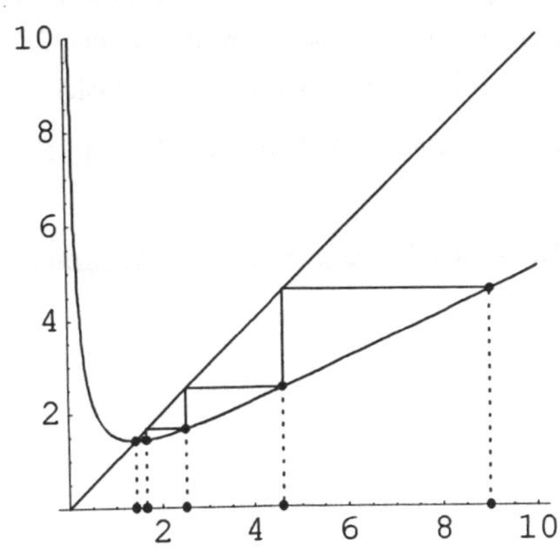

7 Concluding remarks

The example discussed here is richer than may be apparent. Hero's method is a special case of Newton's method for successively approximating roots of a real-valued differentiable function f of a real variable. For those who know some calculus,

Newton's method can serve as an instructive exercise in applying the ideas presented here, or as an alternate path through the same ideas—and more. For Newton's method the function to iterate is $g(x) = x - f(x)/f'(x)$. *Mathematica*'s symbolics may be exploited to derive the formula for $g(x)$, and its graphics employed to display the geometric idea behind Newton's method as well as to visualize, without a cobweb diagram, the iteration process and the approach of successive approximations to their limit. A well-written implementation of Newton's method can even be extended to the case of a multivariate f through one simple change in the definition of g (the Jacobian matrix replaces the derivative); the multivariate case can serve, further, as context for learning about vector algebra and three-dimensional graphics in *Mathematica*. A related exercise, accessible even to learners who know no calculus, is to implement the secant method for root-finding (see Skeel and Keiper [2]). In the secant method it is necessary to keep track of pairs of points through succcessive iterations, so that implementing the method functionally, with `Nest`, can prove to be a bit of a programming challenge.

8 References

[1] Heath, T. A History of Greek Mathematics, vol. II (Chap. 18, pp. 323–326), 1921, Oxford University Press.

[2] Skeel, R. D. and Keiper, J. B. Elementary Numerical Computing with *Mathematica* (sec. 3.5, pp. 97–99), 1993, McGraw-Hill.

[3] Song, K. Visual approach in teaching mathematics and *Mathematica, MathSource* #4286, July 2002, http://library.wolfram.com/infocenter/MathSource/4286; see also Animated *Mathematica* functions, http://documents.wolfram.com/flash).

[4] Wattenberg, F. Calculus in a Real and Complex World, (sec. 1.6, pp. 50–54), 1995, PWS Publishing.

[5] Wolfram, S. The *Mathematica* Book, 4 ed. (page *xv*), 1999, Wolfram Media and Cambridge University Press.

Complex 2D Walks based on Context Independent L-Systems

Erik Jensen
University of California, Santa Cruz
Veikko Keränen
Rovaniemi Polytechnic, School of Technology
veikko.keranen@ramk.fi

1 Abstract

We report here a method of generating 2D walks based on context independent L-systems for two or four letter alphabets. Binary strings (words) are mapped to strings over a four letter alphabet by two separate techniques. 2D walks are generated from the resulting quaternary strings by assigning vectors to the characters as follows: $a \rightarrow (-1,0)$, $b \rightarrow (1,0)$, $c \rightarrow (0,-1)$, $d \rightarrow (0,1)$. In the case of long strings obtained by iterating endomorphisms, one may also use vectors for image words of letters. A walk is generated by adding up the vectors of sequential characters in a word through a regular grid and coloring the cells at which the vectors terminate. Plots of distance from the origin versus string position provide a rapid means for comparison of the strings resulting from different L-systems. Here one may also use matrix multiplication in a powerful way. We examine the 2D walks from a variety of L-systems and provide examples of considerably complex paths from simple L-systems. Square-free strings over four letters can result in both simple and highly complex 2D walks.

2 Introduction

Lindenmayer systems are production systems that are used to model growth by recursively applying parallel rewriting rules to a one dimensional string of symbols (or even to multidimensional structures). In this paper, we consider only one dimensional strings. The rules of such an L-system can specify that symbols (letters) are replaced by longer strings in such a way that the length of the string increases infinitely with the number of iterations. A D0L system is a deterministic Lindenmayer system in which the replacement rules for a given symbol in an alphabet do not depend on the context of the symbol in the string. The rules of a Lindenmayer system are implemented in parallel so that at each step all transformations are applied to the string simultaneously.

Several techniques have been used for graphically displaying L-systems with computers. "Turtle Graphics" were developed by Seymour Papert in 1980. With this method, the graphics of 1D strings are generated by interpreting symbols as instructions typically for moving forward, turning right by a specified angle, and turning left by a specified angle. The paths that result often possess fractal symmetry. The most well known example of this sort of path is the Koch curve first described by Helge von Koch in 1904. The Hilbert curve is another well known example of a path that can be produced with this technique that also shows fractal symmetry. Graphics similar to those used here are often referred to as vectorgrams, and they have previously been used for the analysis of DNA sequences and shift registers of digital computers [3] .

3 Preliminaries

An *alphabet* Σ is a finite non-empty set of abstract symbols called *letters*. A *word* (*string*) over Σ is a finite (unless otherwise indicated) string, or sequence, of letters belonging to Σ. The set of all words [non-empty words] over Σ is denoted by Σ^* [Σ^+]. On Σ^*, the associative binary operation of *catenation* is defined. For words u and v, it is the juxtaposition uv. The *empty word*, which is the neutral element of catenation, is denoted by λ.

Let $w = x_1 \cdots x_m$, $x_i \in \Sigma$. The *length* of the word w, denoted by $|w|$, is the number of occurrences of letters in w, i.e., $|w| = m$. Let $\Sigma = \{a_1, \dots, a_n\}$. The number of occurrences of one letter $x \in \Sigma$ in w is denoted by $|w|_x$, or simply by $|w|_i$ if $x = a_i$. The notation $\psi_\Sigma(w)$ stands for the *Parikh vector* of w, i.e., $\psi_\Sigma(w) = (|w|_1, \dots, |w|_n)$. Usually we will omit the subscript Σ and write simply ψ instead of ψ_Σ.

Let $k \geq 2$ be a given integer. A *k-repetition* [an *abelian k-repetition*] is a non-empty word of the form R^k [$P_1 \cdots P_k$, where $\psi(P_\mu) = \psi(P_\nu)$ for all $1 \leq \mu < \nu \leq k$, i.e., P_i:s are permutations of each other]. Instead of [abelian] 2- and 3-repetitions, the terms [*abelian*] *squares* and [*abelian*] *cubes* are often used. A word or an ω-word (explained below) is called *k-repetition free* [*abelian k-repetition free*], or in short *k-free* [*a-k-free*], if it does not contain any *k*-repetition [abelian *k*-repetition] as a subword.

A *morphism* h is a mapping between free monoids Σ^* and Δ^* with $h(uv) = h(u)h(v)$ for every u and v in Σ^*. Especially, $h(\lambda) = \lambda$. A morphism $h: \Sigma^* \to \Delta^*$, being compatible with the catenation of words, is uniquely defined, if the word $h(x) \in \Delta^*$ is (effectively) given for each $x \in \Sigma$. If $\Delta = \Sigma$, we call h an *endomorphism* (and usually write g instead of h). For a given integer $k \geq 2$, a morphism $h: \Sigma^* \to \Delta^*$ is called *k-free* [*a-k-free*], if $h(w)$ is *k*-free [a-*k*-free] for every k-free [a-k-free] word $w \in \Sigma^*$.

With regards to *L-systems* (Aristid Lindenmayer 1925–1989), we specify the following concepts. A *D0L-system* is a triple $G = (\Sigma, g, \alpha_0)$, where Σ is an alphabet, g: $\Sigma^* \to \Sigma^*$ is an endomorphism, and α_0, called the *axiom*, is a word over Σ. The (*word*) *sequence* $S(G)$ generated by G consists of the words

$\alpha_0 = g^0(\alpha_0)$, $g^1(\alpha_0)$, $g^2(\alpha_0)$, $g^3(\alpha_0)$, ... where $g^i(\alpha_0) = g(g^{i-1}(\alpha_0))$ for $i \geq 1$.

The *language* of G is defined by $L(G) = \{g^i(\alpha_0) \mid i \geq 0\}$. Languages [sequences] defined by a D0L-system are referred to as *D0L-languages* [*D0L-sequences*]. For a thorough discussion of various L-systems we refer the reader to Rozenberg and Salomaa [4].

We will employ two methods of mapping binary strings $w = x_1 \cdots x_m$, $x_i \in \{a,b\}$, to strings over 4 letters. Consider a word (or an ω-word) $\alpha = c_1 c_2 c_3 \ldots$, where each $c_i \in \{a,b\}$. Define a new alphabet Σ_1 by $\Sigma_1 = \{[aa],[ab],[ba],[bb]\}$ and a word (or an ω-word) $\beta = d_1 d_2 d_3 \ldots$ over this alphabet by the condition (A) $d_i = [c_{2i-1} c_{2i}]$ for every $i \geq 1$, or by the condition (B) $d_i = [c_i c_{i+1}]$ for every $i \geq 1$. In what follows, we refer to these procedures as *method A* and *method B* respectively.

4 Generation of L-systems with *Mathematica*

The following code implements a context independent L-system by using replacement rules repeatedly. In the example below, ten iterations of the rewriting rule generate a prefix of the infinite Thue-Morse sequence in a list form.

```
binString = Nest[ReplaceAll[#,
{"a"→{"a", "b"},"b"→{"b","a"}}]&,"a",10]//Flatten;
```

5 Generation of 2D walks

In the following code, a walk is generated by following the vectors of sequential characters in an input word (over a four letter alphabet) through a regular grid and coloring the cells at which the vectors terminate. The input string is referred to here as `z`. A unique color value is assigned to each of the four possible characters.

```
Clear[ourShow];transProductionsForabcd={{-1,0},{1,0},{0,
-1},{0,1}};
rec[{x_,y_}]:=Rectangle[{x,y}-0.45,{x,y}+0.45];
```

```
ourTransForm[z_List,{a_,b_,c_,d_}]:= Module[{x=0,y=0},
  Switch[#,
   "a",{RGBColor[1,.28,.71],rec[{x,y}+=a]},
   "b",{RGBColor[.1,.86,.15],rec[{x,y}+=b]},
   "c",{RGBColor[1,1,0],rec[{x,y}+=c]},
   "d",{RGBColor[.5,.5,1],rec[{x,y}+=d]}]&/@Take[z,
   {1,Length[z]}]];

ourShow[z_List,prod_,ft_,pl_:"",df_:Identity]:=
Show[Graphics[ourTransForm[z,prod]],AspectRatio→1,Axes→
False,Frame-
>True,PlotLabel→pl,DisplayFunction→df,FrameTicks->ft];
```

6 Mapping binary strings to strings over four letters

```
productions = {{"a","a"}→"a",{"a","b"}→"b",{"b","a"}→
            "c",{"b","b"}→"d"};
methodA[x_]:= Partition[x,2]/.productions;
methodB[x_]:= Partition[x,2,1]/.productions;
```

7 2D walks from the Thue-Morse Sequence

Here we will show the 2D walks of the Thue-Morse sequence based on both methods of mapping binary strings to strings over four letters. The paths generated here are from the first 200 characters (method A) or 101 characters (method B) of the Thue-Morse sequence – starting from the letter *a* in the quaternary case. The string over four letters produced by method B is proven to be repetition free (square-free), see [5].

```
Show[GraphicsArray[{ourShow[Take[methodA[binString],100]
,
  transProductionsForabcd,Automatic,"Method A"],
  ourShow[Take[methodB[binString],100],
  transProductionsForabcd,{{0,5,10,15},{0,-5,-10,-15},
  Automatic,Automatic},"Method B"]}]]
```

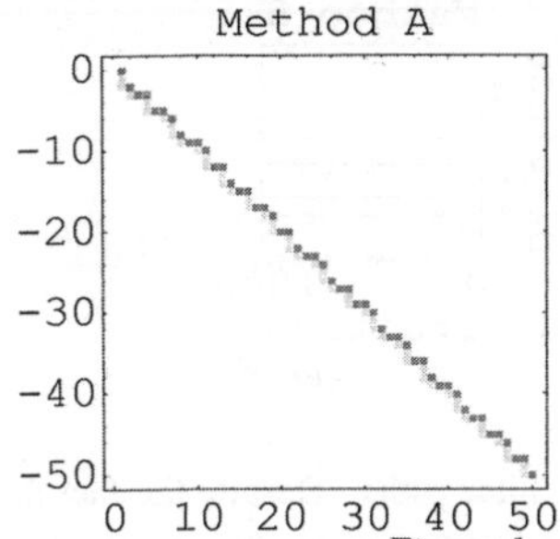

Figure 1

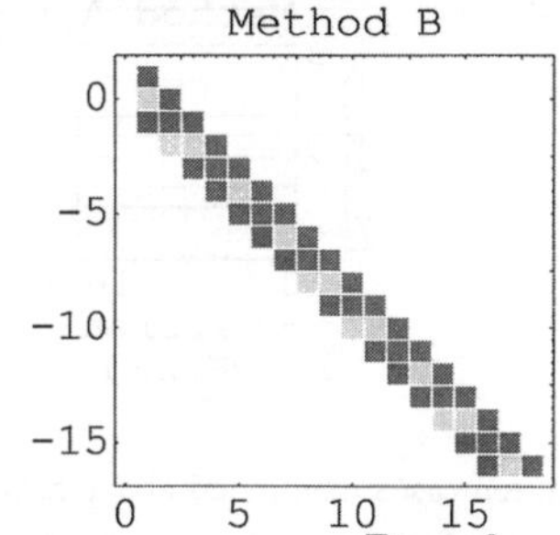

Figure 2

8 Systematically characterizing simple L-systems

We have systematically characterized the 2D walks based on all binary, context independent L-systems with image words of length 3 or less. The vast majority of cases show approximately linear increase in distance from the origin (D) with string position (S). There are 10 cases that result in more complex paths that oscillate irregularly and remain fixed near the origin. Here we show 2D walks and plots of D vs. S for one such case, $g(a) = abb$, $g(b) = baa$, for a total of 1000 steps – starting from a in the quaternary case. This particular morphism is k-free for every integer $k \geq 4$, see [2, p. 49]. It can be shown that transformation with method A results in a square-free ω-word over four letters.

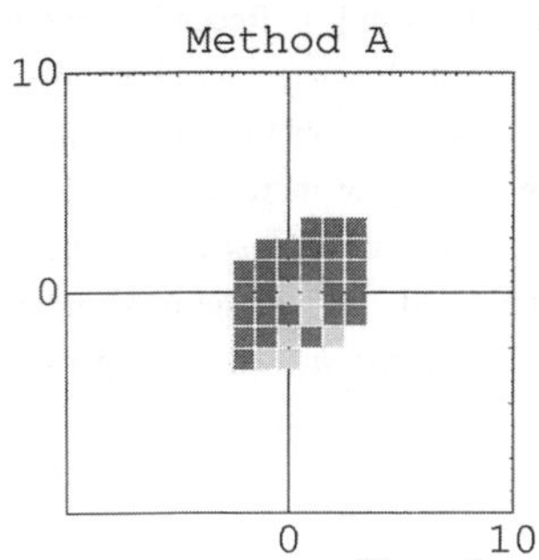

Figure 3

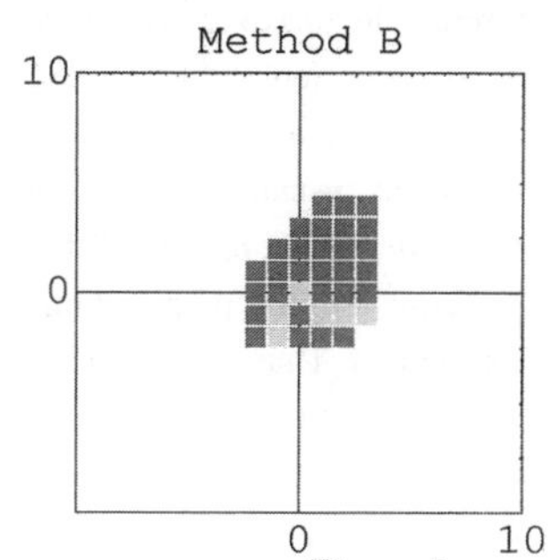

Figure 4

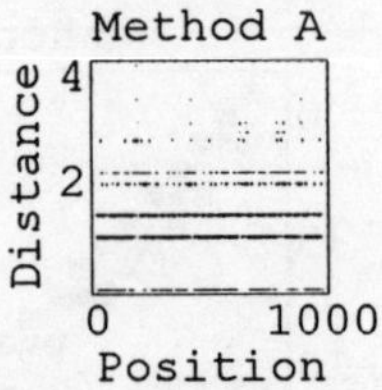

Figure 5

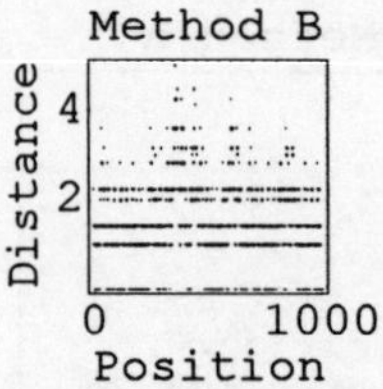

Figure 6

All ten cases mentioned above have similar paths. Despite the fact that there must be simple nesting structure to these strings, plots of D vs. S show seemingly chaotic "jumps" between a small set of possible distances. It seems likely that there is some limit for D as the length of the path approaches infinity in each of these cases. In our systematic characterization of simple binary L-systems, the first 200 characters of each string was tested for repetitions. Repetition-free prefixes resulting from method A always resulted in a 2D walk similar to the attractor type paths shown above. The only strings that had long repetition-free prefixes after transformation with method B were the Thue-Morse string and its complement (exchange of a and b). The existence of other binary omega words that become square-free after transformation with method B is an open problem.

9 2D walks from abelian square-free strings over four letters

In 1991, V. Keränen showed that abelian squares are avoidable on four letters through the discovery of the g_{85} endomorphism [1]. Recently, new infinite a-2-free words over four letters based on a DT0L language has been discovered by V. Keränen. A language of the form $h_k \cdots h_1(w)$, obtained by using compositions of morphims, is called a DT0L-language, see [4] for more details. In the new a-2-free case, all $h_1, \ldots, h_k$ are of the form $\sigma^i(g_{85})$ or $\sigma^i(g_{98})$, $i \in \{0,1,2,3\}$, where σ performs a cyclic permutation of letters, and g_{98} is the new endomorphism over four letters. Thus, one can choose as morphisms h_j all cyclic permutations of g_{85} and g_{98}, but not all permutations in general. Starting with a suitable axiom, the iteration of g_{98} alone produces an a-2-free omega word in spite of the fact that g_{98} does not preserve the a-2-freeness of all words.

Since both endomorphisms g_{85} and g_{98} are over an alphabet of 4 letters, they can be used directly as input for generating a 2D walk by the method we described previously. The following *Mathematica* code defines these endomorphisms. The 2D walks here are based on the first 10000 characters of the strings ${g_{85}}^i(a)$ and ${g_{98}}^i(a)$, $i = 3$. To view longer 2D walks based on these strings we refer the reader to: http://south.rotol.ramk.fi/IMS_IAS_Events.html.

```
cyclicPermutation={"a"->        "b","b"->"c","c"->"d","d"-
>"a"};
g85a="abcacdcbcdcadcdbdabacabadbabcbdbcbacbcdcacbabdabac
adcbcdcacdbcbacbcdcacdcbdcdadbdcbca";
g85b=StringReplace[g85a,cyclicPermutation];
g85c=StringReplace[g85b,cyclicPermutation];
g85d=StringReplace[g85c,cyclicPermutation];
ga="abcacdcbcdcadbdcbdbabcbdcacbabdbabcabdadcdadbdcbdbab
dbcbacbcdbabdcdbdcacdbcbacbcdcacdcbdcdadbdcbca";
gb=StringReplace[ga,cyclicPermutation];
gc=StringReplace[gb,cyclicPermutation];
gd=StringReplace[gc,cyclicPermutation];
g98a = ga; g98b = gd; g98c = gc; g98d = gb;
```

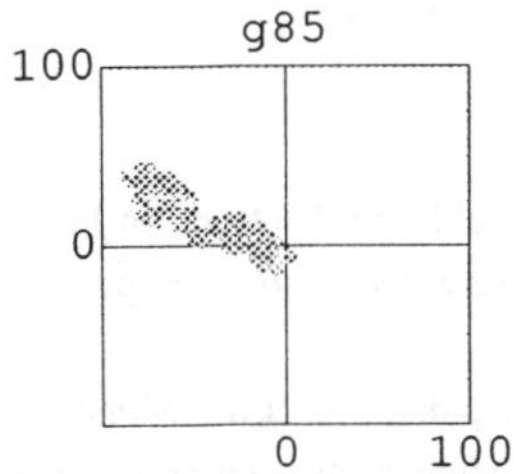

Figure 7

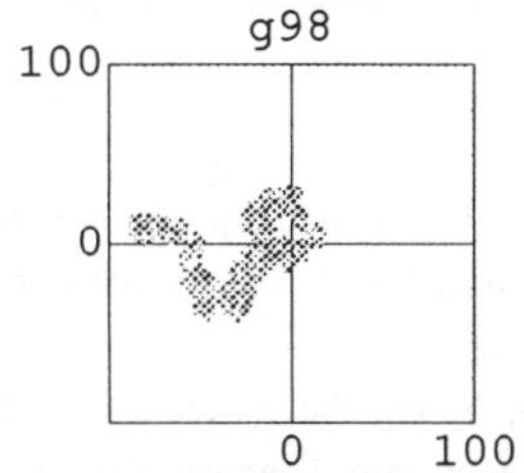

Figure 8

It is possible to very quickly determine the final point in the grid reached by 2D walks based on extremely long words of the form $g^n(a)$. As an example of the case for g_{98} we write a function **m98**, where the columns of the 4×4 – matrix are the Parikh vectors of $g_{98}(a)$, $g_{98}(b)$, $g_{98}(c)$, and $g_{98}(d)$. The final points are computed for $n = 10$ and $n = 11$. Note that the last point has traveled from the second quadrant to the fourth quadrant .

m98[0] = {1, 0, 0, 0}

$$\texttt{m98[n\_] := m98[n] = } \begin{pmatrix} 18 & 28 & 27 & 25 \\ 28 & 27 & 25 & 18 \\ 27 & 25 & 18 & 28 \\ 25 & 18 & 28 & 27 \end{pmatrix} \texttt{.m98[n - 1]}$$

$$\left\{ \begin{pmatrix} -1 & 1 & 0 & 0 \\ 0 & 0 & -1 & 1 \end{pmatrix} \texttt{.m98[10] // MatrixForm,} \right.$$

$$\left. \begin{pmatrix} -1 & 1 & 0 & 0 \\ 0 & 0 & -1 & 1 \end{pmatrix} \texttt{.m98[11] // MatrixForm} \right\}$$

Out[3]= $\left\{\begin{pmatrix} -3489320912 \\ 2415579088 \end{pmatrix}, \begin{pmatrix} 39724367296 \\ -31134432704 \end{pmatrix}\right\}$

10 Regularities in the 2D walks of g_{85} and g_{98}

The 2D walks resulting from g_{85} and g_{98} both have attributes that seem to be highly random. This may to some extent be due to the fact that parts of the path can be "overwritten" by the instructions of subsequent characters of the strings. In the 2D walks of g_{85} and g_{98}, there are necessarily many types of local regularities. If these strings are partitioned into substrings of the same length as $g(a)$ (starting at the beginning of the string), then the only possible blocks in the partitioned string are $g(a)$, $g(b)$, $g(c)$, and $g(d)$. This is due to the fact that these endomorphisms grow uniformly. As a result, for a fixed $i \geq 0$, there must be a set of four "local walks", each with $|g^i(a)|$ steps, that are repeated infinitely many times in different locations of the walks of g_{85} and g_{98}.

References

[1] Keränen, V. Abelian squares are avoidable on 4 letters, in Proc. ICALP '92, Berlin, Germany, 1992. (W. Kuich), Lecture Notes in Comp. Sci., 623:41-52 Springer-Verlag.

[2] Keränen, V. On the k-Freeness of Morphisms on Free Monoids, in Annales Academiæ Scientiarum Fennicæ, Ser. A. I. Mathematica Dissertationes, 1986. (61, O. Lehto) Finnish Science Academy.

[3] Pickover, C. DNA Vectorgrams: Representation of cancer gene sequences as movements along a 2-D lattice, in IBM J. Res. Dev. 31:111-119, 1987, IBM.

[4] Rozenberg, G. and Salomaa, A. The Mathematical Theory of L-systems, 1980, Academic Press.

[5] Salomaa, A. Jewels of Formal Language Theory, 1981, (Chapter 1, Pages 9-10), Computer Science Press.

Generating Closed-Form Formulae that Count Satisfiable Instances of k-SAT

Ronald Monson
University of Western Australia
ronald@maths.uwa.edu.au

1 Abstract

The theory of *NP*-completeness is fundamental to computational complexity and the SATISFIABILITY problem of Boolean Logic is the seminal *NP*-complete problem on which the theory is based. While *NP*-completeness is associated more with the efficiency of deciding whether or not an arbitrary instance is satisfiable, determining the proportion of satisfiable instances of an equivalent problem, 3-SAT, yields important information in regards to this efficiency. Empirical research on the dependence of this proportion on the size of the 3-SAT instance has revealed a "satisfiability" phase transition through which dramatic changes in satisfiability accompany very small changes in problem size. This phenomenon underpins the celebrated threshold conjecture and in this paper we give a new way of attacking the conjecture by describing this transition algebraically. That is, we present and implement an algorithm that, for any given integer n, generates a closed form expression in m, (clause number), and k, (clause size), which counts the proportion of satisfiable instances of 3-SAT. The inherent complexity of the algorithm indicates that generating such a formula for $n \geq 5$ is not feasible, but that, taking the form of a succinct mathematical expression, the algorithm is amenable to further asymptotic analysis. As part of this analysis we show that understanding the threshold amounts to understanding the asymptotics of an alternating sum of k-cylinders. All algorithms and experimentation have been implemented as part of an application package, SATPACK, more details of which appear in an extended version of this paper, (including all references), on the proceedings CD.

2 Generating Closed-Form Formulae

In this section we present the main result, the proof of which is greatly simplified by the following sequence of definitions. With x_i denoting an arbitrary Boolean variable we have the following:

Definition 1 $\left(\mathbb{X}^n, \overline{\mathbb{X}}^n, \mathbb{X}^n_j, \overline{\mathbb{X}}^n_j, \mathbb{A}^n, \mathbb{A}^n_j\right)$

$\mathbb{X}^n = \{x_1, x_2, \ldots, x_n\}$	$\mathbb{X}^n_j = \{J \subseteq \mathbb{X}^n : \lvert J\rvert = j\}$
$\overline{\mathbb{X}}^n = \{\bar{x}_1, \bar{x}_2, \ldots, \bar{x}_n\}$	$\mathbb{X}^n_j = \{J \subseteq \overline{\mathbb{X}}^n : \lvert J\rvert = j\}$
$\mathbb{A}^n = \{\alpha : \alpha : \mathbb{X}^n \to \{1, 0\}\}$	$\mathbb{A}^n_j = \{J \subseteq \mathbb{A}^n : \lvert J \rvert = j\}$

Table 1

Note that $\mathbb{A}^n$ contains all the (2^n) possible ways of assigning either *True*, (1), or *False*, (0), to each boolean variable in $\mathbb{X}^n$. Henceforth, elements of $\mathbb{A}^n$ will be called *assignments*. Next we define k-SAT, the space on which satisfiable formulae will be counted.

Definition 2 ($\mathbb{F}^{k,m}_n$, k **– SAT**)
For n Boolean variables $\{x_1, x_2, \ldots x_n\}$, let $\mathbb{F}^{k,m}_n$ denote the space of formulae of the (Conjunctive Normal) form $C_1 \wedge C_2 \wedge \ldots \wedge C_m$, where:

$C_i = l_1 \vee l_2 \vee \ldots \vee l_k$ for $1 \leq i \leq m$

$l_j \in \mathbb{X}^n \cup \overline{\mathbb{X}}^n$ for $1 \leq j \leq k$

The C_i and l_j defined above are called k-*clauses* and *literals* respectively, while the space, $\mathbb{F}^{k,m}_n$, is commonly referred to as k-*SAT*, with formulae of $\mathbb{F}^{k,m}_n$ being described as *instances* of k-SAT. For $f \in \mathbb{F}^{k,m}_n$ and $\alpha \in \mathbb{A}^n$, we write $\alpha(f) = 1$, (or $\alpha(f) = 0$), and say α *satisfies* f, (or α *falsifies* f), if, after $\alpha(x_i)$ is substituted for each occurrence of x_i in f, the formula evaluates to 1 (or 0) according to the normal rules of Boolean logic. We say that an instance, f, of k-SAT is *satisfiable* if there exists an assignment α such that $\alpha(f) = 1$. Otherwise f is *unsatisfiable*. Given these definitions the *k-SAT problem* can be stated as follows:

Problem (k-SAT) : Given an instance of k-SAT, is this instance satisfiable ?

The SAT problem is similarly stated, except that the arbitrary instance can be *any* propositional formula and, in fact, no generality is lost with this choice. To avoid clutter we will adopt the idiom of having the *doublestruck*, *script* and *gothic* styles refer to, respectively, the set itself, the size of the set and, (where the set is drawn from a universal set), the proportional size of the set. For example, we introduce the following terminology, (in which f denotes formulae):

Definition 3 ($\mathcal{F}_n^{k,m}$, $\mathbb{S}_n^{k,m}$, $\mathcal{S}_n^{k,m}$, $\mathfrak{S}_n^{k,m}$, $\mathbb{U}_n^{k,m}$, $\mathcal{U}_n^{k,m}$, $\mathfrak{U}_n^{k,m}$)

$\mathbb{F}_n^{k,m} =$	$(l_1 \vee \ldots \vee l_k)_1 \wedge \ldots \wedge (l_1 \vee \ldots \vee l_k)_m$	$=$ *Set* of k–CNF f
$\mathbb{S}_n^{k,m} =$	$\{f \in \mathbb{F}_n^{k,m} : \exists v \in \mathbb{A}^n : v(f) = 1\}$	$=$ *Set* of satisfiable f
$\mathbb{U}_n^{k,m} =$	$\{f \in \mathbb{F}_n^{k,m} : \forall v \in \mathbb{A}^n : v(f) = 0\}$	$=$ *Set* of unsatisfiable f
$\mathcal{F}_n^{k,m} =$	$\lvert\mathbb{F}_n^{k,m}\rvert$	$=$ *Number* of k–CNF f
$\mathcal{S}_n^{k,m} =$	$\lvert\mathbb{S}_n^{k,m}\rvert$	$=$ *Number* of satisfiable f
$\mathcal{U}_n^{k,m} =$	$\lvert\mathbb{U}_n^{k,m}\rvert$	$=$ *Number* of unsatisfiable f
$\mathfrak{S}_n^{k,m} =$	$\mathcal{S}_n^{k,m} / \mathcal{F}_n^{k,m}$	$=$ *Proportion* of satisfiable f
$\mathfrak{U}_n^{k,m} =$	$\mathcal{U}_n^{k,m} / \mathcal{F}_n^{k,m}$	$=$ *Proportion* of unsatisfiable f

Table 2

In this paper the emphasis will be on the set of *unsatisfiable* instances, ($\mathbb{U}_n^{k,m}$), since this simplifies the main proof; one could, of course, equally well approach the solvability in terms of *satisfiability* using $\mathcal{F}_n^{k,m} - \mathcal{U}_n^{k,m} = \mathcal{S}_n^{k,m}$. For assignments and sets of assignments as subscripts, we abuse the notations of Definition 3 as follows :

Definition 4 ($\mathbb{S}_v^{k,m}$, $\mathbb{S}_I^{k,m}$, $\mathcal{S}_I^{k,m}$, $\mathbb{U}_v^{k,m}$, $\mathbb{U}_I^{k,m}$, $\mathcal{U}_I^{k,m}$)

$v \in \mathbb{A}^n$	$I \subseteq \mathbb{A}^n$	
$\mathbb{S}_v^{k,m} = \{f \in \mathbb{F}_n^{k,m} : v(f) = 1\}$	$\mathbb{S}_I^{k,m} = \bigcap_{v \in I} \mathbb{S}_v^{k,m}$	$\mathcal{S}_I^{k,m} = \lvert\mathbb{S}_I^{k,m}\rvert$
$\mathbb{U}_v^{k,m} = \{f \in \mathbb{F}_n^{k,m} : v(f) = 0\}$	$\mathbb{U}_I^{k,m} = \bigcap_{v \in I} \mathbb{U}_v^{k,m}$	$\mathcal{U}_I^{k,m} = \lvert\mathbb{U}_I^{k,m}\rvert$

Table 3

Definition 5 (clause-variable ratio)

Rather than refer to m, the number of clauses in a formula, it will often be convenient to consider this number as a multiple of the number of variables. That is, we consider $\mathfrak{U}_n^{k,dn}$ and will refer to d as the *clause-variable ratio*.

Definition 6 ($\mathcal{H}$ – Hamming Closeness)

For $I \subseteq \mathbb{A}^n$, $\mathcal{H}(I) = |\{x \in \mathbf{X}^n : (\forall\, \alpha \in I,\ \alpha(x) = 1) \vee (\forall\, \alpha \in I,\ \alpha(x) = 0)\}|$. By convention, if $I = \{\}$, we will set $\mathcal{H}(I) = 2n$.

Intuitively, $\mathcal{H}(I)$ is the number of variables assigned the same truth value by *all* the assignments of I. We will call this the *Hamming Closeness* of I. We will need to use the *Principle of Inclusion and Exclusion,* the proof of which appears in any discrete math textbook (e.g. [2]). We state it here for reference.

The Principle of Inclusion and Exclusion (PIE)

With index set G, suppose that $(Y_1,\ Y_2,\ \ldots\ Y_{|G|})$ is a family of subsets drawn from a universal set $\mathcal{X}$ and, for $I \subseteq G$, $Y_I = \bigcap_{i \in I} Y_i$ (with, by convention, $Y_\phi = \mathcal{X}$). Then the number of elements of $\mathcal{X}$ which do *not* lie in any of the Y_i is:

$$\sum_{I \subseteq G} (-1)^{|I|}\, |Y_I| \qquad \text{................(PIE)}$$

We are now ready to state the main theorem of the paper.

Theorem 1

Given $\mathbb{A}^n$, the set of 2^n assignments on n Boolean variables, the number of unsatisfiable formulae with m k-clauses is given by:

$$\mathcal{U}_n^{k,m} = \sum_{I \subseteq \mathbb{A}^n} (-1)^{|I|} \left(\sum_{J \subseteq I} (-1)^{|J|} (\mathcal{H}(J))^k \right)^m \qquad \text{................(1)}$$

Proof:

With index set $\mathbb{A}^n = \{v_1, v_2, \ldots v_{2^n}\}$, consider $\{\mathbb{S}_{v_1}^{k,m}, \mathbb{S}_{v_2}^{k,m}, \ldots, \mathbb{S}_{v_{2^n}}^{k,m}\}$, a family of subsets drawn from $\mathbf{F}_n^{k,m}$. The unsatisfiable formulae of $\mathbf{F}_n^{k,m}$ are exactly those formulae which do *not* lie in any of the $\mathbb{S}_{v_i}^{k,m}$ so that, by **PIE**, $\mathcal{U}_n^{k,m} = \sum_{I \subseteq \mathbb{A}^n} (-1)^{|I|}\, S_I^{k,m}$. But $S_I^{k,m} = \left(S_I^{k,1}\right)^m$ since a (conjunctive) formula of $\mathbb{S}_n^{k,m}$ is true iff each of its clauses is true. With index set $I = \{v_1, v_2, \ldots v_{|I|}\}$ and $\{\mathbf{U}_{v_1}^{k,1}, \mathbf{U}_{v_2}^{k,1}, \ldots, \mathbf{U}_{v_{|I|}}^{k,1}\}$, a family of subsets drawn from $\mathbf{F}_n^{k,1}$, we see that those formulae which do *not* lie in any of the $\mathbf{U}_{v_i}^{k,1}$ are exactly those formulae of $\mathbb{S}_I^{k,1}$. Therefore, by **PIE**, $S_I^{k,1} = \sum_{J \subseteq I} (-1)^{|J|}\, \mathcal{U}_J^{k,1}$. Finally, for $J \subseteq \mathbb{A}^n$, a (disjunctive) clause is falsified by all of J's assignments iff all the assignments also falsify all the clause's literals. The variables in such literals must therefore be assigned the *same* truth value by *all* these assignments. There are $\mathcal{H}(J)$ such variables so that $\mathcal{U}_J^{k,1} = (\mathcal{H}(J))^k$. Combining these gives **(1)**. □

SATPACK contains an optimized implementation of $\mathcal{U}_n^{k,m}$ based on **(1)** which can be used to generate formulae for $n = 2, 3$ and 4. Having loaded in this implementation by calling <<SAT`CNFCount`, the following formulae are generated:

In[1]:= $\mathcal{U}_2^{k,m}$

Out[1]= $4^{km} - 4\,(2 - 3\,2^k + 4^k)^m - 4\,(-2^k + 4^k)^m + 2\,(-2^{1+k} + 4^k)^m + 4\,(1 - 2^{1+k} + 4^k)^m + (4 - 2^{2+k} + 4^k)^m$

In[2]:= $\mathcal{U}_3^{k,m}$

Out[2]=

$$6^{km} + (-6 + 3\,2^{2+k} - 8\,3^k + 6^k)^m - 8\,(-3 + 9\,2^k - 7\,3^k + 6^k)^m - 8\,(3 + 3\,2^k - 5\,3^k + 6^k)^m - 24\,(-1 + 5\,2^k - 5\,3^k + 6^k)^m - 24\,(1 + 2^{2+k} - 5\,3^k + 6^k)^m + 2\,(6 - 4\,3^k + 6^k)^m + 32\,(3\,2^k - 4\,3^k + 6^k)^m + 30\,(2 + 2^{1+k} - 4\,3^k + 6^k)^m + 6\,(-1 + 2^{2+k} - 4\,3^k + 6^k)^m + 4\,(-2\,3^k + 6^k)^m + 12\,(1 - 2\,3^k + 6^k)^m + 12\,(2^k - 2\,3^k + 6^k)^m - 8\,(-3^k + 6^k)^m + 12\,(-2 + 7\,2^k - 2\,3^{1+k} + 6^k)^m - 8\,(3 - 3^{1+k} + 6^k)^m - 24\,(1 + 2^k - 3^{1+k} + 6^k)^m - 24\,(2^{1+k} - 3^{1+k} + 6^k)^m + 16\,(6^k + 6\,(2^k - 3^k))^m$$

In[3]:= $\mathcal{U}_4^{k,m}$ **// Shallow[#, {10, 10}] &**

Out[3]//Shallow=

$$8^{km} + 48\,(-10 + 13\,2^k - 3\,2^{1+2k} + 8^k)^m + 8\,(-8 + 3\,2^{2+k} - 3\,2^{1+2k} + 8^k)^m + 8\,(-2^{1+2k} + 8^k)^m + 32\,(1 - 2^{1+2k} + 8^k)^m + 48\,(2^k - 2^{1+2k} + 8^k)^m + 64\,(-5 + 9\,2^k - 3\,2^{1+2k} + 3^k + 8^k)^m + 32\,(-2^{1+2k} + 3^k + 8^k)^m + 288\,(-4 + 7\,2^k - 3\,2^{1+2k} + 2\,3^k + 8^k)^m + 288\,(-2 + 3\,2^{1+k} - 3\,2^{1+2k} + 2\,3^k + 8^k)^m + \ll 127 \gg$$

Note that in the last case, $n = 4$, a "Shallow" wrapper has been used to omit the last 127 terms. Now fix $k = 3$ and express m in terms of a clause-variable ratio x. The next figure illustrates how changes in this clause-variable ratio affect the proportion of unsatisfiable formulae. It is, in some ways, remarkable how the somewhat random and complex formulae generate the smooth curves shown. We discuss this graph in relation to the threshold conjecture in the next section.

```
In[4]:= PlotUCNFCount•[{2, 3, 4}, 3, ShowThreshold → True]
```

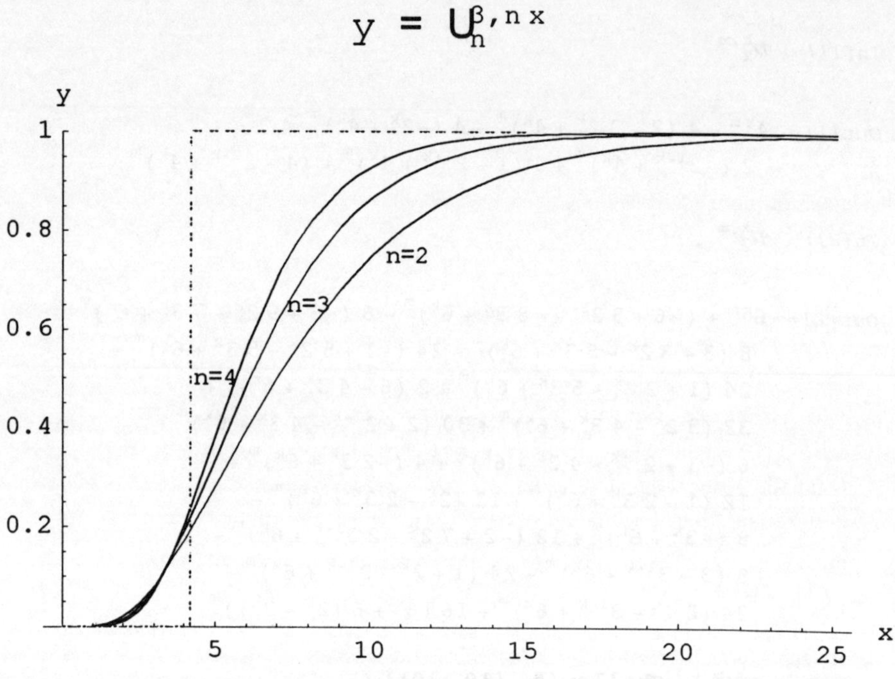

Figure 1

3 The Threshold Conjecture

In *Figure 1* the `ShowThreshold` option was used to generate the dashed line, which represents a step function to which it is conjectured the curves approach in the limit. That is, it represents the curve to which the sequence of functions, $y = \mathcal{U}_n^{k,nx}$, converges as $n \to \infty$. It is intuitively clear that the curves of *Figure 1* should all be increasing functions, since increasing the number of clauses of a random formula increases the probability that any given assignment falsifies at least one of the formula's clauses and therefore the entire formula. Less obvious however, is that most of this increase should take place over a vanishing range for increasing n. During the last decade this is exactly what has been experimentally observed [21][27] with larger simulations for increasing n generating curves that appear to approach the dashed step function illustrated in *Figure 1*. The emergence of this *threshold* or *phase transition*, through which the satisfiability of a random formula changes dramatically, can be stated more precisely as follows:

Threshold Conjecture:

There exists a constant, d_k, such that:

$$\lim_{n\to\infty} \mathbf{U}_n^{k,\,dn} = \begin{cases} 0 & d < d_k \\ 1 & d > d_k \end{cases}. \qquad \text{.................(2)}$$

It is worth noting that this conjecture and related results usually assume a more restricted *standard model* of $\mathsf{F}_n^{k,m}$ than the more general definition previously given here. For large n however, the difference becomes negligible, although this standard model can simplify the asymptotic analysis. It is straightforward to show that $d_0 = d_1 = 0$, while in [18] and [4] it is independently shown that $d_2 = 1$. Theoretical results for d_3, probably due to the *NP*-completeness of 3-SAT, have been harder to come by, with a sequence of improvements in the lower and upper bounds, (surveyed in [10], [1] and [26]), currently placing it somewhere in the interval $3.145 < d_3 < 4.506$ ([1] [8]). It should be pointed out that, since the existence of d_k asserted by the Threshold Conjecture is still an open question for $k \geq 3$, the "bounds" for d_3 are, more correctly, lower and upper bounds for respectively, $sup\{d : \lim_{n\to\infty} \mathbf{U}_n^{3,\,dn} = 0\}$ and $inf\{d : \lim_{n\to\infty} \mathbf{U}_n^{3,\,dn} = 1\}$. Friedgut [14] however, has shown the existence of a sequence, $d_k(n)$, such that $\forall \epsilon > 0$, $\lim_{n\to\infty} \mathbf{U}_n^{k,\,dn} = \begin{cases} 0 & d = d_k(n) - \epsilon \\ 1 & d = d_k(n) + \epsilon \end{cases}$. Hence we know that the curves of *Figure 1* at least approach the shape of a step function in the limit, although it is still unclear whether or not the location of the "step" converges, (in which case $d_k(n) \to d_k$). Extensive experimental evidence, (for example [7] [28]), suggests that $d_3 \sim 4.2$ if such a point does exist.

Prior to their appearance in computational complexity, phase transitions had been observed before with, in particular, striking parallels in random graph theory and certain areas of statistical physics. In a seminal paper Erdös [13] observed threshold behaviour in the evolution of random graphs, including one analogous to k-SAT in which the graph's *chromatic number* undergoes a phase transition around certain vertex/edge ratios. In statistical physics similar analogies exist for phenomena involving percolation, magnetic spin glasses and matter phase changes. In fact, results from graph theory have been used to answer fundamental questions about the threshold of k-SAT [14][26], while techniques from statistical physics have been borrowed to understand the width of the transition [21][27][11]. For a discussion of the background and appearances of phase transitions in general we refer the reader to [11][12].

One of the advantages of having an algorithm for $\mathcal{U}_n^{k,m}$ is that the emergence of this threshold can be more directly studied by analysing the asymptotics of the formula representing the algorithm. The accompanying extended paper in these proceedings gives an example of this process to produce a more succinct expression for $\mathcal{U}_n^{k,m}$ in terms of *h-cylinders* – a characterization of assignments which fix truth values to h variables. More precisely:

Definition 7 (h-cylinder, I^h)

For $J \subseteq \mathbb{A}^n$ call J an h-cylinder if $(\mathcal{H}(J) = h) \wedge (|J| = 2^{n-h})$. For $I \subseteq \mathbb{A}^n$, define $I^h = \{J \subseteq I : J \text{ is an } h\text{-cylinder}\}$.

$$\mathcal{U}_n^{k,m} = \sum_{I \subseteq \mathbb{A}^n} (-1)^{|I|} \, |I^k|^m \qquad \text{.................(7)}$$

Hence "all" that is now needed to prove the threshold conjecture is to use the tantalizingly simple formula of (7) to show that the curves $y = \mathrm{U}_n^{3,nx}$ approach the dashed line of *Figure 1* as n becomes large.

4 Conclusion

Previously, rigorous lower and upper bounds for the threshold phase transition have been obtained through quite disparate methods, (lower bounds having been obtained through complexity analyses of algorithms that solve k-SAT while upper bounds have been attained through sophisticated counting arguments). Now, having a single formula to analyse may provide a more uniform approach to understanding this phase transition, with further progress dependent on understanding more fully the distribution of h-cylinders. A related *Mathematica* application package, SATPACK, has been produced that contains, amongst other things, implementations of the algorithms introduced in this paper (i.e. that count satisfiable solutions), graphing functionality, an inbuilt *SATSolver* that can be used to reproduce previous empirical results and transformations between *NP*-complete problems and their solutions.

UmlaufSatz Movie

Y. Tazawa
Tokyo Denki University
tazawa@cck.dendai.ac.jp

1 Abstract

At IMS2001, I gave a presentation titled "Gauss-Bonnet Theorem by Mathematica." It contained a visualization of the elegant proof of the Umlaufsatz (Theorem of Turning Tangents) by Heinz Hopf, mainly using still figures [1]. In this talk I will show full animations of figures of the detailed proof, incorporated into a movie. Visualizing proofs, and visualizing through computer graphics in particular, gives viewers clear images of the situation, but at the same time it can give some slanted points of view. In this movie I tried to show as many figures as possible to avoid a loss in generality, tracing the figures we imagine in our mind when we follow the proof.

2 Statement of the Theorem

2.1 Polygon on a plane

As is well-known in plane geometry, the sum of the exterior angles α_i of a polygon is 2π

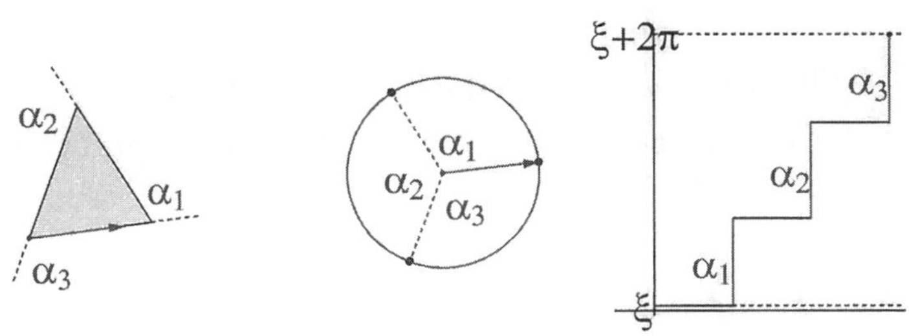

2.1 Umlaufsatz (Theorem of Turning Tangents)

If we replace the edges of the polygon by smooth curves, then we get a simple (no self-intersection) this article. For a closed curve, we have the Theorem of Turning Tangents: the sum of exterior

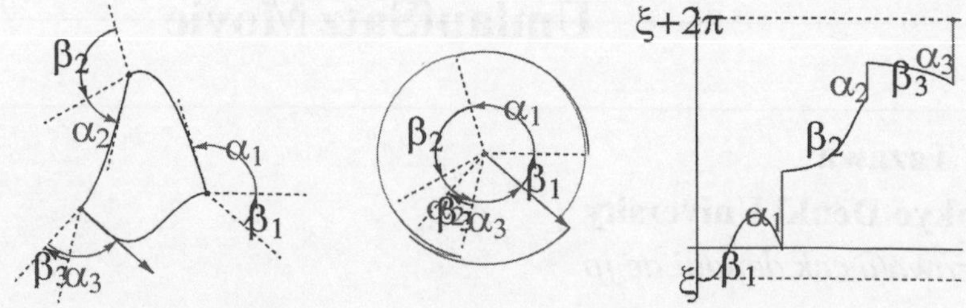

3 Proof of the Theorem

The statement of the theorem above is almost trivial for a closed curve, as shown in Anim more and we need a strict proof.

In the following, I will give the detailed visualization of the proof given by H. Hopf [2] through illustrate various complicated or wild situations that may occur, even if such cases do not happen

3.1 Definition of a map f on a triangle

Hopf defined a nice map f on a triangle into the unit circle, as illustrated by Anim 3, Anim 4

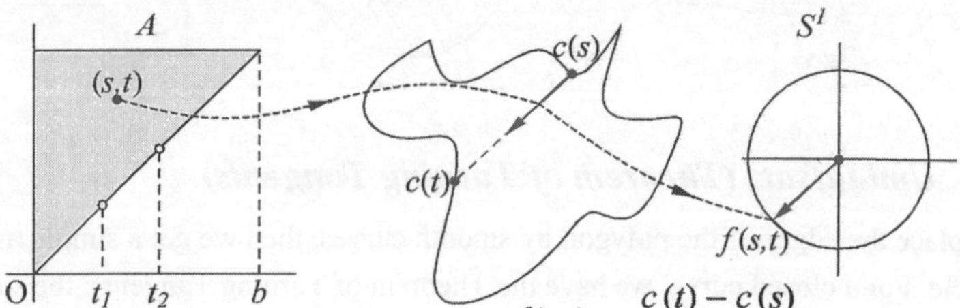

$$f(s,t)=\frac{c(t)-c(s)}{|c(t)-c(s)|}$$

3.2 Continuity of *f*

Continuity of f at $(0, b)$ is shown in: Anim 6.

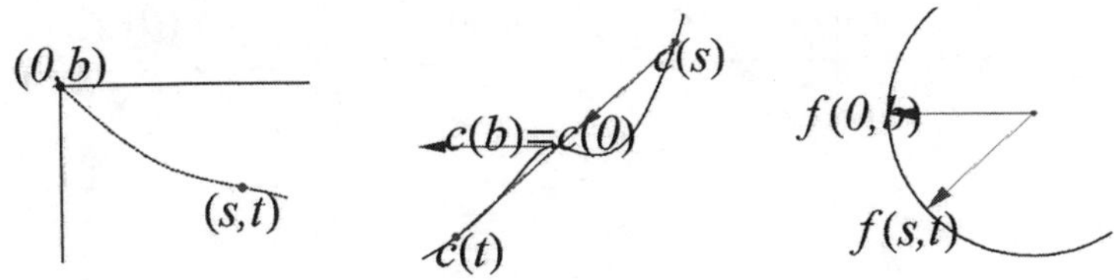

Continuity of f at (u,u) is shown in: Anim 7.

Continuity of f at $(s_0,t_0) \neq (0,b), (u,u)$ is shown in: Anim 8.

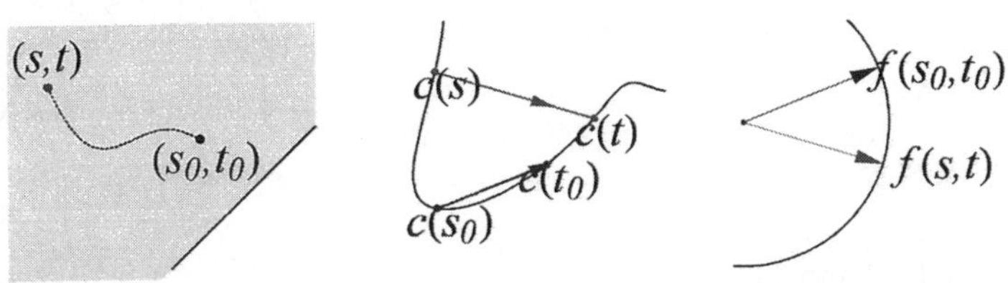

3.3 Extending the map *f* to a function *θ*

We extend the map f to a function θ into $\mathbb{R}$ as illustrated by Anim 9.

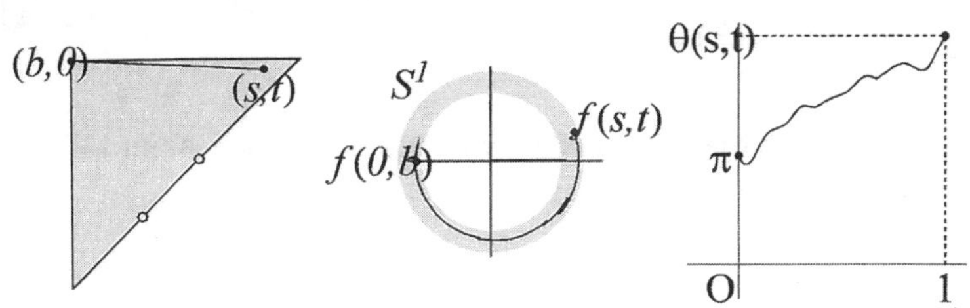

We want to show $\theta(b,b)=2$ by Anim 10.

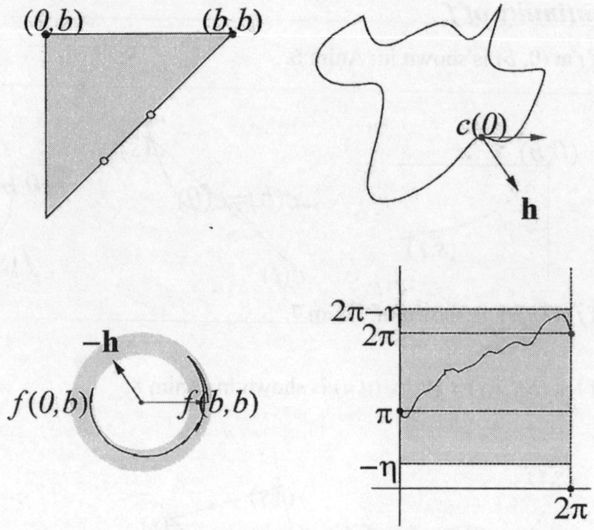

$\theta(0,0)=0$ by Anim 11.

We regard θ as the function into S^1, the covering space of $\mathbb{R}$. Anim 12.

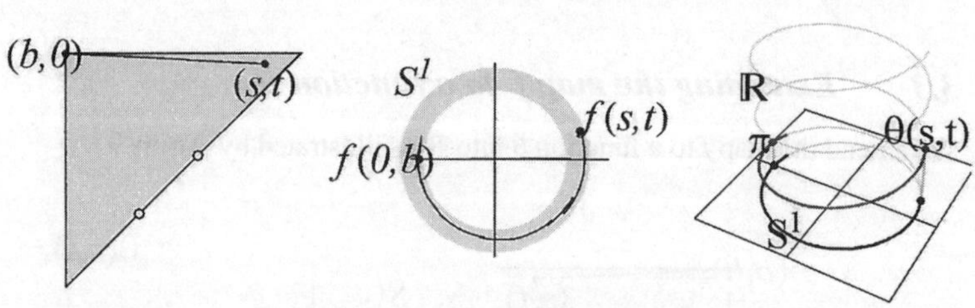

The function θ is continuous as illustrated by Anim 13 and Anim 13B.

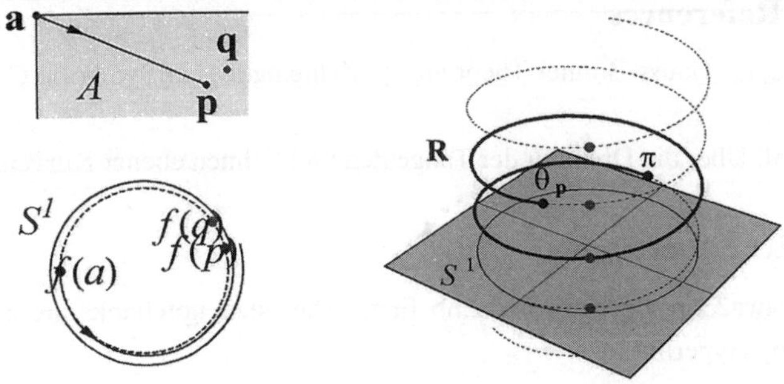

The exterior angle at each vertex is equal to the difference of the limits of the value of θ at (

3.4 *Proof of the theorem*

Anim 16 illustrates the proof of the theorem.

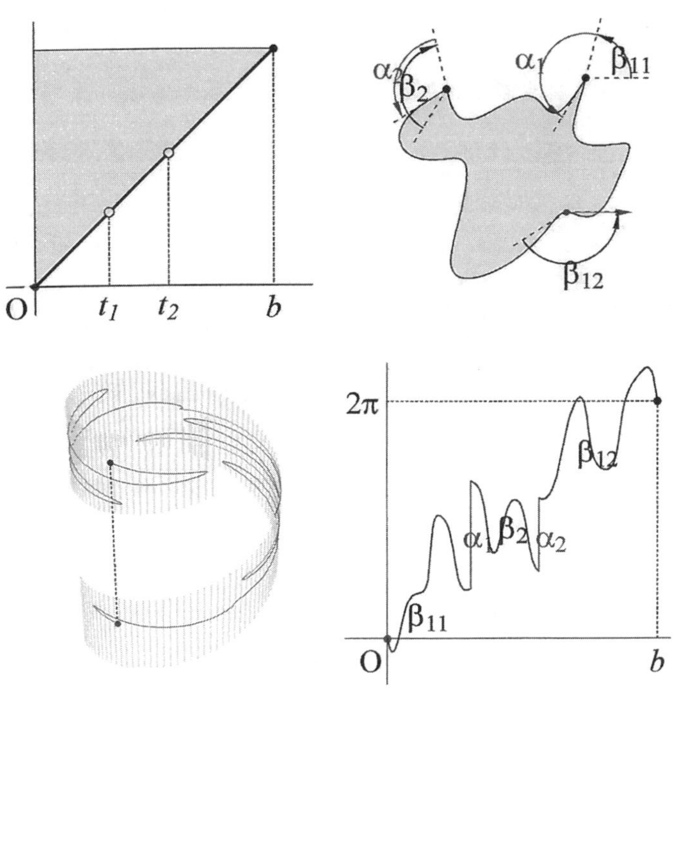

4 References

[1] Tazawa,Y. Gauss-Bonnet Theorem by Mathematica, in Symbolic Computation: New pp485-482.

[2] H. Hopf, Über die Drehung der Tangenten und Sehnen ebener Kurven, Compositio Math.,

Appendix: CD contents

Using Tazawa2.zip, open Tazawa2.nb first. The other notebooks are connected to this notebook by Hyperlink.

Mathematical Interpretation of the Boundary Conditions Phenomenon in the Refined Least Squares Method

R. A. Walentynski
Silesian University of Technology, Gliwice, Poland
rwal@kateko.bud.polsl.gliwice.pl

1 Abstract

The paper presents a mathematical explanation of the phenomenon observed for the Refined Least Squares method. It was found by numerical experiment with *Mathematica* that for some boundary-value or initial value problems it is possible to neglect selected boundary conditions and still obtain a nonsingular system of algebraic equations. Two examples present this phenomenon from the mathematical point of view and I make sense of the solution obtained. This discovery has an important application to ill-conditioned problems of mathematical physics and engineering. The approach makes it possible, in certain cases, to find an approximation which is valid on most of the domain, apart from the boundary layer, or to evaluate a main trend of the function.

2 Introduction

The Refined Least Squares Method applied to initial, boundary and initial-boundary value problems has been implemented within *Mathematica* [1]. The main difference from the classical approach is to consider the boundary or initial conditions as a minimized functional. Only independent approximating functions are allowed. This makes the algorithm much simpler and makes it possible to approximate multidimensional problems with a discontinuous boundary.

It was found that the Refined Least Squares method has a very interesting feature connected with boundary conditions. Under certain circumstances some boundary or initial conditions in the considered boundary-value or initial-value problem described by a differential equation or system of differential equations may be neglected and we still obtain a nonsingular system of algebraic equations (compare [2], [3] and [4]).

Solutions obtained may have a physical sense. Application of this feature of the method is especially useful when searching the approximation of ill-conditioned boundary value problems in the theory of shells [2] and other problems with boundary layers [6]. The phenomenon has been explained physically for shell problems: see [3], [4] and [5]. This contribution presents an explanation of the phenomenon from the mathematical point of view. The mathematical explanation presented here for simple explicit equations may be extended to the problems described with an implicit system of differential equations describing shell problems (see [7]). As this interesting feature of the method has been discovered thanks to *Mathematica* system, it is also explained with its help.

3 Brief description of the method

To focus our attention on the explanation of the phenomenon let us consider an explicit, one dimensional boundary-value (or initial value) problem described with a linear nonhomogenous ordinary differential equation of the function $y(x)$, $e = 0$ with n boundary conditions $b_i = 0$ in the interval $x \in \langle x_a,\ x_b \rangle$:

$$e := \sum_{k=j}^{n} a_k\, y^k[x] - f(x). \qquad \text{................(1)}$$

We can build the following integral for the considered problem.

$$\mathcal{F} := \int_{x_a}^{x_b} (\alpha\, e)^2\, \boldsymbol{d}x + \sum_{i=1}^{n} (\beta_i\, b_i)^2. \qquad \text{................(2)}$$

We assume the expression can be approximated by a linear combination of independent functions.

$$y(x) := \sum_{k=0}^{m} c_k\, u_k(x). \qquad \text{................(3)}$$

In contrast to the classical approach, functions $u_k(x)$ do not have to satisfy boundary or initial conditions as they are already considered in the functional.

The approximation is done by minimization of the functional (2), which is equivalent to $\boldsymbol{m}$ conditions:

$$\partial_{c_i} \mathcal{F} = 0. \qquad \text{................(4)}$$

This produces a system of $\boldsymbol{m}$ linear algebraic equations with a symmetric matrix. An effective computational procedure of the coefficients of the system of equations is presented in [2], [3] and [7], but to explain the considered phenomenon we will apply the method explicitly.

The explanation of the phenomenon connected with boundary conditions will be illustrated with two examples.

4 Example 1

4.1 Equation

Let us consider a boundary value problem in the interval x∈⟨0, 50⟩ described with a differential equation with the following left hand side:

```
In[26]:= e1 := y[x] + y^(4)[x] - (1/2 + x/50 + x (x - 50)^4/25000000) (*==0*)
```

The exact solution of the equation is:

$$y(x) = \frac{x^5-200\,x^4+15000\,x^3-500000\,x^2+6749880\,x+12504800}{25000000} \qquad (5)$$
$$+e^{\frac{x}{\sqrt{2}}}\, c_1 \cos\left(\frac{x}{\sqrt{2}}\right)+e^{-\frac{x}{\sqrt{2}}}\, c_2 \cos\left(\frac{x}{\sqrt{2}}\right)$$
$$+e^{-\frac{x}{\sqrt{2}}}\, c_3 \sin\left(\frac{x}{\sqrt{2}}\right)+e^{\frac{x}{\sqrt{2}}}\, c_4 \sin\left(\frac{x}{\sqrt{2}}\right).$$

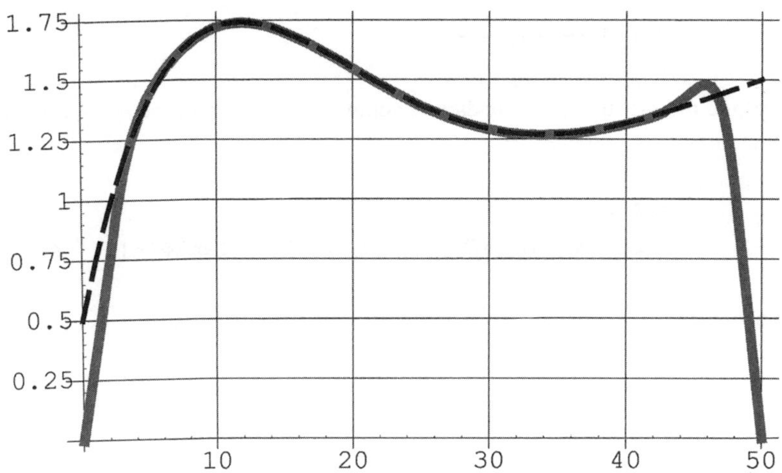

Figure 1

Figure 1 presents the solution of the problem with boundary conditions $y(0) = y(50) = 0$ and $y'(0) = y'(50) = 0$ (grey line), and a comparison to the particular integral of the solution (black dashed line). Note that the particular integral in the considered case is a feasible solution on most of the domain, apart from the boundary layers.

4.2 *Least squares approximation*

We will show that in the considered case all boundary conditions may be neglected in the approximation. Let us assume the interpolation function (3) has the following form. As the particular integral is a polynomial of degree 5 we can set it as follows:

$$\texttt{y[x\_]} = \sum_{\texttt{i=0}}^{5} \texttt{c[i] x}^{\texttt{i}}$$

The functional (2) is defined as an integral with the limits of the given domain.

$$\mathcal{F} = \int_0^{50} \texttt{el}^2 \, d\texttt{x}$$

As we can easily see, the boundary conditions are not considered in it. Differentiating the functional we get a system of linear algebraic equations (4).

$$\texttt{V[i\_]} := \partial_{\texttt{c[i]}} \mathcal{F} == 0$$

Despite the fact that the boundary conditions are neglected, the system is not singular. We can find its solution.

```
sol =
  Solve[Table[V[i], {i, 0, 5}], Table[c[i], {i, 0, 5}]];
```

The functional is equal to zero. It means that the differential equation is satisfied exactly.

```
ℱ /. sol⟦1⟧
```

0

It happens that the method returns a particular integral in the considered case: compare with Eqn. (5).

```
Simplify[y[x] /. sol[[1]]]
```

$$\frac{12504800 + 6749880\,x - 500000\,x^2 + 15000\,x^3 - 200\,x^4 + x^5}{25000000}$$

5 Example 2

5.1 Equation

Let us consider the same interval and boundary conditions as in the previous case but a differential equation with the following left hand side:

```
In[30]:= e2 := 2 z''[x] + z^(4)[x] - x^2 (*==0*)
```

It can be found that the particular integral in this case is a function:

$$\text{In[31]:= } zs[x\_] := -\frac{x^2}{4} + \frac{x^4}{24}$$

5.2 Least squares approximation

As the particular integral is a polynomial of degree 4 we assume an approximation in the following form:

$$\text{In[32]:= } z[x\_] = \sum_{i=0}^{4} d[i]\, x^i$$

The functional, neglecting all boundary conditions, results in a singular algebraic system. It can be easily found, by evaluation of the functional $\mathcal{F}1$, that two coefficients: d[0] and d[1] are absent in it. Therefore two equations of the system of algebraic equations (4) are identities (0=0).

$$\text{In[34]:= } \mathcal{F}1 = \int_0^{50} e2^2\, dx$$

To obtain a non-singular system of algebraic equations we should consider at least two boundary conditions. For example:

$$\text{In[35]:= } \mathcal{F}2 = \int_0^{50} \mathbf{e2}^2\, d\mathbf{x} + \mathbf{z}[0]^2 + \mathbf{z}[50]^2$$

The minimization procedure returns the following function. It satisfies the differential equation and two boundary conditions considered in the functional. It will be called a base solution or a generalized particular integral.

$$-\frac{31175\, x}{6} - \frac{x^2}{4} + \frac{x^4}{24}$$

The choice of boundary conditions is not arbitrary. The following functional leads to the singular system, too.

$$\text{In[36]:= } \mathcal{F}3 = \int_0^{50} \mathbf{e2}^2\, d\mathbf{x} + \mathbf{z}'[0]^2 + \mathbf{z}'[50]^2$$

Figure 2 shows its diagram (black dashed line) compared to the exact solution - with consideration of all boundary conditions (grey line). It shows that the base solution obtained with minimization of the functional $\mathcal{F}2$ represents the main trend of the exact solution. Moreover, this example shows that the method is not limited to problems with a boundary layer.

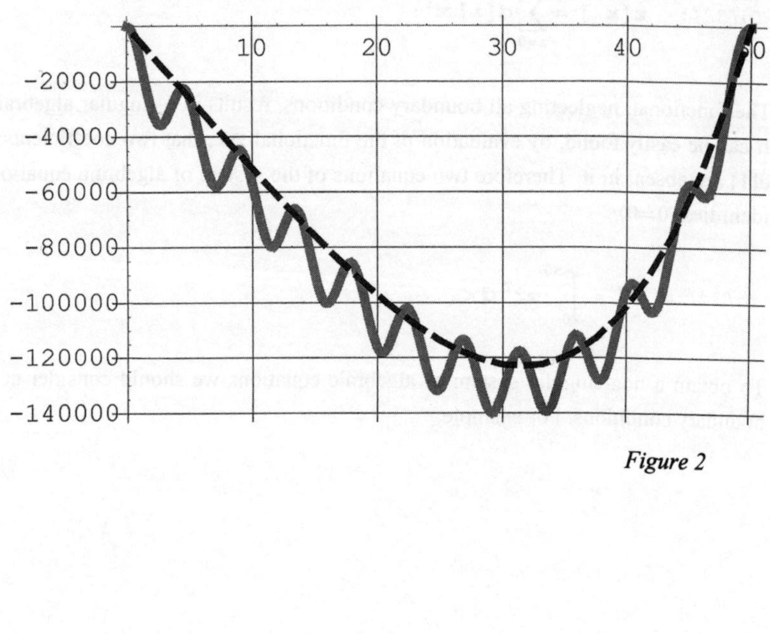

Figure 2

6 Conclusions

The examples lead to the conclusion that the number of boundary conditions that needs to be considered is connected with the lowest order of the function derivative in the differential equation concerned. If the lowest order is zero, as in the first example, we can neglect all boundary conditions. Generally, if the lowest order is n we must consider n boundary or initial conditions. In certain cases the approximation obtained by ignoring some boundary conditions may make sense. The method is especially useful in problems with a boundary layer, as in the first example. Finding the main trend of the solution may also be useful. As already mentioned, the features of the method discovered in this paper can have important engineering applications. This explanation is not a formal mathematical proof, which remains to be done. Nevertheless it must be emphasized that *Mathematica* helped me, as a professional civil engineer, first to discover, then to explore and finally to understand the crucial aspects of the phenomenon.

7 References

[1] Walentynski R.A. A least squares method for solving initial-boundary value problems, in *Mathematics with Vision*, Proceedings of the Second International *Mathematica* Symposium, pp. 483--490, Rovaniemi, Finland, June 1997. (eds. Keränen, V., Mitic, P. and Hietamäki, A), June 1997, Computational Mechanics Publications, Southampton and Boston.

[2] Walentynski R.A. Refined least squares method for shells boundary value problems, in *Symbolic Computations: New Horizons*, Proceedings of the Fourth International *Mathematica* Symposium, pp. 511--518, Tokyo, Japan, June 2001. (eds. Tazawa, Y., Sakakibara, S., Ohashi, S. and Uemura, Y.), June 2001, Tokyo Denki University Press, Tokyo.

[3] Walentynski R.A. Solving symbolic and numerical problems in the theory of shells with *Mathematica*, in *Symbolic and Numerical Scientific Computation*, Proceedings of the Conference SNSC'2001, pp. 40+, Linz-Hagenberg, Austria, September 2001. (eds. Winkler, F. and Langer, U.), 2003, to be printed in series *Lecture Notes in Computer Science* by Springer Verlag, Heildelberg, New York.

[4] Walentynski R.A. Meshless, two steps approach to the boundary layer problems in shells, in *International Scientific Conference*, Proceedings of the 7th Conference ISC'2002, pp.314--317, Kosice, Slovakia, May 2002, (ed. Kristofovic, V.), May 2002, Kosice University of Technology, Faculty of Civil Engineering, Kosice.

[5] Walentynski R.A. Dealing with shells boundary conditions with refined least squares method, in proceedings of the *Fifth World Congress of Computational Mechanics*, http://wccm.tuwien.ac.at/publications/Papers/fp80303.pdf, Vienna, Austria, July 2002. (eds. Mang, H.A., Rammerstorfer, F.G. and Eberhardsteiner, J.), July 2002, Vienna University of Technology, Vienna.

[6] Walentynski R.A. Solving Reissner-Mindlin plates with refined least squares method, in *New Trends in Statics and Dynamics of Buildings*, Proceedings of the International Scientific Conference, pp. 49--54, Bratislava, Slovakia, October 2002. (ed. Kralik, J.) October 2002, Slovak University of Technology, Bratislava.

[7] Walentynski R.A. *Application of Computer Algebra in Symbolic Computations and Boundary-Value Problems of Theory of Shells*, February 2003, submitted to the editor, Monographs, Silesian University of Technology Press, Gliwice.

Constructing Branches of Solutions of Nonlinear PDE ' s

Jay H. Wolkowisky
Mathematics Department
University of Colorado, Boulder, CO
wolkowis@euclid.colorado.edu

1 Abstract

This paper uses a linear partial differential equation (PDE) solver to construct the bifurcating branches of solutions of nonlinear PDE's. This linear PDE solver (developed by the author in a previous paper) uses the Mutigrid algorithm. This algorithm is written in C and the *Mathematica* protocal Mathlink is used to call the C program from *Mathematica*. This combination is very powerful since we can utilize the speed of C together with the symbolic and graphical capabilities of *Mathematica*.

2 The Nonlinear Boundary Value Problem

We will be concerned with the following **nonlinear** boundary value problem(BVP).

$$u_{xx} + u_{yy} + \lambda u + f[x, y, u, u_x, u_y] = 0, \quad 0 < x < \pi, \quad 0 < y < \pi$$
$$u[0, y] = u[x, 0] = u[\pi, y] = u[x, \pi] = 0 \quad \ldots\ldots\ldots\ldots \quad (1)$$

Connected with this problem is the following associated **linear** BVP.

$$u_{xx} + u_{yy} + \lambda u + f_1[x, y] u + f_2[x, y] u_x + f_3[x, y] u_y = g[x, y],$$
$$0 < x < \pi, \quad 0 < y < \pi, \quad u[0, y] = u[x, 0] = u[\pi, y] = u[x, \pi] = 0 \quad \ldots\ldots \quad (2)$$

The reason we wish to be able to solve the linear problem (2) is that we wish to use a generalized Newtons Method to solve the nonlinear problem (1), as follows. Let the linear operator **G[u]** ($v \to$ **G[u]** v) be defined as

```
G[u] v = vxx + vyy + λv +
    fu[x, y, u, ux, uy] v + fux[x, y, u, ux, uy] vx +
    fuy[x, y, u, ux, uy] vy,  0 < x < π, 0 < y < π,
```

where u is a given function. Newton's Method takes the form

$$u_{n+1} = u_n - G[u_n]^{-1}\,\bar{F}[u_n,\lambda]$$

where

$$\bar{F}[u,\lambda] = u_{xx} + u_{yy} + \lambda u + f[x, y, u, u_x, u_y].$$

So in order to calculate $G[u_n]^{-1}\bar{F}[u_n,\lambda]$ we must be able to solve (2) with

```
g[x, y] = F̄[u[x, y], λ];
f1[x, y] =
  fu[x, y, u[x, y], ux[x, y], uy[x, y]];
  f2[x, y] = fux[x, y, u[x, y], ux[x, y], uy[x, y]];
  f3[x, y] = fuy[x, y, u[x, y], ux[x, y], uy[x, y]];
```

3 The Linear Boundary Value Problem

The descretized version of this problem is the following, with mesh size $h = \frac{\pi}{N}$, where N+1 is the number of mesh points in each dimension. The notation `u[i,j]` means the value of the approximate solution at the mesh point `(i,j)`.

```
(uxx + uyy + λu + f1[x, y] u +
    f2[x, y] ux + f3[x, y] uy == g[x, y]) /.

{uxx -> 1/h^2 (u[i + 1, j] - 2 u[i, j] + u[i - 1, j]),

 uyy -> 1/h^2 (u[i, j + 1] - 2 u[i, j] + u[i, j - 1]),

 u -> u[i, j], ux -> (u[i + 1, j] - u[i - 1, j])/(2 h),

 uy -> (u[i, j + 1] - u[i, j - 1])/(2 h), x -> i h, y -> j h}
```

From this we can then write a **Gauss-Seidel** iteration scheme which will form the basis for the **Multigrid** algorithm which is to be used to solve the linear BVP.This method was discussed previously in [1].

4 An Improved Version of the Multigrid Algorithm

The standard Multigrid algorithm uses a 129 x 129 fine-grid and a 3 X 3 coarse-grid. This version of the algorithm works fine as long as λ is not near the eigenvalues of the linear homogeneous BVP. However, we will be interested in the bifurcating branches of solutions (as a function of the parameter λ) of the nonlinear boundary value problem (1). Since the branches of solutions bifurcate from the eigenvalues of the linearized problem, we need to be able to solve the linear problem near its eigenvalues. In order to do this we need to modify the Multigrid algorithm since the 3 X 3 coarse-grid is too crude for convergence of the Multigrid algorithm when λ is near the eigenvalues of the linear problem. Instead of taking the coarsest grid to be 3 X 3 we take it to be 17 X 17. The Multigrid method then requires that we solve the corresponding matrix system of equations "exactly". In order to accomplish this, the coefficient matrix needs to be constructed. This size of this coefficient matrix turns out to be 225 X 225 because of the following. For a 17 X 17 grid the interior grid is 15 X 15, which means there are 225 unknowns and 225 equations. As we shall see, this coefficient matrix is **band diagonal** because of the structure of the **Laplace** operator. In our modified Multigrid algorithm we take advantage of this property by using a very efficient band diagonal matrix solver. We next construct this band diagonal matrix using *Mathematica*'s powerful symbolic commands.

4.1 Construction of the Band Diagonal Coefficient Matrix

We first make some definitions.

```
first = {{-1 / 2, 0, 1 / 2}};
second = {{1, -2, 1}};
laplace = PadLeft[second, {3, 3}, 0, {1, 0}] +
    PadLeft[Transpose[second], {3, 3}, 0, {0, 1}];
MatrixForm[laplace];
```

The first is the stencil for the one-dimensional first order centered difference scheme, the second is the stencil for the one-dimensional second order centered difference scheme, and the third is the stencil for the **Laplace** operator. We can now construct the coefficient matrix for the descretized (finite- difference) version of the operator

$$u_{xx} + u_{yy} + \lambda u + f_1[x, y] u + f_2[x, y] u_x + f_3[x, y] u_y;$$

We define the following for a 17 X 17 grid:

```
U = Table[u[x, y], {x, 2, 16}, {y, 2, 16}];
F1 = Table[f1[x, y], {x, 2, 16}, {y, 2, 16}];
F2 = Table[f2[x, y], {x, 2, 16}, {y, 2, 16}];
F3 = Table[f3[x, y], {x, 2, 16}, {y, 2, 16}];
```

The main tool for the construction of the coefficient matrix is the very powerful ***Mathematica*** command `ListCorrelate.`

```
vder = 1/h^2 ListCorrelate[laplace, U, {2, 2}, 0] +
   λ U + F1 U + 1/h F2 ListCorrelate[
     first, U, {{1, 2}, {-1, 2}}, 0] +
   1/h F3 ListCorrelate[Transpose[first], U, {{2, 1}, {2, -1}}, 0];

vdereqs = Flatten[vder];
length = Length[vdereqs];
veqs = Table[vdereqs[[i]] == 0, {i, length}];
```

Next we pick out the coefficients of the "**u**'s" from the above 225 equations using the *Mathematica* command `LinearEquationsToMatrices`, and then construct the 225 X 225 coefficient matrix.

```
<< LinearAlgebra`MatrixManipulation`
m225 =
   First[LinearEquationsToMatrices[veqs, Flatten[U]]];
MatrixForm[m225];
```

The matrix **m225** has only 5 non-zero "diagonals". It is this " bandedness" that we take advantage of in solving the corresponding system of linear equations.

5 An Example Showing the Construction of a Bifurcating Solution Branch

We will consider the following nonlinear BVP,

$$u_{xx} + u_{yy} + \lambda u + f[x, y, u, u_x, u_y] = 0, \quad 0 < x < \pi,$$
$$0 < y < \pi, \quad u[0, y] = u[x, 0] = u[\pi, y] = u[x, \pi] = 0$$

with the nonlinear term:

```
f[x_, y_, u_, v_, w_] := x y u w^2
```

We define the following:

```
F[x_, y_, u_, v_, w_, λ_] =
  u_xx + u_yy + λ u + f[x, y, u, v, w] ;
f1[x_, y_, u_, v_, w_] = ∂_u f[x, y, u, v, w];
f2[x_, y_, u_, v_, w_] = ∂_v f[x, y, u, v, w];
f3[x_, y_, u_, v_, w_] = ∂_w f[x, y, u, v, w];

f1[u_][i_ /; 1 < i < imax, j_ /; 1 < j < imax] =
  f1[x, y, u, v, w] /. {u -> u[i, j], v -> (u[i + 1, j] - u[i - 1, j])/(2 h1),
    w -> (u[i, j + 1] - u[i, j - 1])/(2 h2), x -> i h1, y -> j h2};

f2[u_][i_ /; 1 < i < imax, j_ /; 1 < j < imax] =
  f2[x, y, u, v, w] /. {u -> u[i, j], v -> (u[i + 1, j] - u[i - 1, j])/(2 h1),
    w -> (u[i, j + 1] - u[i, j - 1])/(2 h2), x -> i h1, y -> j h2};

f3[u_][i_ /; 1 < i < imax, j_ /; 1 < j < imax] =
  f3[x, y, u, v, w] /. {u -> u[i, j], v -> (u[i + 1, j] - u[i - 1, j])/(2 h1),
    w -> (u[i, j + 1] - u[i, j - 1])/(2 h2), x -> i h1, y -> j h2};

FF[u_, λ_][i_ /; 1 < i < imax, j_ /; 1 < j < imax] =
  F[x, y, u, v, w, λ] /.

  {u_xx -> 1/h1^2 (u[i + 1, j] - 2 u[i, j] + u[i - 1, j]),

   u_yy -> 1/h2^2 (u[i, j + 1] - 2 u[i, j] + u[i, j - 1]),

   u -> u[i, j], v -> (u[i + 1, j] - u[i - 1, j])/(2 h1),

   w -> (u[i, j + 1] - u[i, j - 1])/(2 h2),

   x -> i h1, y -> j h2};
```

We now use Newton's Method to solve the descritized version of equation (2) with the coefficients **f1, f2, f3** defined by the above corresponding descritized functions, and the right hand side **g[x,y]** defined by the above function **FF**. The solution of this descritized linear problem is accomplished by our improved Multigrid algorithm with the coarse grid of 16 X 16 and the fine grid of 128 X 128. The C program, working through *MathLink*, solves over 16,000 linear equations. This procedure is as follows.

```
mmm[m_, λ_] := PDEmlmod[100, 3, 3, λ, rhs[m, λ],
   ff1[m], ff2[m], ff3[m], m, a[m, N[π/16], λ]];
```

Here **PDEmlmod** is the *MathLink* -created function and is similar to the corresponding function discussed in [1]. It uses the definitions:

```
ff1[m_] := Table[f1[r][k, l] /. r[i_, j_] :> m[[i, j]],
   {k, 129}, {l, 129}];
ff2[m_] := Table[f2[r][k, l] /. r[i_, j_] :> m[[i, j]],
   {k, 129}, {l, 129}];
ff3[m_] := Table[f3[r][k, l] /. r[i_, j_] :> m[[i, j]],
   {k, 129}, {l, 129}];
rhs[m_, λ_] := Table[FF[r, λ][k, l] /.
    r[i_, j_] :> m[[i, j]], {k, 129}, {l, 129}];
```

Thus, Newton's method takes the form:

```
newtonIteration[m_, λ_, ε_] :=
  (l = 0; FixedPoint[(l++; newton[#, λ]) &, m,
    SameTest -> ((MatrixNorm[#1 - #2]) < ε &)]);
newton[m_, λ_] := m - mmm[m, λ];
```

The construction of a branch of solutions of the nonlinear BVP (1) bifurcating from an eigenvalue of the linear homogenous BVP is accomplished by a continuation process. We start by getting a non-trival solution for **λ** near an eagenvalue of the linear problem. This is done by using the eigenfunctins of the linear problem which can be solved explicitly. Once we have such a solution of the nonlinear problem for **λ** near an eigenvalue, we can increment **λ** by a small amount and use Newton's method to converge to a solution for this new value of **λ**. We proceed in this manner, constructing a branch of solutions as a function of **λ**. This procedure is executed by the following function:

```
idbranch2[m0_, λ0_, dλ_, jm_, ε_] :=
 Module[{idb}, idb[0] := m0;
  idb[1] := newtonIteration[idb[0], λ0, ε];
      idb[i_] := (idb[i] =
      newtonIteration[(2 idb[i - 1] - idb[i - 2]),
       λ0 + (i - 1) dλ, ε]) /; i >= 2 ;
   Table[me[k] := idb[k], {k, jm}];  Do[
  Print[{λ0 + (j - 1) dλ, MatrixNorm[idb[j]], 1}],
  {j, jm}]]
```

In this function, **m0** is a starting solution, usually taken as the zero matrix, **λ0** is the starting value of **λ**, **dλ** is the increment in **λ** to be taken, **jm** is how many increments to do, and **ε** is the error parameter which tells Newton's method when to stop. As an example we construct a branch bifurcating from the double eigenvalue **λ= 5,** as follows.

```
branch =
  idbranch2[m0, 4.993013080573769`, -.001, 40, .0001];
ListPlot[Table[{branch[[i, 1]], branch[[i, 2]]},
  {i, Length[branch]}], PlotJoined → True,
 AxesLabel → {λ, Norm}, PlotStyle → Hue[.1]]
```

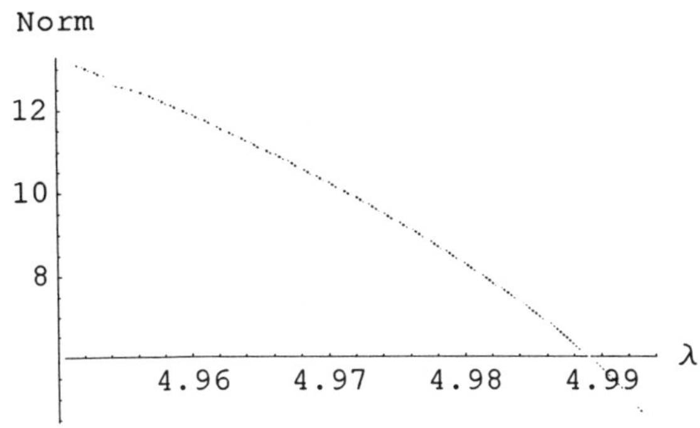

6. References

[1] Wolkowisky, Jay H., A Full Multi-Grid Package Using *MathLink* for Solving Elliptic PDE's; Proceedings of the Fourth International *Mathematica* Symposium; Chiba, Japan; June 2001. (eds.Tazawa, Y., Sakakibara, S., Ohashi, S., Uemura, Y.), Tokyo Denki University Press.

Bayesian Statistical Models for Financial Audits

Karl Heiner
State University of New York at New Paltz, USA
kwheiner@aol.com
Anthony O'Hagan and David J. Laws
Univeristy of Sheffield, UK
a.ohagan@sheffield.ac.uk and d.j.laws@sheffield.ac.uk

1 Abstract

Bayesian Statistical Models provide statisticians and auditors with a set of powerful tools for attacking financial audit problems. Some of these statistical auditing problems may be solved analytically. For these, functions for manipulating equations and standard packages make *Mathematica* a natural environment for work. However, most Bayesian statistical models suitable for attacking financial audit problems are not accessible in closed form. For these problems *Mathematica's* symbolic programming language, efficient array handling functions, and numerical optimization routines form the base for building large-scale simulations. The beta-binomial model for estimating the proportion of transactions in error is described. Models for multilocation audits are discussed briefly, as are models for sequential multiple location audits. Why *Mathematica* is an important tool in this type of data analysis is shown.

2 Introduction

Large medical insurers (e.g. private insurance companies, state and federal governments) depend on statistical audits to resolve financial disputes between the medical providers and the payers. In the U.S., it is not uncommon for courts and administrative law judges to base fiscal determinations on statistical audits. When fraud is alleged, prosecutors depend on these procedures to establish that financial criteria are met when returning felony indictments. In the U.S. and U.K. the administration of large governmental programs are audited at various levels of government. Laws and O'Hagan[1] describe such audits conducted by the British Audit Agency. In the U.S., the Food Stamps program (a program in which state and local governments distribute coupons redeemable in grocery stores and markets for food) is an example of a large program audited by the federal government and in some cases by state governments.

In some audits, a conjugate model (e.g. the beta-binomial-beta model) is useful for combining prior information and data into a probability model for a parameter of interest, e.g. error rate. Here, the posterior probability distributions are known to *Mathematica*, which also provides cumulative distribution functions, and the ability to make draws from these distributions. When sampling from finite populations, we need to sample from the predictive distribution. In the simple beta-binomial model, the predictive distribution is the Polya distribution, a distribution not generally provided in software. The relationship between the results produced using the beta-binomial predictive distribution and Polya urn model sampling is interesting and endorses the use of both.

Frequently, governmental audit programs consider several administrative locations and systems. Here, the prior, likelihood, and posterior probability models are multivariate (e.g. multinomial, Dirichlet). Also, the hierarchal nature of the administrative units and subjective opinions of experts may be incorporated into the models using Bayes Theorem. In these applications, within location data are often sparse. Laws and O'Hagan[1] describe such multilocation audits conducted in the UK. We will describe a multilocation audit in New York State where the ratio of error payments to book value is of interest for all counties in New York State. In some of these counties, very little data are available. The problem may be further complicated and the estimates made more precise where these audits are performed periodically.

3 Models

When applying Bayesian statistical models, we are interested in estimating some parameter θ (or vector of parameters, $\boldsymbol{\theta}$) associated with some probability distribution. In order to estimate θ, we rely on an assumed statistical model, prior information about θ, and data, D. Bayes theorem tells us that the posterior probability distribution of θ given the data D is the prior probability distribution of θ, updated by multiplying by the likelihood of the data $P(D \mid \theta)$ and normalized by dividing by a constant c. The normalizing constant forces the posterior probabilities to sum to one. Thus,

$$\text{Posterior probability } P(\theta \mid D) = \frac{\text{Prior probability } P(\theta) \ x \text{ Likelihood } P(D|\theta)}{c}$$
$$\propto \text{Prior probability } P(\theta) \ x \text{ Likelihood } P(D \mid \theta).$$

In Bayesian models, both the parameter of interest θ and the data ,denoted Y, are considered random variables. So we have $p(\theta \mid Y) = \frac{p(\theta)\, p(Y|\theta)}{\sum_{\theta} p(\theta)\, p(Y|\theta)}$ for discrete random variables, or $p(\theta \mid Y) = \frac{p(\theta)\, p(Y|\theta)}{\int p(\theta)\, p(Y|\theta)\, d\theta}$ for continuous quantities.

3.1 *Single Centre Model*

Suppose that from a sample of size n transactions, we wish to estimate the proportion of the population in error. The number of errors in the sample follows the binomial probability mass function. From a frequentists point of view, this probability mass function tells us the probability of X successes given a specified probability of success, say θ. From the Bayesian point of view, the probability mass function tells us the likelihood of the possible values of θ given that the number of errors in the sample of size n is known. The binomial likelihood is $f(x \mid \theta) = \binom{n}{x} \theta^{x}(1-\theta)^{n-x}$.

The beta distribution is a continuous family of distributions on the interval (0,1). With $\alpha > 0$, $\beta > 0$, and with $B(\alpha,\beta)$ and $\Gamma(\alpha)$ referring to Euler's Beta and Gamma functions, respectively, the probability density function (pdf) of the beta distribution is $f(y \mid \alpha, \beta) = \frac{1}{B(\alpha,\beta)} y^{\alpha-1}(1-y)^{\beta-1} = \frac{\Gamma(\alpha+\beta)}{\Gamma(\alpha)\Gamma(\beta)} y^{\alpha-1}(1-y)^{\beta-1}$. The beta distribution is a convenient prior distribution for the binomial likelihood because it is defined on the range of values assumed by the binomial parameter θ. The family can assume a variety of shapes which makes it convenient for subjective priors; and it is the conjugate prior for the binomial likelihood, i.e. a beta posterior will result from updating a beta prior with a binomial likelihood. Thus, following Bayes theorem

$$p(\theta \mid X) =$$

$$\frac{p(\theta)\, p(X \mid \theta)}{\int p(\theta)\, p(X \mid \theta)\, d\theta} = \frac{1}{c}\left[\frac{\Gamma(\alpha+\beta)}{\Gamma(\alpha)\Gamma(\beta)}\theta^{\alpha-1}(1-\theta)^{\beta-1}\right]\left[\binom{n}{x}(\theta^{x}(1-\theta))^{n-x}\right] =$$

$$\frac{\binom{n}{x}\frac{\Gamma(\alpha+\beta)}{\Gamma(\alpha)\Gamma(\beta)}\theta^{x+\alpha-1}(1-\theta)^{n-x+\beta-1}}{\int_0^1 \binom{n}{x}\frac{\Gamma(\alpha+\beta)}{\Gamma(\alpha)\Gamma(\beta)}\theta^{x+\alpha-1}(1-\theta)^{n-x+\beta-1}\, d\theta} =$$

$$\frac{\binom{n}{x}\frac{\Gamma(\alpha+\beta)}{\Gamma(\alpha)\Gamma(\beta)}\theta^{x+\alpha-1}(1-\theta)^{n-x+\beta-1}}{\binom{n}{x}\frac{\Gamma(\alpha+\beta)}{\Gamma(\alpha)\Gamma(\beta)}\frac{\Gamma(x+\alpha)\Gamma(n-x+\beta)}{\Gamma(n+\alpha+\beta)}}$$

The denominator of the final expression is known as the beta-binomial, or Polya distribution. So, $p(\theta \mid X) = \frac{\Gamma(n+\alpha+\beta)}{\Gamma(x+\alpha)\Gamma(n-x+\beta)} \theta^{x+\alpha-1} (1-\theta)^{n-x+\beta-1}$, the beta distribution with parameters $x+\alpha$ and $n-x+\beta$.

Thus, if we have prior beliefs about the error rate θ, we may specify α and β, incorporate our data (sample size n and error count x), and estimate θ with the mean, mode, or median of the appropriate beta distribution. A probability interval for θ is constructed by finding small and large quantiles of the beta distribution. Functions for accomplishing these tasks are included in *Mathematica*'s Standard Packages. Alternatively, we could have simulated the descriptive measures of the posterior distribution of θ from the beta distribution, or from the product of the prior and the likelihood.

When population sizes are small, to estimate the total number of errors, we may predict the number of errors, y, in the unseen portion of the population, m, and add these numbers to the observed sample. The predictive distribution is the integral of the product of the likelihood and the posterior distribution, $\int_0^1 \left[\frac{\Gamma(n+\alpha+\beta)}{\Gamma(x+\alpha)\Gamma(n-x+\beta)} \theta^{x+\alpha-1} (1-\theta)^{n-x+\beta-1} \right] \left[\binom{m}{y} (\theta^y (1-\theta))^{m-y} \right] d\theta$ $= \binom{m}{y} \frac{\Gamma(n+\alpha+\beta)}{\Gamma(x+\alpha)\Gamma(n-x+\beta)} \frac{\Gamma(y+x+\alpha)\Gamma(m-y+n-x+\beta)}{\Gamma(m+n+\alpha+\beta)}$, a beta-binomial or Polya pdf. From this we make predictions and develop prediction intervals for the number of errors, y, in the unobserved m transactions, to which we add the observed x and n. The Polya distribution is not included in *Mathematica*'s Standard Packages, but we may easily construct the needed functions. We would like to define a discrete probability distribution and work with it as if it were defined in the *Mathematica* standard add-on packages. For the Polya distribution, we know the probability mass function, a function of three parameters, m, α, and β, as well as its recurrence formula. We define betaBinomial[m, α, β] so that PDF[betaBinomial[m, α, β], y] returns the above pdf, and so that Quantile[betaBinomial[m, α, β], q] would return the qth quantile of this distribution.

The relationship between the results produced using the beta-binomial predictive distribution and Polya urn model sampling is interesting and endorses the use of both. Meeden and Varderman [1] recommend simulating the number of errors in each of a finite number of categories from the data in the observed sample. When the number of categories is two, this is equivalent to the Polya urn model, see Feller [3]. A ball is randomly drawn from an urn containing r red and b black balls. The drawing is *with replacement*, but in addition to replacing the selected ball, c balls of the same color are added to the urn, as well as d balls of the opposite color. Now, a new ball is selected from the urn which now contains $r+b+c+d$ balls, and replaced along with the additional $c+d$ balls, etc. We will be concerned with the Polya model where $c = 1$. We will draw from an urn containing x red and $n - x$ black balls. The drawing will be with replacement and an additional ball, identical in color to the selected ball, will be added. We will repeat this process for $m = N - n$ trials. Thus, the original size of the urn will be n, our sample size, while the final size of the urn will be $N = n + m$, the size of the finite population. In order to calculate the probability that we are left with $x + y$ red balls and $(n - x) + (m - y)$ black balls after the m Polya draws we start by calculating the probability of a sequence y reds followed by $m - y$ blacks. The probability of this sequence is $P(r \text{ on } 1^{\text{st}})\ P(r \text{ on } 2^{\text{nd}} \mid r \text{ on } 1^{\text{st}})\ \ P(r \text{ on } 3^{\text{rd}} \mid r \text{ on } 1^{\text{st}} \text{ two}) \ldots P(r \text{ on } y^{\text{th}} \mid r \text{ on } 1^{\text{st}} \{y-1\})$ $P(b \text{ on } \{y+1\}^{\text{st}} \mid r \text{ on } 1^{\text{st}}\ y)$, etc. But, there are $\binom{m}{y}$ posssible sequences. So, the probability of y red balls in m Polya draws is

$$\binom{m}{y} \frac{\dfrac{\Gamma(\text{total } r\text{'s})\,\Gamma(\text{total } b\text{'s})}{\Gamma(\text{final urn size})}}{\dfrac{\Gamma(\text{original } r\text{'s})\,\Gamma(\text{original } b\text{'s})}{\Gamma(\text{original urn size})}},$$

the Polya predictive distribution when the prior is Beta[0,0].

In the prior distribution of the error rate, we may provide the model with information. The prior information may be subjective, perhaps elicited from experts, or may be more objective, based on previously obtained data. This approach is used in the sequential models. Prior information and our belief about its strength is introduced to the model by specifying the parameters of the prior distribution. Here α and β are the parameters of the beta distribution. If prior knowledge about α and β is unavailable, our prior distribution will be uninformative. Uninformative priors will be weak, i.e. non committal about θ and easily "swamped" by the data introduced to the model though the likelihood. In the case of the beta distribution, the sum of the parameters α and β may be thought of as the "sample size" associated with the prior. A beta distribution with $\alpha = 1$ and $\beta = 1$ is a uniform distribution on the range [0,1], flat and intuitively uninformative, and having an effective sample size easily overwhelmed by the data in the likelihood. In a recent study on the behavior of various methods for obtaining interval estimates for the binomial parameter.

Brown, Cai, and DasGupta [4] showed that the beta-binomial model with the Beta [1,1] prior considerably out-performed more commonly used classical frequentist approaches. Beta [1/2,1/2] performed even better. Meeden and Varderman [1] proposed a simulation approach which can be shown to be equivalent to using a Beta [0,0] prior. A Beta [0,0] is not a prior probability density, but we find this relationship to simulation to be interesting.Suppose we specify the flat prior, Beta [1,1]. The posterior distribution of θ will be Beta $[x+1, n-x+1]$ and we use *Mathematica* to easily obtain a point and interval estimate for θ, e.g.

Mean[BetaDistribution[$1+x$, $1+n-x$]] and Quantile[BetaDistribution [$1+x$, $1+n-x$], 0.025], Quantile[BetaDistribution[$1+x$, $1+n-x$], 0.975]. *Table 1* below shows the point and interval estimates for the 2002 upstate New York Food Stamps audit where 44 of 389 contained errors and for a county where one of twelve contained errors. Alternative, we may construct an ordered list of randomly generated θ's, and find the appropriate statistics empirically.

In *Table 1*, for the State we find little difference between classical frequentist's techniques and Bayesian estimates using various proposed uninformative prior distributions for θ. This is due to the fact that with a sample of size 389, the asymptotic normality assumptions of the classical approach are appropriate, and the data are swamping the uninformative priors in the Bayesian method. However, for county A with a sample of 12, the lower confidence limit of classical symmetric confidence interval is less than zero, and we can expect that the upper confidence interval is also too low due the positive skew in the data. The Bayesian intervals are more realistic and according to Brown, Cai, and DasGupta [4] have superior coverage. The last row of *Table 1* shows the results of using an informative prior based on combined information from other counties. Notice that the raw proportion, 0.083 is shrunk toward the State proportion 0.11, while the precision has been measurably increased.

	State			CoA		
Method	Mean	LCL	UCL	Mean	LCL	UCL
Classical	0.113	0.082	0.145	0.083	−0.07	0.240
Beta[0, 0]	0.113	0.084	0.146	0.083	0.002	0.285
Beta[0.5, 0.5]	0.114	0.085	0.147	0.115	0.009	0.328
Beta[1, 1]	0.115	0.085	0.148	0.143	0.019	0.360
Beta[$k\ \theta_0$]				0.103	0.028	0.220

Table 1. Point and Interval Estimates for θ Using Five Methods

3.2 Multiple Centre Model and Sequential Multi Centre Model

When we have several locations within a single audit (e.g. counties within a state), data in some locations may be sparse or non existent. If location error rates are of interest, Laws and O'Hagan[1] describe how the posterior distribution for the collection of locations may be used as the prior for each location. As in the sequential model, this method involves increasing the variance (i.e. reducing the effective sample size) to specify the prior distribution. In *Table 1* we see how County A has borrowed information from the State to yield an error rate that has been shrunk toward the State rate, and thus we have traded bias for increased precision.

We may rely on results of the previous audit to provide a prior to the beta-binomial model. If we used the posterior distribution from the first audit as the prior for the second we would be assigning equal importance to the two audits, the posterior distribution would be Beta $[1+x_1+x_2,\ 1+(n_1-x_1)+(n_2-x_2)]$. If we believe that the probability distribution associated with the error rate may change over time, we have a dynamic system. Thus, our Bayesian model should allow for evolution and if the system is evolving, the information gathered about the system loses value with time. A dynamic model (see Pole and Harrison [5]) will reflect this dilution of information while letting the previous audit provide information through the prior. We reduced the strength of the prior by increasing its variance, i.e. reducing the effective sample size. For the beta-binomial model, we may proportionally down weight each parameter of the initial posterior distribution, thereby retaining the mean of that distribution while increasing the variance.

When multilocation audits are conducted periodically, we may establish the collective posterior distribution for each period and apply the multilocation model described by Laws and O'Hagan[1], who also show how to handle more than two error types and taints. This is accomplished by successive updating and projection steps, based on Bayesian conjugate analysis and precision decay, respectively. The *Mathematica* platform faciltates the definition of functions to perform and combine these steps.

3.3 Models with More Than Two Finding Types and Taints

One is sometimes interested in more than the two findings, error or correct. In this case our likelihood will be multinomial. The Dirichlet distribution is the conjugate prior for the multinomial likelihood. So if we adopt the Dirichlet distribution as the form of our prior, we have the posterior distribution of $\theta=\{\theta_1,\theta_2,\ \ldots\ ,\theta_J\}$, $\boldsymbol{\theta}_t\,|\,\boldsymbol{\theta}_{t-1}\sim\mathrm{Di}[k_{t-1}\,\boldsymbol{m}_{t-1}]=\mathrm{Di}[\boldsymbol{a}_t]$. $m_{tj}=E[\theta_{tj}]=a_{tj}/\sum_j a_{tj}$, which may also be obtained by simulation. For functions of the θ_{tj}'s we will use simulation and construct a function dirichletDraw[ ] for sampling from a Dirichlet distribution. Draws from the multinomial are handled with standard *Mathematica* functions. Frequently, error rates are defined to be the proportion total payments in error, e.g. pound in error/ pounds paid. This is incorporated into the models by defining taint to be the ratio of error payments to total payments. In each category of error, we observe payments and errors and make draws from either an empirical distribution of payments and errors in each error type, or from some fitted model of their relationship.

4 Conclusions and Functions

Bayesian models and *Mathematica*'s environment for implementing them provide statisticians with powerful tools for analyzing data resulting from statistical audits. These techniques have been used by insurers, national and regional governmental agencies, and courts for in making decisions involing considerable sums. Many new models are being developed and tested in all aspects of applied statistics including statistical auditing. Functions involving the Polya and Dirichlet are available by e-mailing kwheiner@aol.com.

5 References

[1] Meeden, G. and Varderman, S. A noninformative bayesian approach to interval estimation in finite population sampling, in JASA, 1991, American Statistical Assn.

[2] Laws, D. J. and O'Hagan, A. A hierarchical Bayes model for multilocation auditing, in JRSS(D), 2002, Blackwell.

[3] Feller, W. An Introduction to Probability and Its Applications, 1968, Wiley.

[4] Brown, L.D., Cai, T.T. and DasGupta, A. Interval estimation for a binomial proportion, in Statistical Science, 2001, Institute of Mathematical Statistics.
Brown, Cai, and DasGupta

[5] Pole, A., West, M., and Harrison, J. Applied Bayesian Forcasting and Time Series Analysis, 1994, Chapman & Hall.

Symbolic and symbolic-numeric calculations in applied mathematics, mechanics and ecology

I.E.Poloskov
Perm State University
Igor.Poloskov@psu.ru

1 Abstract

It is known that a lot of mathematical calculations must be processed during the practical use of exact and approximate methods that are designed to analyze different deterministic and random phenomena in dynamical systems. The present paper describes some schemes for such analysis, each requiring an application of the computer algebra package *Mathematica* for long-term symbolic and symbolic-numeric computations. Here we consider tasks mainly connected with the stochastic analysis of various effects such as: the estimation of sensitivity; water pollution; the calculation of some numeric characteristics of multi-dimensional random vectors; and tasks concerned with finding stochastic potentials.

2 Introduction

We first introduce some notation. We usually assume that a random n-dimensional phase vector $x(t)$ satisfies a system of stochastic differential equations (SD Eqs.) in the Stratonovich sense [1], [2]

$$\dot{x}(t) = f(x,t) + G(x,t)\,\xi(t),$$

where ξ is a m-dimensional vector of independent Gaussian white noises with unit intensities; f is a n-dimensional deterministic vector; and G is a non-random $n{\times}m$-matrix. The probability density function $p(x,t)$, the main characteristic of the vector x, satisfies the Fokker-Planck-Kolmogorov equation (FPK Eq.)

$$\frac{\partial p}{\partial t} = \mathcal{L}\,p, \quad p(x, t_0) = p_0(x),$$

where the operator $\mathcal{L}$ has the form

$$\mathcal{L}\,u = -\sum_{i=1}^{n} \frac{\partial}{\partial x_i}\,[a_i(x, t)\,u] + \frac{1}{2}\sum_{i,j=1}^{n} \frac{\partial^2}{\partial x_i\,\partial x_j}[b_{ij}(x, t)\,u],$$

$$a_i(x,t) = f_i(x,t) + \tfrac{1}{2}\sum_{j=1}^{n}\sum_{k=1}^{m} \frac{\partial g_{ik}(x,t)}{\partial x_j}\, g_{jk}(x,t),$$
$$B(x,t) = G(x,t)\, G^{\mathrm{T}}(x,t).$$

3 Package ProbRel

Very often the application of probability methods demands the use of numerical characteristics of random variables such as the cumulants (semi-invariants) λ_α, the quasi-moments, the initial moments $m_\alpha = \mathbf{M}[x^\alpha]$, and the central moments $\mu_\alpha = \mathbf{M}[(x-m)^\alpha]$. Among the two last-named, we further distinguish the first moments such as the mean values $m_i \equiv m_{e_i} = \mathbf{M}[x_i]$, the variances $\sigma_i^2 \equiv \mu_{2e_i} = \mathbf{M}[(x_i - m_i)^2]$, the mixed initial moments of second order $m_{ij} \equiv m_{e_{ij}} = \mathbf{M}[x_i\, x_j]$, and the covariances $\mu_{ij} \equiv \mu_{e_{ij}} = \mathbf{M}[(x_i - m_i)(x_j - m_j)]$. Here $\mathbf{M}$ is a symbol of mathematical expectation, α is a multi-index, $e_i = \{\delta_{ij}\}$, $j = 1, 2, ..., n$, $e_{ij} = e_i + e_j$, and δ_{ij} is a Kronecker delta. For example, some of the characteristics, mentioned above, are used to exclude infinite sets of ordinary differential equations (ODEs). for initial moments during an exploration of nonlinear dynamical systems affected by random fluctuations (here, the quasi-Gaussian or cumulant closure is exploited).

To use the relations derived in [3], a *Mathematica* [4] code package *ProbRels* has been implemented. The aims of this package include the following: (i) to calculate indefinite integrals, where the integrands are products of multi-dimensional Gaussian probability densities and polynomials, in symbolic mode by computer; (ii) to derive relations for the quasi-Gaussian and cumulant closure. Below we demonstrate the use of the package, in the form of calls to package procedures and the results of calculations.

Example 4.1. If we need to find the mean value of the expression $3\, x_1{}^2 x_2 + 4\, x_1 x_2 x_3 - x_1{}^5 x_3$, where $\{x_1, x_2, x_3\}$ is a vector with a Gaussian distribution, then we may call procedure *IntGauss* as follows:

```
IntGauss[ {x1, x2, x3}, 3 x1^2 x2 + 4 x1 x2 x3 - x1^5 x3,
   {m1, m2, m3},
  {{μ11, μ12, μ13}, {μ21, μ22, μ23}, {μ31, μ32, μ33}}]
```

where the third parameter of the procedure is the vector of mean values and the fourth one is the matrix of covariances. The result is

$$3\, m_1^2\, m_2 - m_1^5\, m_3 + 4\, m_1\, m_2\, m_3 + 3\, m_2\, \mu_{11} -$$
$$10\, m_1^3\, m_3\, \mu_{11} - 15\, m_1\, m_3\, \mu_{11}^2 + 6\, m_1\, \mu_{12} + 4\, m_3\, \mu_{12} -$$
$$5\, m_1^4\, \mu_{13} + 4\, m_2\, \mu_{13} - 30\, m_1^2\, \mu_{11}\, \mu_{13} - 15\, \mu_{11}^2\, \mu_{13} + 4\, m_1\, \mu_{23}$$

Note we have taken into account here that the matrix of covariances is a symmetric one.

Example 4.2. If we need to produce the Gaussian closure of the expression $3\, x_1{}^2 x_2 + 4\, x_1 x_2 x_3 - x_1{}^5 x_3$ to exclude moments with orders more than 3, where $\{x_1, x_2, x_3\}$ is a vector with the vector α of mean values and the matrix $\{\mu_{ij}\}$ of covariances, then we may call procedure *QuasiGauss* as follows:

```
QuasiGauss[ {x1, x2, x3}, 3 x1^2 x2 + 4 x1 x2 x3 - x1^5 x3, {m1, m2, m3},
  {{μ11, μ12, μ13}, {μ21, μ22, μ23}, {μ31, μ32, μ33}}, 3, "α"]
```

The result is

$$-2\, m_1^3\, m_3\, \mu_{11} - 8\, m_1\, m_3\, \mu_{11}^2 - m_1^4\, \mu_{13} - 10\, m_1^2\, \mu_{11}\, \mu_{13} - 15\, \mu_{11}^2\, \mu_{13} + 4\, \alpha_{1,1,1} - m_1^3\, \alpha_{2,0,1} - 7\, m_1\, \mu_{11}\, \alpha_{2,0,1} + 3\, \alpha_{2,1,0} - 2\, m_1\, \mu_{13}\, \alpha_{3,0,0}$$

Example 4.3. If we need to produce the cumulant closure of the expression $3\, x_1{}^2 x_2 + 4\, x_1 x_2 x_3 - x_1{}^5 x_3$ to exclude moments with orders more than 3, where $\{x_1, x_2, x_3\}$ is a vector with the vector α of mean values and the matrix $\{\mu_{ij}\}$ of covariances, then we may call procedure *CumulClosure* as follows:

```
CumulClosure[{x1, x2, x3}, 3 x1^2 x2 + 4 x1 x2 x3 - x1^5 x3, 3, "α", "λ"]
```

The result is

$$24\, \alpha_{0,0,1}\, \alpha_{1,0,0}^5 - 30\, \alpha_{1,0,0}^4\, \alpha_{1,0,1} + 4\, \alpha_{1,1,1} - 60\, \alpha_{0,0,1}\, \alpha_{1,0,0}^3\, \alpha_{2,0,0} + 60\, \alpha_{1,0,0}^2\, \alpha_{1,0,1}\, \alpha_{2,0,0} + 30\, \alpha_{0,0,1}\, \alpha_{1,0,0}\, \alpha_{2,0,0}^2 - 15\, \alpha_{1,0,1}\, \alpha_{2,0,0}^2 + 3\, \alpha_{2,1,0} - 10\, \alpha_{2,0,1}\, \alpha_{3,0,0}$$

Example 4.4. If we need to express moments of a random vector in terms of cumulants, then we write

```
ColumnForm[MomCum[3 , 3, "α", "λ"] ]
```

Here the first number is a dimension of this random vector and the second value is the major order of moments considered. A part of the result is the following

$$\alpha_{1,1,1} == \lambda_{0,0,1}\, \lambda_{0,1,0}\, \lambda_{1,0,0} + \lambda_{0,1,1}\, \lambda_{1,0,0} + \lambda_{0,1,0}\, \lambda_{1,0,1} + (\lambda_{0,0,1}\, \lambda_{1,1,0} + \lambda_{1,1,1})$$

$$\alpha_{2,0,1} == \lambda_{0,0,1}\, \lambda_{1,0,0}^2 + 2\, \lambda_{1,0,0}\, \lambda_{1,0,1} + \lambda_{0,0,1}\, \lambda_{2,0,0} + \lambda_{2,0,1}$$

$$\alpha_{2,1,0} == \lambda_{0,1,0}\, \lambda_{1,0,0}^2 + 2\, \lambda_{1,0,0}\, \lambda_{1,1,0} + \lambda_{0,1,0}\, \lambda_{2,0,0} + \lambda_{2,1,0}$$

$$\alpha_{3,0,0} == \lambda_{1,0,0}^3 + 3\, \lambda_{1,0,0}\, \lambda_{2,0,0} + \lambda_{3,0,0}$$

Using procedure *CumMom*, we can express cumulants of a random vector in terms of moments in the same way.

4 Sensitivity of the probability density function

Let us consider the stochastic differential equation

$$\dot{x} = \nu\, x + \beta_1\, \xi, \quad x(0) = x_o.$$

Here x_o has the Gaussian distribution $N(x;\beta_2,\beta_3^2)$ with the mean value β_2 and the variance β_3^2; the random parameter v is uniformly distributed on the segment $[\beta_4, \beta_5]$. Suppose x_o and v be independent random variables. The task is to estimate the sensitivity of the probability density function $p(x,t;\beta)$ to perturbation of the parameters β_i, $i = 1, 2, ..., 5$ in a neighbourhood of the vector $\beta_o = \{0,1;5;1;1;2\}^T$.

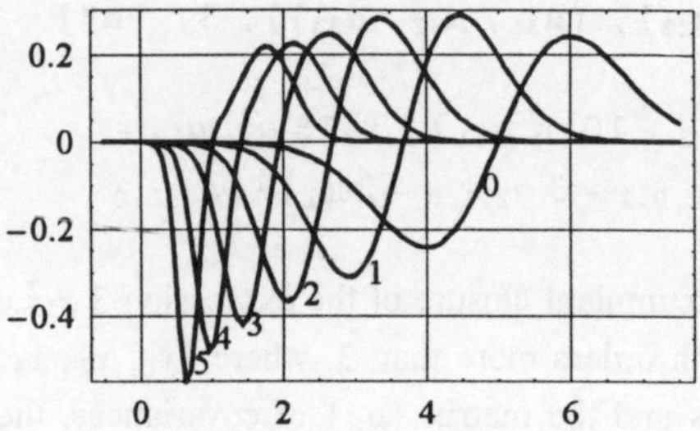

Figure 1

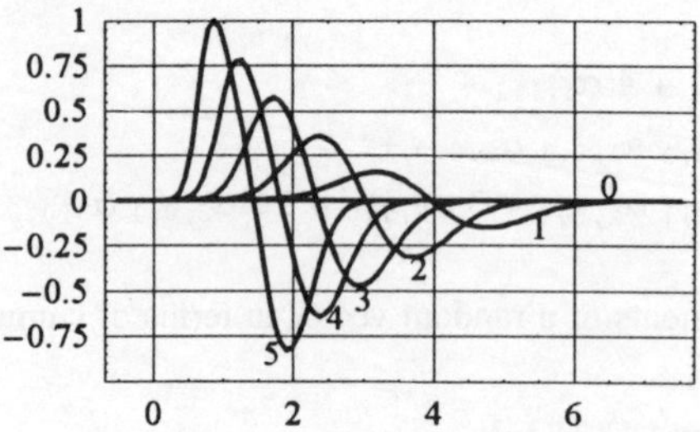

Figure 2

To solve this problem, let us take into account the linearity of the last equation with respect to x and write the FPK Equation satsified by the conditional probability density $\pi = \pi(x,t|v;\beta)$:

$$\frac{\partial \pi}{\partial t} = \frac{\beta_1^2}{2}\frac{\partial^2 \pi}{\partial x^2} + \frac{\partial (v x \pi)}{\partial x}, \qquad \pi(x,0|v;\beta) = p_0(x;\beta).$$

This density can easily be calculated as $\pi = N(x;m,D)$, where the functions m and D satisfy the following ODEs:

$$\dot{m} = -v\,m, \quad m(0) = \beta_2, \quad \dot{D} = -2\,v\,D + \beta_1^2, \quad D(0) = \beta_3^2.$$

The solutions of these equations look like

$$m = \beta_2 e^{-vt}, \quad D = (\beta_3^2 - \beta_1^2/2v)\,e^{-2vt} + \beta_1^2/2v.$$

Then the probability density function $p(x, t; \beta)$ takes the form:

$$p(x,t;\beta) = \frac{1}{\beta_5 - \beta_4}\int_{\beta_4}^{\beta_5} N(x;m,D)\,dv$$

and sensitivity functions $h_q(x, t)$ can be defined as follows:

$$h_q(x, t) = \frac{\partial p(x,t;\beta)}{\partial \beta_q}\Big|_{\beta=\beta_0}, \quad q = 1, 2, ..., 5.$$

These functions were tabulated with the help of a *Mathematica* code program. Graphs of the functions $h_q(x, t)$, $q = 2, 4$, are shown on Fig.1 and 2, where the digits 0,1,...,5 mark profiles of $h_q(x, t)$ at different time values.

5 Monitoring of water pollution

Below we consider a task concerned with environmental problems caused by the increase of pollutant loads discharged into natural water bodies. Here a water pollution in a river is modelled by the system of linear stochastic differential-difference Eqs. with multiple delays in the form [5]:

$$\begin{pmatrix} \dot{y}_k(t) \\ \dot{q}_k(t) \end{pmatrix} = \begin{pmatrix} -1.32 & 0 \\ -0.32 & -1.29 \end{pmatrix}\begin{pmatrix} y_k(t) \\ q_k(t) \end{pmatrix} + \begin{pmatrix} 0.1 \\ 0 \end{pmatrix} r_k + \begin{pmatrix} 0.9\, y^*_{k-1}(t) + 4.19 \\ 0.9\, q^*_{k-1}(t) + 1.9 \end{pmatrix} +$$

$$\begin{pmatrix} g_k \\ 0 \end{pmatrix} \xi_k(t),$$

where $k = 1, 2, 3$;

$$y^*_{k-1}(t) = \begin{cases} y^*_0, & k = 1, \\ a_1\, y_{k-1}(t) + a_2\, y_{k-1}(t-\tau) + a_3\, y_{k-1}(t-2\tau), & k > 1; \end{cases}$$

$$q^*_{k-1}(t) = \begin{cases} q^*_0, & k = 1, \\ a_1\, q_{k-1}(t) + a_2\, q_{k-1}(t-\tau) + a_3\, q_{k-1}(t-2\tau), & k > 1; \end{cases}$$

$$\tau = 0.5; \quad y^*_0 = 0; \quad q^*_0 = 10;$$

$$a_1 = 0.15; \quad a_2 = 0.70; \quad a_3 = 0.15;$$

$$r_k, g_k = \text{const}; \quad z = \{y_1, q_1, y_2, q_2, y_3, q_3\}^{\mathrm{T}}.$$

In this model the river is considered as a sequence of three bodies. Water of the river flows through all bodies and changes a level of its quality estimated by the vector z.

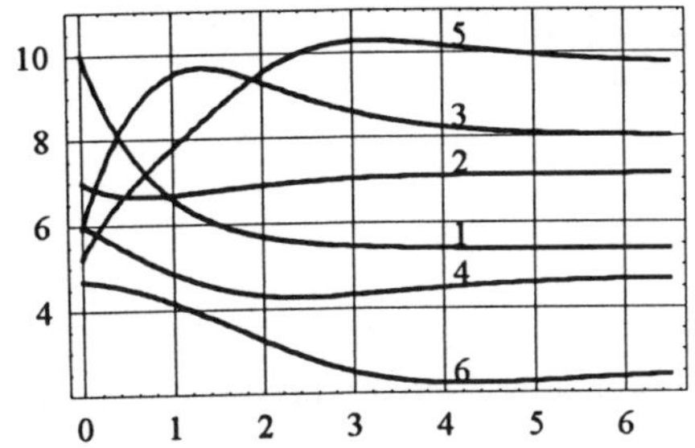

Figure 3

To study a random regime of this change, we used a modification of our method of a phase space extension [6]. The aim of our calculations was to compute the first moments, i.e. the vector of mean values and the matrix of covariances, for the vector z with a Gaussian distribution. To achieve the result, we have realized a new multi-step algorithm with the help of *Mathematica*. This algorithm is based on the classical method of steps for deterministic ODEs. with one delay. The following actions are produced on each step of the algorithm: (i) extension of the previous phase vector; (ii) building a list of new first moments, which correspond to the present phase vector; (iii) symbolic derivation of an ODE system for the vector of mean values and the matrix of covariances; (iv) preparation of initial conditions; (v) numerical integration of the system; (vi) remembering of results of computations at the end of step for further calculations and for a final visualization. We show graphs of the mean values for z_i, i = 1, 2, ..., 6 in Fig.3; graphs of σ_i^2, i = 1, 3, 5 in Fig.4; and graphs of σ_i^2, i = 2, 4, 6 in Fig.5.

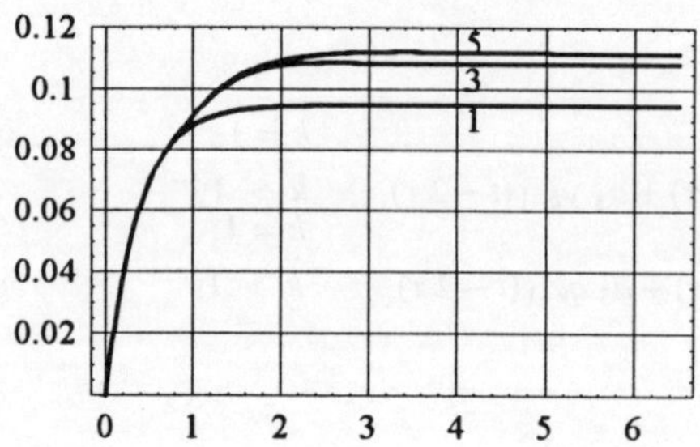

Figure 4

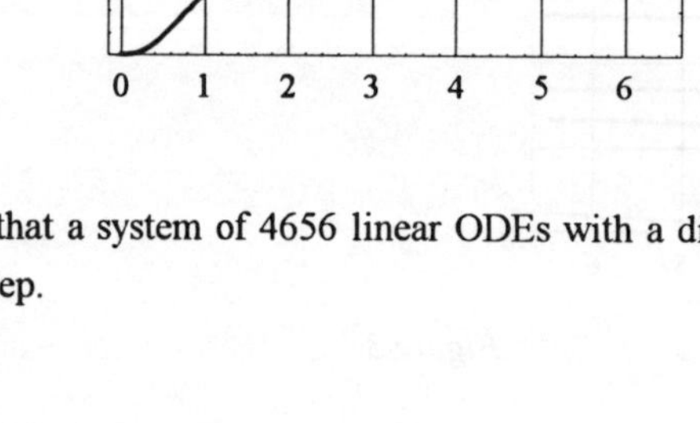

Figure 5

Note that a system of 4656 linear ODEs with a discharged matrix was integrated at the last step.

6 Polynomial Stochastic Potentials

Stochastic systems with polynomial coefficients arise in a lot of problems. A calculation of the steady-state probability density function $p_s(x)$ for the system of such type can be reduced to determination of the steady-state stochastic potential $\phi(x) \equiv -\ln p_s(x)$. This potential satisfies the modified stationary FPK Eq.[7]:

$$\frac{1}{2}\sum_{i,j=1}^{n} \mathbb{A}_{ij}\left(\frac{\partial^2 \phi}{\partial x_i \partial x_j} - \frac{\partial \phi}{\partial x_i}\frac{\partial \phi}{\partial x_j}\right) + \sum_{i=1}^{n} \mathbb{B}_i \frac{\partial \phi}{\partial x_i} - \mathbb{C} = 0,$$

where

$$\mathbb{A}_{ij} = b_{ij}, \quad \mathbb{B}_i = -a_i + \sum_{j=1}^{n} \frac{\partial b_{ij}}{\partial x_j},$$

$$\mathbb{C} = \sum_{i=1}^{n} \frac{\partial}{\partial x_i}\left(-a_i + \frac{1}{2}\sum_{j=1}^{n} \frac{\partial b_{ij}}{\partial x_j}\right).$$

In a number of cases the function $\phi(x)$ can be represented by a polynomial in the variables $x_1, x_2, \ldots, x_n$. Usually it is very hard to obtain an explicit form of such a polynomial because the exact solution of a large system of quadratic algebraic equations is necessary.

In [7] we built a relation connecting the degrees N_s of the function $\phi(x)$ in x_s and the degrees of the modified FPK Eq. coefficients. This relation has the form:

$$2M_s^* + N_s = K_s^* + 1, \quad s = 1, 2, \ldots, n,$$

and allows to close a formally constructed infinite system of algebraic equations satisfied by coefficients of Maclaurin-series expansion for the potential $\phi(x)$ and to simplify tasks under consideration. The relation can be used for solution of inverse problems too.

Example 6.1. Let us construct a stochastic system with the potential $\phi(x) = (x_1^2 + x_2^2)^2 - (x_1^2 + x_2^2)$ if there are only additive noises. Here $n = 2$, $N_1 = N_2 = 4$. Suppose $m = 2$, the matrix G is a diagonal one, $G = \mathrm{diag}\,(g_1, g_2)$. Then $M_1^* = M_2^* = 0$ and we obtain $K_1^* = K_2^* = 3$ from the relation, i.e.,

$$f_i = \sum_{p=0}^{3} \sum_{q=0}^{3-p} f_{ipq}\, x_1^p x_2^q, \quad i = 1, 2.$$

The result of calculations with the help of a *Mathematica* code program is as follows:

$$f_{100} = f_{120} = f_{200} = f_{220} = f_{102} = f_{202} = f_{111} = f_{211} = 0,$$
$$f_{130} = f_{112} = -2g_1^2, \quad f_{203} = f_{221} = -2g_2^2, \quad f_{110} = g_1^2, \quad f_{201} = g_2^2,$$
$$f_{101} + f_{210} = 0, \quad f_{103} + f_{230} = 0, \quad f_{121} + f_{230} = 0, \quad f_{212} - f_{230} = 0.$$

We see that some coefficients are indefinite. Let C_1 and C_2 be arbitrary constant values. Then the functions f_i take the form

$$f_1 = g_1^2 x_1 - C_1 x_2 - 2\, g_1^2 x_1 (x_1^2 + x_2^2) + C_2\, (x_1^2 x_2 + x_2^3),$$
$$f_2 = C_1 x_1 + g_2^2 x_2 - 2\, g_2^2 x_2 (x_1^2 + x_2^2) - C_2\, (x_1^3 + x_1 x_2^2).$$

7 Acknowledgements

This work was partly funded by the RFBR (grants "Ural-2002" No. 02-01-96406 and No. 02-01-96409) and by the Ministry of education (No. §¦02-1.0-151).

8 References

[1] Gardiner, C.W. Handbook of Stochastic Methods for Phisics, Chemistry and the Natural Sciences. 2nd edn., 1985, Springer-Verlag, Berlin Heidelberg New York.

[2] Malanin, V.V. and Poloskov, I.E. Random Processes in Nonlinear Dynamic Systems. Analytical and Numerical Methods of Analysis, 2001, Regular and Chaotic Dynamics, Ijevsk (in Russian).

[3] Poloskov, I.E. Package ProbRel for Symbolic Probabilistic Calculations, in Bulletin of Perm University. Information Systems and Technologies, 2002 (be printed in Russian).

[4] Wolfram, S. The Mathematica Book, 4th edn., 1999, University Press, Cambridge.

[5] Tamura, H. A Discrete Dynamical Model with Distributed Transport Delays and its Hierarhical Optimisation to Preserve Stream Quality, in IEEE Trans., SMC, 1974, V.4, 424-429.

[6] Poloskov, I.E. Extension of Phase Space in Analysis of Differential-Difference Systems Effected by Random Fluctuations, in Automatica & Telemechanica, 2002, N 9, 58-73 (in Russian).

[7] Poloskov, I.E. Method of Semi-Inverse Task in Analysis of Some Stochastic Systems, in Problems of Mechanics and Control, Nonlinear Dynamical Systems, 2001, 113-125, Perm State University, Perm (in Russian).

Derivative and Integration on Time Scale with *Mathematica*

Ahmet Yantir
Department of Mathematics, Izmir Institute of Mathematics, Turkey
ahmetyantir@iyte.edu.tr

1 Abstract

Mathematical modelling of time dependent systems is always interesting for applied mathematicians. First continuous and then discrete mathematical models were built in the mathematical development from ancient to modern times. With the discovery of time scale, the problem of irregular systems was solved in the 1990s. In this paper we explain the derivative and integral of functions of time scales and the solution of some basic calculus problems using *Mathematica*.

2 Introduction

In section 2, we define time scale and the forward and backward jump operators which are used in the definition of time scale and derivative. We also give *Mathematica* commands which find the values of forward operators at given points. In section 3, we give the definition and some important theorems on the time scale derivative. In this paper we deal with only the forward difference derivative on time scales. We give *Mathematica* commands in two cases: first for given time scales which are the combination of discrete and continuous sets, and then for discrete cases with a rule given by the user. In section 4, we present the integration on time scales with *Mathematica* in two different Modules, one for purely discrete cases and one for a combination of discrete and continuous cases.

3 What is Time Scale?

Let T be any closed subset of the real numbers, R. T is called a time scale.

Example 3.1

The real numbers, the integers, the natural numbers, and nonnegative integers are examples of time scales, as are

```
In[1]:= T1 = 0 ≤ x ≤ 3 || x == 4 || x == 9;
        T2 = Table[2^n, {n, 0, 10}]
```

The rationals, the irrationals, complex numbers, and the open intervals of real numbers are not time scales.

In order to define the derivative on time scale we need to define backward and forward jump operators which are used in the definition of time scale derivative. [2]

Definition 3.1

Let T be a time scale. For $t \in T$; $t <$ MaxT; where MaxT is a real number, we define the forward jump operator σ: $T \rightarrow T$ by $\sigma(t)=\text{Inf}\{ s: s\in t; s > t \}$.

We define this operator in *Mathematica* as follows [2,3,4]:

```
In[3]:= << Algebra`InequalitySolve`
        Clear[T1];
        inf[exp_] :=
          Min[Select[Level[exp, 2], Element[#, Reals] &]];
        T1 = 5 <= x <= 10 || x == 12 || x == 18;
        sigma[t_] := inf[InequalitySolve[T1 && (x > t), x]];
        sigma[5]; sigma[10]
```

If T = R, then $\sigma(t)=t$. If T = Z (the integers), then $\sigma(t)=t+1$. Also $\sigma(\text{MaxT}) = \text{MaxT}$.

Definition 3.2

t is called a right-dense point if $\sigma(t)=t$, and right-scattered if $\sigma(t)>t$.

Definition 3.3

Let T be a time scale. For $t \in T$; $t >$ MinT; we define backward jump operator ϱ: $T \rightarrow T$ by $\varrho(t)=\text{Sup}\{ s: s\in t; s < t \}$.

t is called a left-dense point if $\varrho(t)=t$, and left-scattered if $\varrho(t)<t$.

4 Derivative on Time Scale

In order to define the derivative on time scales we need the set below which is derived from the time scale T as follows:

Definition 4.1

$$T^k = \begin{cases} T - \text{MaxT} & \text{if MaxT} < \infty \text{ and MaxT is right-scattered} \\ T & \text{otherwise} \end{cases}$$

T^k is called the region of differentiability.

Definition 4.2

Let f :T→R be a function and t ∈ T^k. If there exist a neighborhood U_t such that

| f(σ(t)) - f(s) - a [σ(t) - s]| ≤ ϵ |σ(t) - t|

is satisfied for all s ∈ U_t, ϵ > 0 then f is Δ differentiable at the point t. a is called Δ derivative of f at the point t.

a=$f^{\Delta}(t) = \lim_{s \to t} \frac{f(\sigma(t)) - f(s)}{\sigma(t) - s}$.

We give an example of the Δ derivative in *Mathematica* as follows:

Example 4.1

Let f(x)=x^2 and g(x)=$x^3 + x^2$+1. Let T4=[5,10]∪{12,18}∪[20,22]. Find $\left(\frac{f}{g}\right)^{\Delta}(10)$.

Solution:

By definition 4.2

```
In[9]:= Clear[f, g, x, s, T1];
        T1 = 5 <= x <= 10 || x == 18 || 20 <= x <= 22;
        f[x_] = x^2; g[x_] = x^3 + 1 / x;
        sigma[t_] := inf[InequalitySolve[T1 && (x > t), x]];
        s = sigma[10];
        Limit[(f[s] / g[s] - f[t] / g[t]) / (s - t), t -> 10] // N
```

So we get $\left(\frac{f}{g}\right)^{\Delta}$(10)=-0.00555437 [2,3,4]

Theorem 4.1

Let f, g : T → R be functions and t ∈ T^k. If f and g have a Δ derivative at t then
1) f g has a Δ derivative at t and

$$(f\,g)^{\Delta}(t) = f^{\Delta}(t)\,g(t) + f(\sigma(t))\,g^{\Delta}(t) = f(t)\,g^{\Delta}(t) + f^{\Delta}(t)\,g(\sigma(t))$$

2) If g(t) g(σ(t)) ≠0 then $\frac{f}{g}$ has a Δ derivative at t and

$$\left(\frac{f}{g}\right)^{\Delta}(t) = \frac{f^{\Delta}(t)\,g(t) - f(t)\,g^{\Delta}(t)}{g(t)\,g(\sigma(t))}.$$

Example 4.2

[1] has a 'paper' proof of this theorem. We found the value of the derivative of f/g in example 4.1 by using definition 4.2. We can use Theorem 4.1 to compute the derivative of f/g [2,3,4]:

```
In[15]:= fd = Limit[(f[s] - f[t]) / (s - t), t -> 10];
         gd = Limit[(g[s] - g[t]) / (s - t), t -> 10];
         result = (fd * g[10] - gd * f[10]) / (g[s] * g[10]) // N
```

We define the derivative for purely discrete time scales in *Mathematica* as follows [3,4]:

```
In[18]:= Clear[f, g, t, T, s, i]
         Tderivative[f_, t_, i_] :=
           Module[{T, n, s, point1, lim1},
            T = Table[i, {a, 10}];
            n = Position[T, t]; n = Flatten[n];
            point1 = T[[n + 1]];
            lim1 = Limit[(f /. x -> T[[n + 1]]) -
                (f /. x -> s) / (T[[n + 1]] - s), s -> t];
            Print["The function is= ", f,
             "\n\nThe time scale is=", Table[i, {a, 10}],
             "\n\nThe point is= ", t];
            Print["The next point is=",
             point1, "\n\nThe derivative is=", lim1]];
         Tderivative[x^2 + 1 / x, 4, a / 2]
```

5 Integration

Definition 5.1

A function f: T→R is called regulated if it has finite right-sided limits at all right-dense points in T, and has finite left-sided limits at all left-dense points in T. A function f:T→R is called rd-continuous if it is continuous at all right-dense points in T and its left-sided limits exist (and is finite) at all left-dense points in T. The set of rd-continuous functions in T is denoted by C_{rd}.

If $F^{\Delta}(t)$=f(t) is satisfied for all t∈T^k, then F is called the anti-derivative of f.

Theorem 5.1

Let a, b ∈ T and f ∈ C_{rd}. If [a,b] consists of only isolated points, then

$$\int_a^b f(t)\,\Delta t = \begin{cases} \sum_{t\in[a,b)} (\sigma(t)-t)\,f(t) & \text{if } a<b \\ 0 & \text{if } a=b \\ -\sum_{t\in[b,a)} (\sigma(t)-t)\,f(t) & \text{if } a>b \end{cases}$$

In particular $\int_t^{\sigma(t)} f(\tau)\,\Delta\tau = (\sigma(t)-t)\,f(t)$

See [1] for a 'paper' proof of this theorem. We define the definite integral for purely discrete time scales in *Mathematica* as follows [3,4]:

```
In[21]:= Clear[f, g, l, u];
         Tintegrate1[f_, g_, l_, u_] :=
          Module[{T, s}, T = Table[f, {i, l, u}];
             Print["Time Scale is= ", T];
             Print["The function is= ", g];
             s = Sum[(g /. x -> T[[i]]) * (T[[i + 1]] - T[[i]]),
             {i, 1, u - 1}];
             Print["The value of Integral= ", s]]
         Tintegrate1[i / 2, x^3, 1, 10]
```

Here f denotes the discrete time scale rule, g denotes the integrand, l denotes the lower bound of the integral, u denotes the upper bound of the integral.

If T is a union of continuous and discrete sets we use another Module to solve the problem.

Example 5.1

Let T =[1,3]⋃{5}⋃[6,8]⋃{12,13} and f(t)=t^3 . Evaluate $\int_2^{12} f(t)\,\Delta t$.

Solution:

We can solve this problem with *Mathematica* as follows [3,4]:

```
In[24]:= A = Interval[{1, 3}, {5, 5}, {6, 8}, {12, 13}];
         Tintegrate2[f_, a_, b_] :=
          Module[{B, i, j, Sum1, Sum2},
           B = IntervalIntersection[A,
              Interval[{a * 1., b * 1.}]]; i = 1;
           Print["The time scale is = ", A];
           Print["The Integral Interval=", B];
           While[B[[i]][[2]] ≠ Max[B], i = i + 1];
            Sum1 = 0; Sum2 = 0; j = 1;
           While[j ≤ i,
           Sum1 = Sum1 +
               Integrate[f, {x, B[[j]][[1]], B[[j]][[2]]}];
           Print["Integration from ", B[[j]][[1]],
              "  to ", B[[j]][[2]]]; j++;];
           Print["the value of integral=", Sum1]; j = 1;
           While[j < i,
           Sum2 = Sum2 + f /. x ->
               B[[j]][[2]] * (B[[j + 1]][[1]] - B[[j]][[2]]);
           Print["Discrete integral from=", B[[j]][[2]],
              " to ", B[[j + 1]][[1]]]; j++;];
           Print["The value of discrete integral=", Sum2];
           Print["The result=", Sum1 + Sum2 ];]
         Tintegrate2[x^2, 2, 6]
```

References

[1] M. Bohner-A. Peterson, Dynamic Equations on Time Scales, 2001, Birkhauser Boston.

[2] A. Yantir-U. Ufuktepe, Basic Calculus on Time Scale, 2003, Lecture Notes on Computer Science (will be published).

[3] J. W. Gray, Mastering *Mathematica,* 1997, Academic Press.

[4] S. Wolfram, The *Mathematica* Book, 1996, Cambridge University Press.

Image, video, and 3D visualization extensions in the Digital Image Processing application package.

A. Mariusz Jankowski
University of Southern Maine
mjkcc@usm.maine.edu

1 Abstract

This paper presents current efforts in extending the Digital Image Processing application package by linking it to external libraries using J/Link and Java. Native *Mathematica* code has been augmented with Java Advanced Imaging, a 2D image processing library, and Java Media Framework, a Java package for manipulation of time-based media such as audio and video. 3D image processing and visualization capabilities were also added to DIP by using the Visualization Toolkit, which is an open-source, object-oriented software system that supports 3D/2D graphics, visualization, image processing and volume rendering.

2 Introduction

Mathematica is a general-purpose scientific computing and visualization system. It has a flexible and powerful programming language and extensive functionality in all areas of symbolic and numeric computing. In recent years, with the addition of the Digital Image Processing (DIP) application package [1] and significant performance improvements to the *Mathematica* kernel, it has become an excellent platform for prototyping imaging solutions [2]. Additionally, the inclusion of a Java runtime and JLink has made it possible to quickly and conveniently add existing public-domain Java-based functionality to *Mathematica* and DIP. Examples of relevant Java libraries include the Java Advanced Imaging (JAI) and Java Media Framework (JMF) APIs from Sun Microsystems and The Visualization Toolkit (VTK) from Kitware Inc.

3 2D image processing with Java Advanced Imaging API

The Java Advanced Imaging API provides a set of object-oriented interfaces that support a simple, high-level programming model which allows developers to manipulate images easily [3]. Examples of image processing techniques available in JAI range from simple operations such as contrast enhancement, cropping, and scaling to more complex operations such as geometric warping and frequency domain processing. There are approximately 80 operators, many with native support for optimized performance. The fundamental image data object in Java, introduced in JDK 1.2, is defined by the `BufferedImage` class [4]. It plays the same role as the new `ImageData` object introduced in the *Digital Image Processing* package, encapsulating information about pixel values, data organization, and color spaces. Two-way translation between *Mathematica* and Java image objects for several Java data types, color spaces and compatible pixel-interleaving schemes has been implemented in a new BufferedImage.m package.

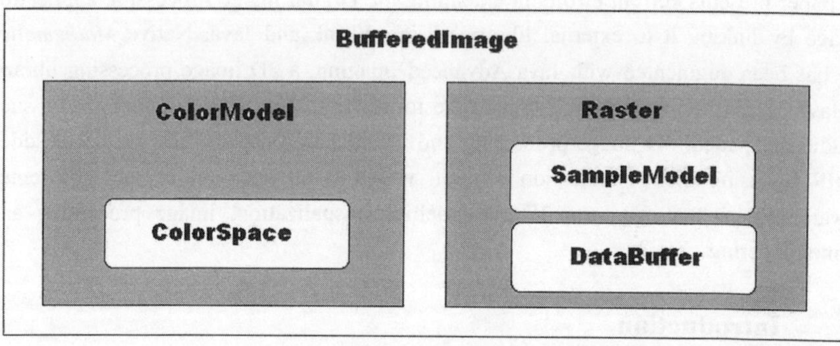

Figure 1. BufferedImage class hierarchy.

The `BufferedImage` class consists of two objects: `ColorModel` and `Raster`. The `Raster` class encapsulates raster image data as a rectangular grid of pixels. `Raster` is in turn made up of two objects, `SampleModel` and `DataBuffer`, that manage how the pixel data is organized in memory. The `DataBuffer` object contains the actual raw image data stored in a manner defined by `SampleModel`. The `ColorModel` class is needed to interpret how the raw pixel values are mapped to a specific color.

Here is a typical image processing example using JAI operators. I use the "convolve" operator to implement a smoothing operation using a so-called moving average filter. This loads the `ImageProcessing` and J/Link packages and launches the Java runtime.

```
In[1]:= << ImageProcessing`;
        << JLink`;
        InstallJava[];
```

Here, I read an example image and convert to a `BufferedImage` object.

```
In[4]:= img = ImageRead["cameraman.tif"];
        src = ImageDataToBufferedImage[img];
```

Here, I load the JAI interface, define a parameter block, and add the required operands. Most other JAI image operators may be implemented in a similar manner. Utility functions have been added to the BufferedImage.m package that simplify this process for the *Mathematica* user.

```
In[6]:= jai = JavaNew["javax.media.jai.JAI"];
        ker = JavaNew["javax.media.jai.KernelJAI",
           11, 11, Flatten[BoxFilter[11, 11]]];
        pb = JavaNew["java.awt.image.
             renderable.ParameterBlock"];
        pb@addSource[src];
        pb@add[ker];
        dst = jai@create["convolve", pb];
```

This translates the Java object `dst` back to a *Mathematica* `ImageData` expression and displays the result.

```
In[12]:= Show[GraphicsArray[{Graphics[img],
            Graphics[BufferedImageToImageData[dst]]}]];
```

With one exception (color representation using ICC profiles), the functionality of the JAI API is a subset of the *Digital Image Processing* package. Therefore, the primary reason for adding the JAI package to DIP would be to gain a performance advantage. The speed of a representative selection of image processing operators scripted in Java was measured and compared to the DIP implementations. This comparison reveals significant performance benefits from Java implementations of user-defined high-level *Mathematica* code. The opposite is true for kernel-based *Mathematica* operators.

Out[12]//TableForm=

Operation	Mathematica	JAI	JAI+
scalar times	0.0372	0.1	0.62
add	0.0264	0.06	0.58
3x3 convolve	0.312	0.12	0.64
3x3 median	2.353	0.5	1.02
dilate	8.172	0.28	0.8
rotate	8.75	0.58	1.1

Table 1. Timing comparison between Mathematica and Java implementations of selected image processing operators.

JAI+ timing data includes the time needed to transfer the image data from the *Mathematica* to the Java side and back. The data was measured using a 512×512 grayscale image on a 500 MHz Pentium III computer (all numbers in seconds).

4 Video processing with Java Media Framework API

The ever-increasing availability of digital video cameras is having a profound impact on experimental data gathering in many areas of research, in particular in the biological and medical sciences. The analysis of experimental data encoded in the form of a video stream, at the most basic level requires access the video on a frame-by-frame basis. It is therefore of interest to extend the *Digital Image Processing* package by adding functionality needed to decode individual frames of a digital video stream and to advance a frame pointer to any desired position in the video sequence.

A prototype `MReader`, using Java Media Framework is currently under development[5]. The Java™ Media Framework API (JMF) enables audio, video and other time-based media to be added to Java applications and applets [6]. This optional package, which can capture, playback, stream and transcode multiple media formats, extends the multimedia capabilities on the JSE platform, and gives multimedia developers a powerful toolkit to develop scalable, cross-platform technology. `MReader` is written in Java and accesses these capabilities directly from *Mathematica*, providing an easy interface that is capable of opening a number of different formats. Presently, Quicktime and AVI formats have the best support. Functionality should continue to expand as Sun continues to develop JMF.

Here is an example of a video frame grabbing operation using Java Media Framework. Each frame of a movie is translated to an `ImageData` object for further processing and analysis with *Mathematica*. `MReader` just provides a link between *Mathematica* and JMF. It can be scripted in *Mathematica* or coded directly in Java. The Java code is quite straightforward. Here, I instantiate a new reader and then open a movie file in Microsoft's avi or Quicktime formats.

```
In[13]:= movie = JavaNew["MReader"];
         movie@
           Mopen["file://c:/Documents and Settings/mariusz/
              my documents/ims2003/cinepak.mov"];
```

The following line gets the 5th frame of the movie and returns it as a `BufferedImage`.

```
In[15]:= img = movie@MgetFrame[5];
```

The data can be translated to a *Mathematica* expression for further processing. Here I simply display the frame using Show.

```
In[16]:= Show[Graphics[BufferedImageToImageData[img],
         ImageSize -> {160, 120}]];
```

5 3D data visualization using The Visualization ToolKit

The Visualization ToolKit (VTK) is an open source, freely available software system for 3D computer graphics, image processing and visualization used by thousands of researchers and developers around the world [7]. VTK consists of a C++ class library, and several interpreted interface layers including Tcl/Tk, Python, and most notably Java. VTK supports a wide variety of visualization algorithms including scalar, vector, tensor, texture, and volumetric methods; and advanced modeling techniques such as implicit modelling, polygon reduction, mesh smoothing, cutting, contouring, and Delaunay triangulation. In addition, dozens of imaging algorithms have been directly integrated to allow the user to mix 2D imaging / 3D graphics algorithms and data.

The VTK Java interface permits, as in the case of JAI and JMF libraries, immediate access to all functions and operators via JLink. Here is a simple example of rendering a cone object. I begin by creating instances of vtkPanel and vtkConeSource and set one of its properties.

```
In[17]:= JavaNew["vtkPanel"];
         cone = JavaNew["vtk.vtkConeSource"];
         cone@SetResolution[100];
```

Next, in a typical sequence I create an instances mapper (vtkPolyDataMapper), actor (vtkActor), and renderer (vtkRenderer). A renderer is like a viewport. It is part or all of a window on the screen and it is responsible for drawing the actors it has.

```
In[20]:= coneMapper = JavaNew["vtk.vtkPolyDataMapper"];
         coneMapper@SetInput[cone@GetOutput[]];
         coneActor = JavaNew["vtk.vtkActor"];
         coneActor@SetMapper[coneMapper];
         ren1 = JavaNew["vtk.vtkRenderer"];
         ren1@AddActor[coneActor];
```

Finally I create the render window which will show up on the screen and put our renderer into the render window using AddRenderer. I render the image and rotate the active camera to obtain a non-default view of the object.

```
In[26]:= renWin = JavaNew["vtk.vtkRenderWindow"];
         renWin@AddRenderer[ren1];
         ren1@GetActiveCamera[]@Azimuth[35];
         renWin@Render[];
```

This results in the pop-up window as shown in Figure 2 below.

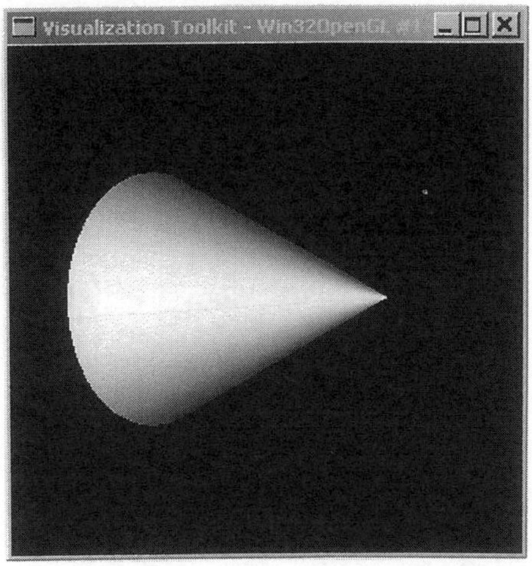

Figure 2. 3D cone object rendered to screen using VTK.

6 References

[1] http://www.wolfram.com/products/applications/digitalimage/

[2] Jankowski, M., "Comet Assay – an introduction to the Digital Image Processing application package," in Proc. 2nd European Medical & Biological Engineering Conference, Vienna, Austria, December 2002.

[3] http://java.sun.com/products/java-media/jai/

[4] Rodrigues, L., "Building imaging applications with Java technology : using AWT Imaging, Java 2D, and Java Advanced Imaging (JAI)," (Chapter 7, page 271), 2001, Addison-Wesley.

[5] Martell, J., "Movie Reader for *Mathematica,*" presented at *Mathematica* Developer Conference, April 2003, Champaign, IL, USA.

[6] http://java.sun.com/products/java-media/jmf/

[7] The Visualization Toolkit User's Guide, 2001, Kitware, Inc.

Volume rendering with MathGL3d

J.-P. Kuska
Department for Computer Graphics and Image Processing, University of Leipzig, Germany
kuska@informatik.uni-leipzig.de

1 Abstract

MathGL3d is an OpenGL based *Mathematica* application of advanced scientific visualization. The interactive visualization of scalar functions of three variables or of three dimensional volume data is a frequent task in many fields of research. Medical data from computer tomography (CT) or magnet resonance tompgaphie (MRT) are especially a real challenge for a visualization application. The paper will demonstrate the build-in functions of MathGl3d for volume data visualization.

2 Volume rendering with isosurfaces

The volume data passed to MathGL3d [1] via **`MVListContourPlot3D[]`** [2] , [3] , [4] as a 3d list of reals. This has the advantage that the data can preprocessed by Mathematica. One disadvantage of this method is, that the Mathematica kernel ***and*** MathGL3d each have to store the data in their own memories. Volume data can often consume large amounts of memory and for the pure visualization storing the data twice should be avoided. **`MVListContourPlot3D[]`** also accepts compressed volume data files as the first argument. I will use this notation here because the size of the volume data can range form 64 — 512 MByte and it is desirable to have only one copy of these data in the memory. The **`MVWriteVolumeData[]`** and **`MVReadVolumeData[]`** functions can be used to read and write the volume data from/into the *Mathematica* kernel. The C++ source code of these functions comes with MathGL3d and it should be easy to integrate direct creation of the volume data into other applications. The volume data format used an adaptive arithmetic coder [5] for lossless compression of volume data. Typically it obtain compression rations between 2 — 8.

One of the most common ways to visualize a function $f(x, y, z)$ with $\{f : \mathbb{R}^3 \mapsto \mathbb{R}\}$ is to draw one or more surfaces with $f(x, y, z) = c_i$. Here is an example of a CT scan of a lobster using three transparent surfaces:

```
Get["OpenGLViewer`"]

MVListContourPlot3D["lobster.v3d",
  Contours → {32., 96., 128.0},
  ContourStyle -> {RGBColor[1, 1, 1],
   RGBColor[1, 0, 0], RGBColor[0, 0, 1]},
  MVAlpha -> 0.5,
  ViewPoint -> {0.158293, -1.53617, 1.68329},
  ViewCenter -> {0.5, 0.5, 0.343},
  MVFlipFace → True, MVNewScene → True, Boxed → False,
  MVReturnValue -> None];
MVPasteGraphics[];
```

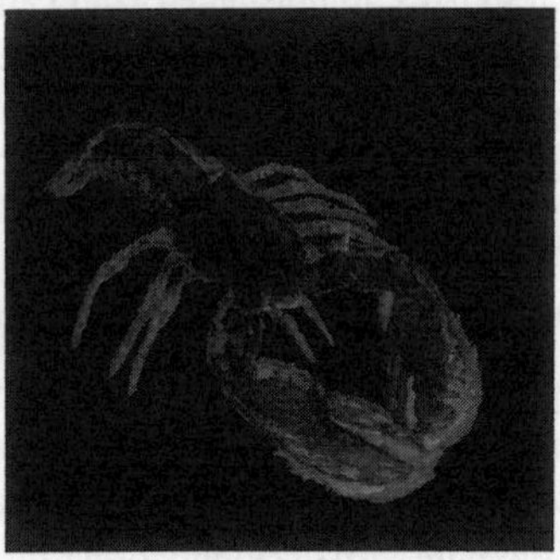

The lobster data set has 320×320×34 voxel. With the growing resolution of modern computer tomographs more slices with a resolution of 256 × 256 to 512 × 512 become available. Here is an example of a CT scan of a bonsai tree with 256×256×256 voxel. The surface is build up with 380490 polygons.

```
bonsaiGr =
  MVListContourPlot3D["bonsai.v3d", Contours → {32.},
   ViewPoint -> {1.43923, -2.65703, 0.76027},
   MVFlipFace → True, MVNewScene → True,
   Boxed → False, MVReturnValue -> None];
MVPasteGraphics[];
```

On an 1200 MHz Athlon *Mathematica* (without OpenGL) needs nearly an half hour to display the bonsai while MathGL3d need two minutes to compute the polygons of the surface and OpenGL (on a GeForce4 graphics card) needs an additional four seconds to show the surface. The timing in *Mathematica* makes it obvious that medical volume data sets with a resolution higher than 128^3 cannot be visualized with *Mathematica* alone in a reasonable time.

The number of polygons needed for the visualization of the isosurfaces in such volume data sets quickly exceeds the available memory. The polygon reduction implemented in MathGl3d works well for smooth surfaces but for complex surfaces of biological objects the effect of the polygon reduction will reduce the polygon count only by 10—20 %. There are many applications that only need visualization of the volume data set and not an explicit polygonal description of the surface. In these cases direct volume rendering may be used.

3 Direct volume rendering

The basic idea of direct volume rendering [6] is the assignment of a transparence value α for every function value $f(x, y, z)$. For medical data this is especially a very natural description, because for the x-rays of a computer tomography the human body is a semi-transparent object. The assignment of a simple function for $\alpha(f(x, y, z))$ produce images very similar to traditional x-ray images. But the 3d volume data set allows it to generate several different renderings without an additional radiation exposure to the patient. MathGL3d allow also the usage of color and transparence $\{r(f), g(f), b(f), \alpha(f)\}$ to emphasize different function values. Since *Mathematica* has no graphics primitive to specify the transparence, the **`GrayLevel[]`** function is used to define the opacity of the volume and the pair **`{RGBColor[r,g,b],GrayLevel[α]}`** is used to assign color and transparence. Here is a rendering of a CT scan of the human heart (courtesy of the

Department of Diagnostic Radiology, University Tübingen, Germany) with a resolution of $512 \times 512 \times 213$ voxel. A polygonal representation of a implicit surface would generate 1.7×10^6 polygons.

```
CardioColor[d_?NumericQ] :=
 Which[d < 0.2,   {d, d, d, 0.0},
       d < 0.25,  {2 d, 0, 0, 0.05},
       d < 0.29,  {2 d, 2 d, 0, 0.75},
       d < 0.325, {0, 0, 2 d, 1.0},
      True,       {1 - d, 1 - d, 1 - d, 0.0}] /.
   {r_, g_, b_, α_} :>
     {RGBColor[r, g, b], GrayLevel[α]}

MVListDensityPlot3D["smoothcardio_40.v3d",
  ColorFunction → CardioColor,
  MVTextureSize → {512, 512, 412},
  ViewPoint → {2.68309, -0.501043, 2.0},
  MVNewScene → True, MVGrayBackground → True];
```

For direct volume rendering, MathGl3d computes semi-transparent textures and maps the textures onto planes perpendicular to the view direction. When these surfaces are blended form back to front, one obtains an approximate solution of the radiation transport equation in a semi-transparent volume with a glowing material. The **MVTextureSize → {r** option specifies the resolution r_x and r_y of the textures used for blending and the number of textures in view direction n_z. Due to the limitations of the texture size in OpenGL, r_x and r_y must be powers of two. The textures should fit into the memory of the graphics card. For smooth functions a low texture resolution and a low number of planes in the view direction are sufficient. Volume data with large gradients need more planes in the view direction and higher resolved textures to give pleasant results.

While **`MVContourPlot3D[]`** returns polygons to *Mathematica* by default, there is no way to display several semi-transparent polygons with the help of the PostScript interpreter of *Mathematica* (and with PostScript in general). By default **`MVDensityPlot3D[]`** and **`MVListDensityPlot3D[]`** return the bitmap image rendered by OpenGL to *Mathematica*.

One problem with this simple form of volume rendering is that surfaces like the vessels in the last picture appear flat, because the underling model of light transmission includes no scattering effects.

4 Shaded volume rendering

Modern graphics hardware like a GeForce 3/4/FX or a Radeon 9700 allow lighting calculation per pixel. Per pixel lighting allow the inclusion of scattering effects into the volume rendering [7], [8], [9]. The gradient of the function $\{\partial_x f, \partial_y f, \partial_z f\}$ can be used to compute diffuse and specular lighting for every pixel to modulate the pixel color.

The **`MVShade`** option of **`MVListContourPlot3D[]`** controls the shading of volume rendering. Since the shaded volume rendering works only on modern graphics cards, it is switched off by default and ignored when the OpenGL library does not allow per pixel lighting effects.

Here is an example of a gorilla skull (200^3 voxel) with different lighting effects. The upper left image shows the pure diffuse lighting, the upper right pure specular lighting and the lower left image diffuse and specular lighting. The lower right image show the rendering of the 134702 polygons in *Mathematica.* The volume rendering use only 443 textured polygons and it is faster to render these 433 polygons than the rendering of the 134702 polygons in the isosurface.

```
Block[{$DisplayFunction = Identity, opts},
  opts = {ViewPoint → {1.71396, -0.86, 1.4052},
    MVNewScene → True};
  monkeyGr = MVListContourPlot3D[
    "monkey.v3d", Contours → {64.},
    MVFlipFace → True, Sequence @@ opts];
  monkeyShade = MVListDensityPlot3D["monkey.v3d",
      ColorFunction → (GrayLevel[#^2] &),
      MVTextureSize → {256, 256, Automatic},
      MVShadeVolume → #,
      MVGrayBackground → True, Sequence @@ opts] & /@
    {DiffuseShading, SpecularShading, True};
];
Show[GraphicsArray[
    Partition[Append[monkeyShade, monkeyGr], 2]]];
```

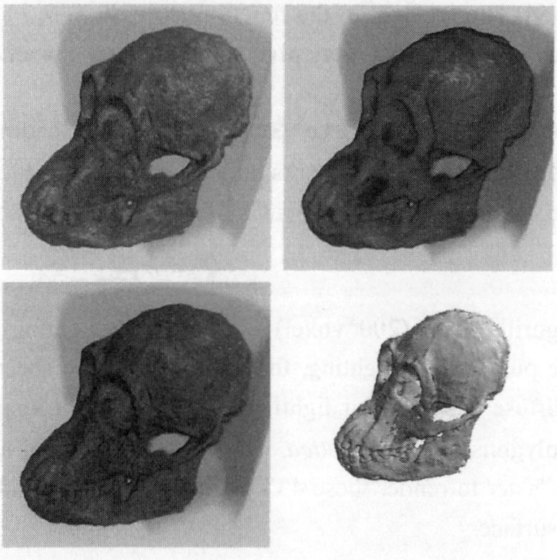

When the resolution of the volume data becomes too large, shaded volume rendering is the only possibility for obtaining high quality pictures of data. Here is a shaded volume rendering of the CT scan of a human heart with 412 polygons and 512^2 textures.

```
MVListDensityPlot3D["smoothcardio_40.v3d",
 ColorFunction → CardioColor,
 MVTextureSize → {512, 512, 412},
 ViewPoint -> {2.68309, -0.501043, 2.0},
 MVShadeVolume -> True,
 MVNewScene → True, MVGrayBackground → True];
```

5 Conclusion and Outlook

MahtGL3d offers two functions to visualize large volume data sets. While the isosurface computation offers a polygonal description of the surface, direct volume rendering can be used to visualize the volume data when the polygon count of the isosurface exceeds the available memory. Shaded volume rendering can generate renderings with nearly the same quality as for polygon-based images. MathGL3d allows the *Mathematica* user to have instant access to the latest development in modern 3d graphics hardware.

The current direct volume algorithms used in MathGL3d use 2d textures for volume visualization. Future versions will extend the rendering methods to 3d volume textures and more sophisticated shading models and better image quality. The new possibilities of free programable graphics hardware for shading (OpenGL fragment programs) will be included, when graphic cards with OpenGL 2.0 become available.

6 References

[1] Kuska, J.-P. MathGL3d, http://phong.informatik.uni-leipzig.de/~kuska/mathgl3dv3/.

[2] William Lorensen and Harvey Cline, Marching Cubes:a High Resolution 3D Surface Construction Algorithm,Proceedings of SIGGRAPH 87 (Anaheim CA,Jul.27-31,1987).In Computer Graphics,21,4 (Aug.1987) ACM SIGGRAPH,New York,1987,pp.163-169.

[3] Bloomenthal, J. An Implicit Surface Polygonizer, Graphics Gems IV (P. Heckbert, ed.), Academic Press, New York, 1994.

[4] Douglas Moore and Joe Warren,Compact Isocontours from Sampled Data,Graphics Gems III,David Kirk,editor,Academic Press,New York,1992.

[5] Kuska, J.-P. MathGL3d, Volume data processing with MathGL3d and *Mathematica* for medical applications, in Proc. 2nd European Medical and Biological Engineering Conference EMBEC '02, December 2002, Verlag der Technischen Universität Graz, pp. 1648—1653

[6] Wilson, O. and Van Gelder, A. and Wilhelms, J. Direct Volume Rendering via 3D Textures, Tech. Rep of the University of California, UCSC-CRL-94-19, 1994

[7] Wilson, O. and Van Gelder, A. and Wilhelms, J. Direct Volume Rendering with Shading via Three-Dimensional Textures, Tech. Rep of the University of California, UCSC-CRL-96-16, 1996

[8] Michael Meißner, M. and Guthe, S. and Straßer, W. Higher Quality Volume Rendering on PC Graphics Hardware, Technicla Rep. of the Wilhelm Schickard Institute for Computer Science,Graphical-Interactive Systems (WSI/GRIS),University of Tübingen,WSI-2001-12

[9] Meißner, M. and Guthe, S. Interactive Lighting Models and Pre-Integration for Volume Rendering on PC Graphics Accelerators, in Graphics Interface, 2002,Canadian Information Processing Society

Creation of multi-channel stereo-images by Mathematica

I. Relke
4D-Vision GmbH, Germany

relke@4d-vision.de

1 Abstract

The problem of creation of a stereo-image for a 3D display from a set of the 2D images is especially important. Using Mathematica's ability we are able to test the different models of 2D image distribution for the evaluation of one stereo-image. For example, we show that the color stereo-image on our 3D display may be obtained also from a special set of a 2D gray-scaled images. The different types of 2D image encoding, corresponding to a definite number of the presented perspectives are studied. General relations, written in Mathematica's notation, allows us to evaluate combined stereo-image from any set of 2D images. Thus, Mathematica's power helps to create new multi-channel systems and improvments in well-known ones. The last our results allows us to present images with a large number of perspectives for very good volumetric perception.

2 Introduction

Recently 3D displays became standard in all large international exhibitions. Moreover, now there are on the market a lot of different models of 3D display and some of them have a very attractive price. A large problem in 3D activity now is the existence of only a small quantity of 3D images and animations, because more than 90% of the all visual information is presented in 2D. Not only are 3D images very impressive for the observer, but they also present much more useful information than ordinary 2D images. So it is important to demonstrate how people may already present their results in 3D form. In this paper we present programs for *Mathematica* users which allow to present their results in stereo-form. The power of *Mathematica* [1] allows us to test many different models for 3D display in a short time to create our own high quality 3D displays reaching the leading world positions in 3D display technologies. This is due to the fact that we have used *Mathematica*'s ability not only to evaluate parameters of different 3D displays, but also to calculate the stereo-image using a set of the ordinary 2D images.

Our company produces autostereoscopic displays, which produce multi-channel stereo-images on a TFT, plasma and projection displays of almost any size. In our 3D display we are able to demonstrate *Mathematica*'s 3D objects and animations with a very high quality and large depth. Moreover, these *Mathematica* objects may be observed on our autostereoscopic displays without the use of any aids such as glasses, helmets or other eye-devices and for many observers simultaneously. Optical filters developed in our company process and decode given multi-channel information. Decoding depends of the position of the observer i.e. each eye of each observer automatically obtains the information which corresponds to his position with respect to the display's surface. This means that each eye of each observer watches different 2D-images, corresponding to different perspectives. Only a human brain is capable of volumetric perception, corresponding to the observed stereoscopic image.

3 S-matrix

The stereo-image from 4D-Vision consists of the set of ordinary 2D images, corresponding to the different perspectives of the considered 3D scene. In this section we consider the matrix S_{ij}, which defines the distribution of the array of 2D-images in the stereo-image and can help to reproduce the stereo-image from the set of the 2D images.

The properties and concrete values of this matrix S_{ij} will be presented below. In this paper we consider only those forms of the matrix S_{ij} which represent a high-quality stereo-impression on the 3D display. The *Mathematica* equation for evaluating a simple S matrix was presented by Relke [2]. A new and more complicated form of the S matrix describing the distribution of 2D-images, depending on the total number of perspectives used, Np, is defined by

$$
\begin{aligned}
&S[i\_, j\_] := \\
&\sum_{m=1}\left(\mathrm{Mod}\left[\left(i + \frac{\mathrm{Np}}{\mathrm{tH}}\,(\mathrm{tH} - m)\right), \mathrm{Np}, 1\right] * \mathrm{KroneckerDelta}[\right. \\
&\quad i - m, j] + \mathrm{Mod}\left[\left(i + \frac{\mathrm{Np}}{\mathrm{tH}}\,(m - 1)\right), \mathrm{Np}, 1\right] * \\
&\qquad \left.\mathrm{KroneckerDelta}[i, j - m + 1]\right)
\end{aligned}
\tag{1}
$$

This *Mathematica* expression allows us to calculate the 2D image distribution in one stereo-image. The matrix presented here corresponds to the definite model of the encoder of the stereo 3D-image by means of the set of 2D-images.

The variable tH defines the period of the structure of the optical filter in the horizontal direction. The optimum value for this variable is equal to 8 and therefore every value for this variable below 8 will be assumed equal to 8. For the concrete realization of equation (1) let us substitute the values of the parameters {imaxima=15; jmaxima=15; Np=24;} in Eq. (1). Then we can obtain, with help of the *Mathematica,* the following table:

1	4	7	10	13	16	19
23	2	5	8	11	14	17
21	24	3	6	9	12	15
19	22	1	4	7	10	13
17	20	23	2	5	8	11
15	18	21	24	3	6	9
13	16	19	22	1	4	7
11	14	17	20	23	2	5
9	12	15	18	21	24	3
7	10	13	16	19	22	1
5	8	11	14	17	20	23
3	6	9	12	15	18	21

Table 1

This table defines the positions and values of the subpixels in the stereo-image. Every number here corresponds to the appropriate perspective. The value of the subpixel in every position (i, j) in this table is taken from the value of the subpixel in the same position of the appropriate 2D-image. As a matter of fact, Table 1 is already used now in 3D displays manufactured by our company 4D-Vision. The matrix $s_{i,j}$ presented here has the obvious translational properties of symmetry:

$$s_{i,j} = s_{i+12\,n,\,j+8\,m} \tag{2}$$

where n and m are integers. Users can examine these properties with the help of *Mathematica,* using Eqs. (1, 2). Note that Eq. (2) defines the dimension of the basic matrix, which in this case is equal to 12x8. Knowledge of the basic matrix has not only theoretical but also the significant practical interest, allowing computer calculations to reproduce the necessary stereo image quickly. Taking into account these properties of symmetry, we'll present everywhere below the $S_{i,j}$ matrixes in form of the basic matrix.

Moreover in our latest realizations we are able to present 3D displays with 40 used channels. If we set Np = 40 in Eq. (1), the interested user may obtain the image distribution for 40 channels. It should be noted that in this case the dimension of the basic matrix is smaller, although the number of the used channels is almost twice as much. Eventually it allows us to calculate the 3D images much more quickly. 40 channels image distribution is already implemented in some new 3D display models from 4D-vision.

Now we present the other possibility for 3D image creation, where only 8 channels are used, with the same optical filter and consequently the same 3D display from 4D-vision.

$$\begin{aligned}&\texttt{Table}\Bigg[\sum_{m=1}^{40}\Big(\texttt{KroneckerDelta}[i-m,\ j]\ *\\&\texttt{Mod}\Big[\texttt{IntegerPart}\Big[-m+8\left(-1+\frac{i}{\texttt{Np}}\right)\Big],\ 8,\ 1\Big]+\\&\texttt{KroneckerDelta}[i,\ j-m+1]\ *\\&\texttt{Mod}\Big[\texttt{IntegerPart}\Big[m+\frac{8\,(-1+i)}{\texttt{Np}}\Big],\ 8,\ 1\Big]\Big),\\&\{i,\ 1,\ \texttt{imaxima}\},\ \{j,\ 1,\ \texttt{jmaxima}\}\Bigg]\ //\\&\texttt{TableForm}\end{aligned} \qquad (3)$$

Substituting the following values of the parameters {imaxima=12; jmaxima=9; Np=24;} in the Eq. (3) we obtain Table 2.

1	2	3	4	5	6	7	8	1
8	1	2	3	4	5	6	7	8
7	8	1	2	3	4	5	6	7
7	8	1	2	3	4	5	6	7
6	7	8	1	2	3	4	5	6
5	6	7	8	1	2	3	4	5
5	6	7	8	1	2	3	4	5
4	5	6	7	8	1	2	3	4
3	4	5	6	7	8	1	2	3
3	4	5	6	7	8	1	2	3
2	3	4	5	6	7	8	1	2
1	2	3	4	5	6	7	8	1

Table 2

As one can see from Tables. (1, 2) the translational properties are the same, but the number of channels used in the second case in 3 times less. This permits a reduction in the time for stereo-image production.

4 Creation of the stereo-image

All examples presented above determine the values of the matrix $S_{i,j}$ for the distribution of 2D images in one stereo-image. Now we show how using *Mathematica* with the matrix $S_{i,j}$ one can calculate, from Eq. (3), the values of the subpixels in one stereo-image.

```
twoD[k_] :=
Partition[Array[k#2,#3 &, {12, 5, 3}] // Flatten, 15];        (4)
```

Here every k-th 2D-image is described in abstract form by the matrix twoD[k]. Every perspective k here is a function of the azimuthal angle, because every perspective differs from others only by the angle under which the corresponding two-dimensional snapshot is produced. In the case considered here there are 8 processed 2D perspectives. The indices i and j define the position of every subpixel in any 2D or stereo-image. For an XGA display (1024x768) the limits for i and j are 1 to 768 and 1 to 3072 respectively, corresponding to the columns and rows of the subpixels in every 2D or stereo-image. The results of the evaluations are in Table 3.

$$\texttt{stereo[i\_, j\_] :=} \sum_{k=1}^{24} (\texttt{KroneckerDelta}[k - s[i, j]] * \texttt{twoD}[k][[i, j]]) \qquad (5)$$

```
test1 =
 Table[stereo[i, j], {i, 1, 12}, {j, 1, 6}] // TableForm
```

$1_{1,1}$	$2_{1,2}$	$3_{1,3}$	$4_{2,1}$	$5_{2,2}$	$6_{2,3}$
$8_{1,1}$	$1_{1,2}$	$2_{1,3}$	$3_{2,1}$	$4_{2,2}$	$5_{2,3}$
$7_{1,1}$	$8_{1,2}$	$1_{1,3}$	$2_{2,1}$	$3_{2,2}$	$4_{2,3}$
$7_{1,1}$	$8_{1,2}$	$1_{1,3}$	$2_{2,1}$	$3_{2,2}$	$4_{2,3}$
$6_{1,1}$	$7_{1,2}$	$8_{1,3}$	$1_{2,1}$	$2_{2,2}$	$3_{2,3}$
$5_{1,1}$	$6_{1,2}$	$7_{1,3}$	$8_{2,1}$	$1_{2,2}$	$2_{2,3}$
$5_{1,1}$	$6_{1,2}$	$7_{1,3}$	$8_{2,1}$	$1_{2,2}$	$2_{2,3}$
$4_{1,1}$	$5_{1,2}$	$6_{1,3}$	$7_{2,1}$	$8_{2,2}$	$1_{2,3}$
$3_{1,1}$	$4_{1,2}$	$5_{1,3}$	$6_{2,1}$	$7_{2,2}$	$8_{2,3}$
$3_{1,1}$	$4_{1,2}$	$5_{1,3}$	$6_{2,1}$	$7_{2,2}$	$8_{2,3}$
$2_{1,1}$	$3_{1,2}$	$4_{1,3}$	$5_{2,1}$	$6_{2,2}$	$7_{2,3}$
$1_{1,1}$	$2_{1,2}$	$3_{1,3}$	$4_{2,1}$	$5_{2,2}$	$6_{2,3}$

Table 3

The equations presented above allow us, in principle, to calculate the stereo-image from the set of the 2D-images. However, the calculations using these formulae is very slow. So for concrete calculations we have used another more effective algorithm. See the CD version of this paper. A similar program was used in our company for the first calculation of the stereo-images.

There is indeed an infinity of different variants for production of a stereo-image from any set of 2D-images. Here we'll consider one of this possible variants. Using this method one can produce the stereo-image with a very large number of perspectives. Let us define the azimuthal angle corresponding to k-th perspective by $\varphi(k)$. Note that the Table 3, which defines the values of the subpixes in one stereo-image, was obtained from Eq. (5) using the 2D images from Eq. (4) and the S matrix from Eq. (3). Repeating this procedure 8 times with the other 2D images we can obtain another stereo-image. If the azimuthal angle between the first and the 8-th new stereo-images is of the same order as $\varphi(8)$-$\varphi(1)$, then we can already obtain a stereo-image with 64 channels. This procedure may be repeated. It means that in the case considered here one can also build the stereo-image with a multiple number of channels.

5 Gray-scale 2D images

Here we show how to calculate the stereo-image from a set of gray-scale 2D images. In Table (2), we use only part of the information from each 2D image in the stereo-image. For example, in the stereo-image consisting of 8 images, only 1/8 of the information in the original 2D image is used. This fact allows us to decrease the size of the original 2D image. To illustrate this, consider Table 2. In this case the distance in the horizontal direction between two subpixels with the same color and belonging to the same image, is 24. Therefore, out of every 8 pixels, only 3 belong to the certain perspective, and the distance between these subpixels is more than 1 pixel. This allows us to create a gray scale image where we have all the necessary information for the color stereo-image. There are also different variants for creation of gray-scale images. The corresponding program is on the CD version of this paper.

Since the true information about the perspective in every gray-scale image is situated on the same places as in the original color 2D image, we can decode this true information by means of an optical filter from 4D-Vision. This property of the gray-scale image is used to align the position of the optical filter as well as to calibrate the 3D display from 4D-Vision.

The results presented here have noticeable practical interest, because they demonstrate which definite part of information is used for production of the stereo-image. On one hand this fact may be used for rendition programs, allowing us to render only a definite part of the 2D perspective from the 3D scene, and that may lead to a decrease of the rendition time in some cases. On the other hand the results presented may help to create the necessary algorithm for scaling the stereo images. Since there are 3D displays with different resolutions, the process of scaling the stereo-image from one resolution to another without loss of quality may be not simple. We are working on creation of new mathematical algorithms for scaling stereo images with minimum loss of quality and information.

6 Conclusion

We have demonstrated how to calculate a multi-channel stereo-image from a set of 2D images. To visualize the stereo-images, you need 4D-Vision's 3D display. *Mathematica* allows us to obtain significant results in 3D activity. Conversely, our displays and software developments may help to visualize *Mathematica's* 3D objects on stereo-screens. In this paper we demonstrated the solution of only a small number of 3D problems. A tremendous growth of 3D activity and the necessity of solving new 3D problems requires new *Mathematica* functions to be created, allowing further graphical possibilities for 2D and 3D object processing and visualization.

7 References

1. Wolfram, S. Das Mathematica Buch, 1977, 3 Auflage, Addison Wesley Longman Verlag GmbH.
2. Relke, I. The use of Computer Algebra System for the 3-D image processing in 3-D Display, in Proc. 4th IMS, Tokyo Denki University, Japan, (ed. Tazawa, Y.), June 2001, Tokyo Denki University Press.

Appendix: New since IMS2001 and CD contents

The following are new since IMS2001.

1. In our first attempts to create 3D displays we used an optical filter with a primitive distribution of eight 2D images. Unfortunately the optical filter corresponding to the distribution of these 2D images creates Moire bands which cause disturbances that prevent viewers from obtaining high-quality stereo-image. We can now create a high-quality 3D display and reach leading world positions. (see the image distribution with 24 channels: see eqs 1-3, and Tables 1-2).
2. Translational properties of the basic matrix (Eq. 2)
3. Mathematica programs to evaluating the dimensions of basic matrix (CD chapter 3) and to calculate a colour stereo-image from a set of gray-scale images (CD Chapter 4)
4. All procedures for calculating a stereo image with the new 2D image distribution (see Eqs. (3-5) and Table 3.).

Improved Java Photo Editor And Digital Image Processing

Junzo SATO
University Of Kumamoto, Japan
jsato@fc.kuh.kumamoto-u.ac.jp
Mariusz JANKOWSKI
University Of Southern Maine, USA
mjkcc@usm.maine.edu
Kiseong KIM
Laxtha Inc., Korea
sola_scriptura7@hotmail.com
Chikara MIYAJI
Japan Institute Of Sports Science, Japan
Chikara.MIYAJI@jiss.ntgk.go.jp
Akira TAKADA
University Of Kumamoto, Japan
takada@fc.kuh.kumamoto-u.ac.jp
Hiroyuki YOSHIHARA
University Of Kyoto, Japan
yoshi@fc.kuh.kumamoto-u.ac.jp

1 Abstract

Over the years, the authors have developed the interactive photo editing application Java Photo Editor (JPE) to supplement the *Mathematica* Frontend [1] when image process operations are done using *Mathematica* and the Digital Image Processing (DIP) package, which provides plentiful scientific knowledge of this field. The JPE can be executed simply as the standalone graphics painting program because it has been implemented as a Java application based on the Java classes defined by J/Link. As the result, sufficient performances of both graphics manipulations usability with a mouse and transfer rate of the image data between Java and *Mathematica* were maintained successfully. Data conversion between Java's BufferedImage and DIP's ImageData were effectively done by DIP's fine tuned algorithm. JPE's technique has been applied for development of other applications related to Java or *Mathematica*. The clinical chart eDolphin and the field simulator MathFields were constructed using JPE's resources.

2 Java Image And *Mathematica* Graphics

Image data conversion between Java and *Mathematica* is one of the most basic requirements when building the photo editing application for *Mathematica*. The JLinkProgramming`JavaSwing` package, which is a wrapper for JFC Swing classes, was developed for convenience of handling each Java class from *Mathematica*. Construction of the Java Image object from *Mathematica* is done through the javax.swing.ImageIcon class. Because the java.awt.image.PixelGrabber can retrieve each pixel data in the form of 4-byte ARGB integers, the Java Image can be decomposed into a matrix of {R, G, B} integers. Converting this matrix of colors into the RasterArray[] or Raster[] enables the user to show this image as the *Mathematica* Graphics object.Two functions GrabPixels[] and ShowJavaImage[] were defined in the JLinkProgramming`JavaPhotoEditor` package for these purposes. On the other hand, the function ToJavaImage[] was defined to convert the *Mathematica* Graphics into the Java Image object.

3 Java Photo Editor

As a simple Java graphics painting application, the Java Photo Editor, which is archived in the file "javaphotoeditor.jar", can be launched by doubleclicking it, because the manifest text is setup and stored in the archive. When the user wants to use it from the *Mathematica* Frontend, the main class javaphotoeditor.JPEFrame should be loaded. This class automatically generates a javaphotoeditor.JPEPanel object to hold the image data. The javaphotoeditor was defined based on J/Link's com.wolfram.jlink.MathGraphicsJPanel. The method loadImage was defined in JPEFrame so that it can acquire image data from the location of the data specified by java.awt.Image, java.io.URL, java.io.File, or java.lang.String. As the result of using JPEPanel, it is also possible to draw the result of the evaluation of the *Mathematica* command by calling the setMathCommand method, which is J/Link's built-in functionality.

4 Tools And Menus

The image region of interest (ROI) and ROI bounds can be selected using a mouse and the area selection tool and the marquee rect tool. Information about these tools may be obtained from *Mathematica* in the form of a list of mouse trajectory points or {left, top, width, height} integers. Painting graphics and text tools also have information about their shapes. Because JPE itself was designed as a supplemental GUI tool for the *Mathematica* Frontend, only a few menu commands were added to define a default Java application. However, with the help of the JLinkProgramming`JavaSwing` package, the user can customize menus freely. Each menu command handler is described in the form of a *Mathematica* expression. The class JLinkProgramming.JavaSwing.ThreadMathActionListener was defined for time-consuming calculations to avoid thread-blocking in the GUI.

5 Working With Digital Image Processing

For best integration of the DIP package with the Java environment, finely-tuned algorithms were developed to convert DIP's ImageData format from/to the java.awt.image.BufferedImage.

6 Integration With the Clinical Chart System - eDolphin

The electronic chart system eDolphin [2] for clinics has been developed as a joint project between the Miyazaki Health Welfare Network [3] and the Kumamoto Regional Health Welfare Open Network. eDolphin was written in Java, and part of JPE's source code has been used to add the functionality to medical image painting.

7 A Field Simulator - MathFields

One of us, Kiseong Kim, contructed a field simulator for development of NMR Gradient Coil imaging using the target field method. All code was written in *Mathematica*. The JLinkProgramming`JavaSwing` package was adopted to build the MathFields' GUI.

8 Screenshots

The following figure shows that the ROI is selected and the image filter defined by *Mathematica* is to be applied by the customized menu command. As a result of the definition of the menu handler, the image data bounded by the marquee rect tool is retrieved by *Mathematica* in the form of the DIP's ImageData.

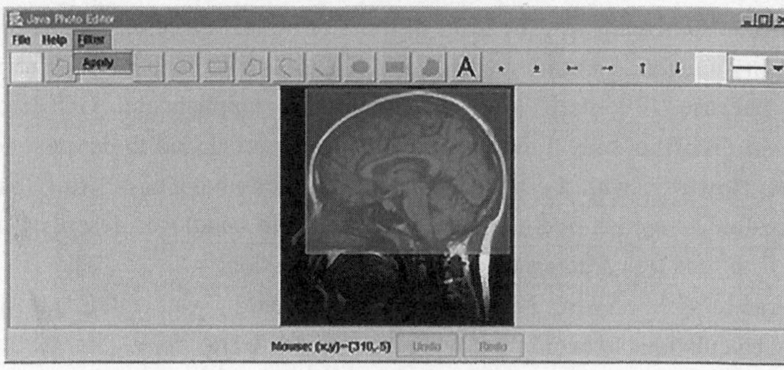

Figure 1

Then, the filtered image is shown in a new *Mathematica* notebook.

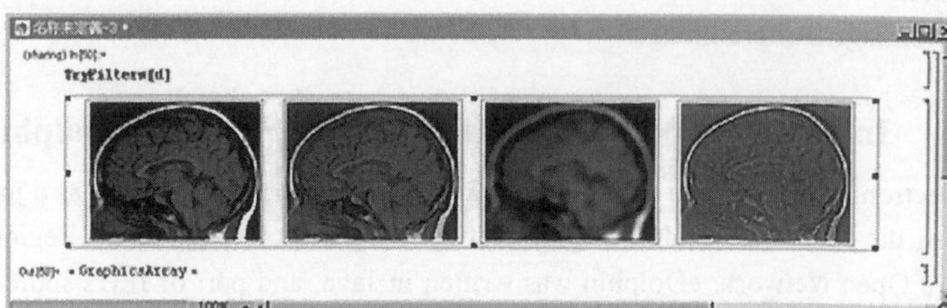

Figure 2

The clicked points are drawn as a polyline using the mouse tool.

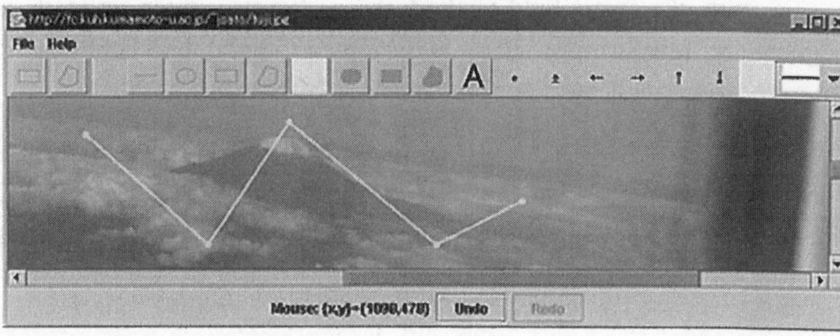

Figure 3

Using DIP's ShowRegionOfInterest[] and ShowLineProfile[], the pixel data on the line path is analyzed.

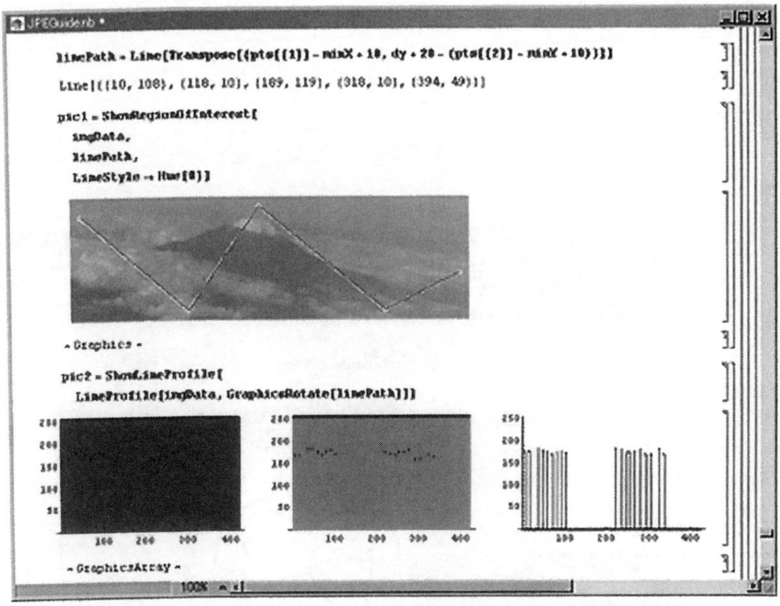

Figure 4

eDolphin has a built-in graphics painting application based on the JPE's source code.

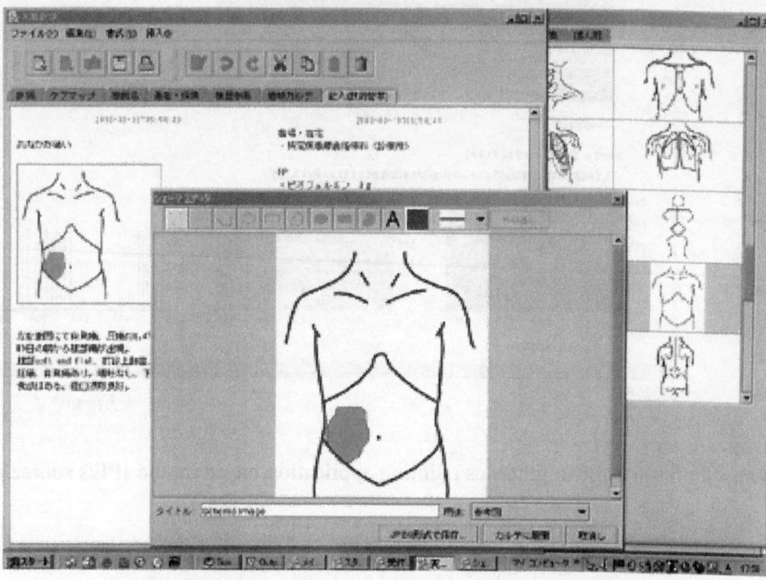

Figure 5

MathFields was constructed using the JLinkProgramming`JavaSwing` package.

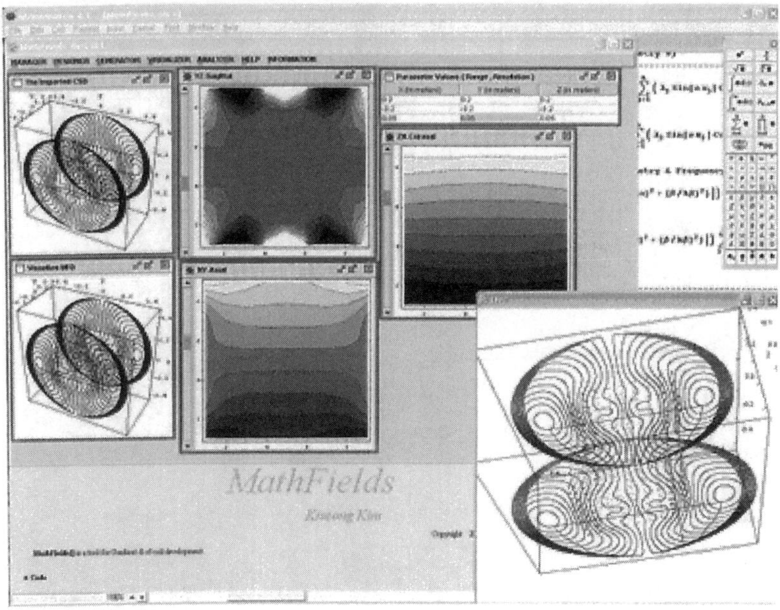

Figure 6

9 References

[1] Sato, J., Miyaji, C., and Jankowski, M., Implementation of Java Photo Editor, *Mathematica* Developer Conference 2001

[2] The Dolphin Project, http://www.kuh.kumamoto-u.ac.jp/dolphin/

[3] Miyazaki Haniwa Network, http://www.haniwa-net.jp/

[4] Java Photo Editor, http://www2.kuh.kumamoto-u.ac.jp/jsato/mathematica.htm

10 Acknowledgement

One of authors, Junzo Sato expresses his gratitude to Mr. Kazushi Minagawa, the president of Digital Globe, Inc., for successive work on eDolphin.

Visualization of Real Projective Algebraic Curves on Models of the Real Projective Plane

Stefan Welke
spwelke@aol.com

1 Abstract

A CAS can compute and plot nearly every curve that can be represented by formulae, and packages like GeomView, MathLive, or MathGL3D allow a 3D representation of curved surfaces which can be rotated in real-time. These are used to discuss three models of the real projective plane: two-sheeted covering by the sphere, the disk-model, and the Roman surface of J. Steiner. Examples of projective curves on these surfaces will be presented.

2 The real projective plane

A common definition of the real projective plane, denoted by $\mathbb{PR}^2$, reads as follows: $\mathbb{PR}^2$ is the set of all straight lines in $\mathbb{R}^3$ that pass through the origin. Since each line through the origin intersects the unit sphere $S^2 \subset \mathbb{R}^3$ in two antipodal points, we immediately get an equivalent geometric definition: $\mathbb{PR}^2$ is equivalent to the set of all pairs of antipodal points on the unit sphere. Thus a first attempt at visualization is to take one half of a sphere. If we look at the earth as a sphere, for example, either hemisphere serves as a model for $\mathbb{PR}^2$. Since antipodal points on the equator have to be identified, we must add a half circle of the equator with one endpoint, as in Figure 1.

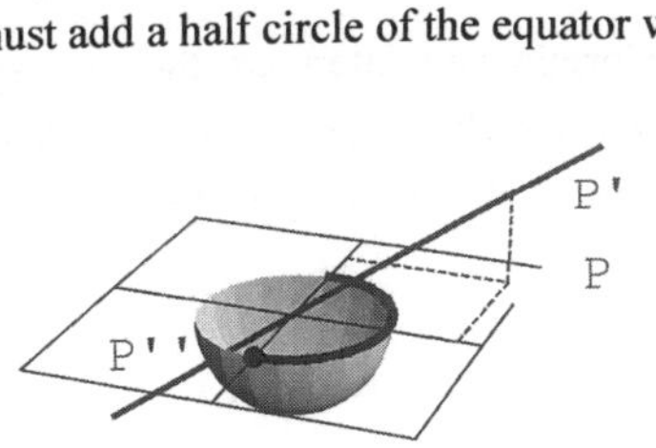

There has been no reason, until now, to call this half of a sphere a plane, projective or not. The connection with a plane is established as follows: Consider the two-dimensional Euclidean plane $\mathbb{R}^2$ embedded in the Euclidean space $\mathbb{R}^3$ in the usual way, i.e. a point $(x, y) \in \mathbb{R}^2$ is mapped at the point $(x, y, 0) \in \mathbb{R}^3$. There corresponds to each point P of the $x-y-$plane a line l through the point $P'(x, y, 1)$ and the origin $(0, 0, 0)$. This line intersects the southern hemisphere in a point P''. We have found a one-to-one correspondence between all points in the Euclidean plane and all points in the southern hemisphere except the points on the equator. This is demonstrated in figure 1 above.

To each point of the half-equator there is a corresponding line in the $x-y-$plane. We can understand such a line as an additional ideal point in $\mathbb{R}^2$ infinitely distant from the origin. Both directions of that line correspond to the same ideal point. This is the classical idea of the projective plane as an extension of the Euclidean plane. We call this model the *hemisphere model of* $\mathbb{P}\mathbb{R}^2$.

3 Plane algebraic curves

Now we want to draw real algebraic curves in the plane and on the southern hemisphere. Consider a parabola with equation $p(x, y) = x^2 - y + 1 = 0$. We extend the corresponding polynomial p of order two to a homogenous polynomial p^* of order two with a third variable z: $p^*(x, y, z) := x^2 - y \cdot z + z^2$. The equation $p^*(x, y, z) = 0$ defines implicitely an algebraic surface S. Since $p^*(tx, ty, tz) = t^2 \cdot p^*(x, y, z)$ we find that with each point (x_0, y_0, z_0) that satisfies $p^*(x_0, y_0, z_0) = 0$, all points of the line $t \cdot (x_0, y_0, z_0)$, $t \in \mathbb{R}$, satisfy the same equation. Thus the surface S consists of straight lines with common point $(0, 0, 0)$. Therefore S is a *ruled surface,* i.e. a surface generated by a moving straight line.
This surface is symmetric with respect to the origin and intersects the unit sphere symmetrically. The intersection $S \cap S^2$ defines a symmetric curve γ on the sphere. Its restriction to the southern hemisphere is the image of the parabola in the hemisphere-model.

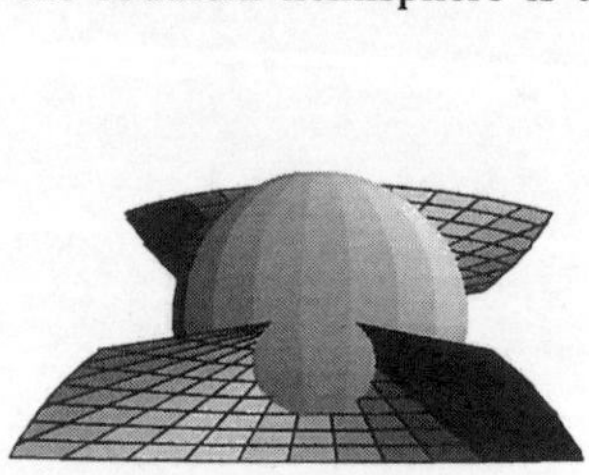

Figure 2

Additionally, in $\mathbb{R}^3$ all points $(0, y, 0)$ satisfy the equation $p^*(x, y, z) = 0$. So the y-axis intersects S^2 in two points on the equator. This pair of points defines a point in the projective space that belongs to the curve γ and which has no counterpart in the Euclidean plane. We recognize that the parabola is a closed curve in $\mathbb{PR}^2$. This is in striking contrast to the case we can see in $\mathbb{R}^2$. We can extend every real plane algebraic curve in this way to $\mathbb{PR}^2$, which is the natural space for this kind of curve.

4 Complete visualization of algebraic curves in the projective plane

Figure 2 suggests how to visualize algebraic curves in $\mathbb{PR}^2$ completely. We must compute the intersection of the sphere and a ruled surface. [4] describes a way to plot the intersection-curve of a sphere and an algberaic surface with *Mathematica* using Groebner Bases. We can avoid the use of Groebner Bases in this special case. In order to achieve this we have to introduce another model of $\mathbb{PR}^2$, the so called *disk-model.*
Stereographic projection, which is known in geometry and cartography, maps the sphere, with the exception of one point, the (north) pole, onto the Euclidean plane in such a way that the southern and northern hemispheres correspond to the interior and exterior respectively of the unit circle. Figure 3 demonstrates this projection:

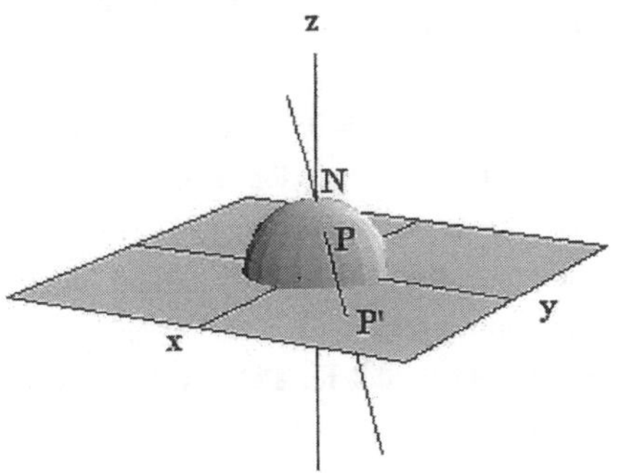

Figure 3

The point P on the sphere is connected by a straight line with the north pole N. This line intersects the $x-y-$plane at the point P', which is the image of P under stereographic projection. It is clear that this is a one-to-one map of $S^2 \setminus \{N\}$ to $\mathbb{R}^2$. We can extend this map to a one-to-one map $s : S^2 \rightarrow \mathbb{R}^2 \cup \{\infty\}$ in such a way that N corresponds to ∞. If (x, y, z) are the coordinates of a point P on S^2, then the coordinates of $P' = s(P)$ are given by (see [4]):

$$s : (x, y, z) \to \left(\frac{x}{1-z}, \frac{y}{1-z} \right) \tag{1}$$

The formula for the inverse map $s^{-1} : \mathbb{R}^2 \cup \{\infty\} \to S^2$ is given by:

$$s^{-1} : (x, y) \to \left(\frac{2\,x}{1 + x^2 + y^2}, \frac{2\,y}{1 + x^2 + y^2}, \frac{-1 + x^2 + y^2}{1 + x^2 + y^2} \right) \tag{2}$$

Now let $p(x, y)$ be any polynomial defining a real algebraic curve and let $p^*(x, y, z)$ be the corresponding homogenous polynomial. Let $\deg(p)$ denote the degree of p. We define:

$$p^{**}(x, y) := p^*\left(\frac{2\,x}{1 + x^2 + y^2}, \frac{2\,y}{1 + x^2 + y^2}, \frac{-1 + x^2 + y^2}{1 + x^2 + y^2} \right) \cdot (1 + x^2 + y^2)^{\deg(p)} \tag{3}$$

p^{**} is clearly a rational function. Since p^* is homogenous, the common denominator is cancelled by the second factor. Thus p^{**} is a polynomial. We see that $1 + x^2 + y^2 \geq 1$, so we have the following equivalence:

$$p^*\left(\frac{2\,x}{1 + x^2 + y^2}, \frac{2\,y}{1 + x^2 + y^2}, \frac{-1 + x^2 + y^2}{1 + x^2 + y^2} \right) = 0 \iff p^{**}(x, y) = 0 \tag{4}$$

We conclude that the algebraic curve defined by $p^{**}(x, y) = 0$ is the stereographic image of the real projective curve defined by $p^*(x, y, z) = 0$. Since the image of the southern hemisphere with the equator is the unit disk, we have caught the real projective plane together with the infinitely distant ideal points within the unit disk. The ideal points correspond to pairs of opposite points on the unit circle. This is the so called unit-disk model of the projective space $\mathbb{P}\mathbb{R}^2$. An equivalent model for $\mathbb{P}\mathbb{R}^3$ is described in [3]. Now we can draw a complete image of a real projective curve.

The parabola $p(x, y) = x^2 - y^2 + 1 = 0$ may serve as an example. We find as above: $p^*(x, y, z) = x^2 - y \cdot z + z^2$. Let *Mathematica* compute $p^{**}(x, y)$:

```
p1[x_, y_, z_] := x^2 - y*z + z^2;

p1[2x/(1 + x^2 + y^2), 2y/(1 + x^2 + y^2), (-1 + x^2 + y^2)/(1 + x^2 + y^2)] // Together //
  Numerator
```

```
1 + 2 x^2 + x^4 + 2 y - 2 x^2 y - 2 y^2 + 2 x^2 y^2 - 2 y^3 + y^4
```

The advantage is, that we can use the *Mathematica* function `ImplicitPlot[..` draw the curve:

```
Needs["Graphics`ImplicitPlot`"]

ImplicitPlot[% == 0, {x, -1, 1},
  Axes -> True, AspectRatio -> Automatic,
  Epilog -> {{Dashing[{.05}], Circle[{0, 0}, 1]}},
  PlotRange -> {{-1, 1}, {-1, 1}},
  Ticks -> {{-1, -.5, .5, 1}, {-1, -.5, .5, 1}}]
```

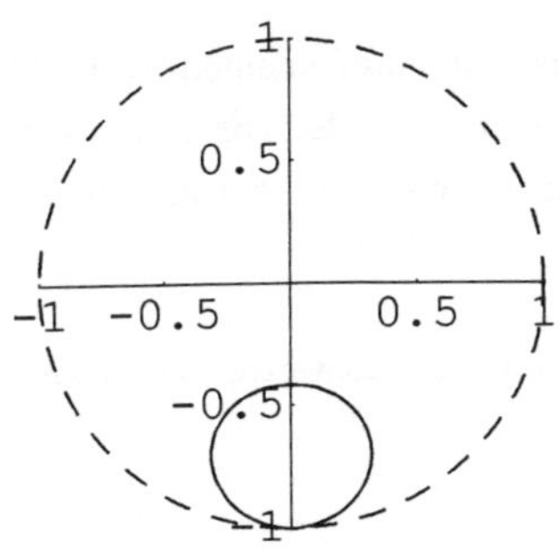

Figure 4

We notice that a parabola is a closed curve in the projective plane.

5 3D-Visualization of projective curves

Once we have plotted a curve in the unit-disk model we can extract the data of this curve from the graphics output. Then we can map each point of the curve with inverse stereographic projections onto both hemispheres. The same construction holds for the inverse of the north-map, which is the inverse stereographic projection as described above, and the inverse of the south-map, except that the line through P passes through the antipodal point of N, the south pole. Both parts together represent the intersection of S^2 and the ruled surface defined by the equation $p^*(x, y, z) = 0$. Figure 5 shows the projective version of Neill's parabola $y^2 - x^3 = 0$, i.e. the intersection of $y^2 \cdot z - x^3 = 0$ and the unit sphere.

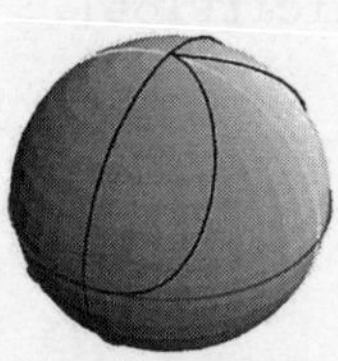

Figure 5

6 A *Mathematica* example

All the calculations we have done so far can be implemented with *Mathematica* functions. We take the so-called *devil's curve* as an example. The code for the function definitions is very technical, so it is omitted in order to save space. Some examples of functions are presented here:

```
devil = y^4 - x^4 - 1.1 y^2 + x^2 ; homogenise[devil, z]
```

$$-x^4 + y^4 + x^2 z^2 - 1.1 y^2 z^2$$

```
northMap[devil]
```

$$0.275\, x^2 - 1.55\, x^4 + 0.275\, x^6 - \frac{y^2}{4} - 0.05\, x^2 y^2 + 0.3\, x^4 y^2 + \frac{3\, y^4}{2} - 0.225\, x^2 y^4 - \frac{y^6}{4}$$

```
GraphicsArray[{UnitDiskModel[devil],
    Show[mapOntoSphere[devil], Boxed -> False]}] // Show
```

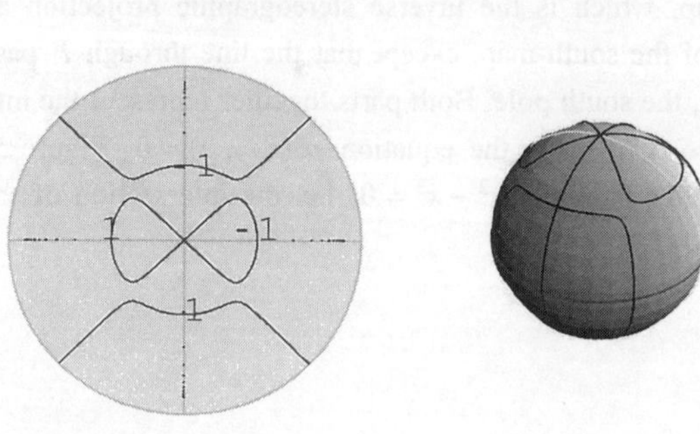

7 Dynamic 3-D visualization

The sphere S^2 is, in topological terms, a two-sheeted covering of the real projective line $\mathbb{PR}^2$. Each pair of antipodal points on S^2 corresponds to one point in $\mathbb{PR}^2$. However, if we use orthogonal projection for displaying 3D graphics, we can only see at most one hemisphere. So we have a look at just one sheet of that covering.
It is possible, due to the existence of dynamic 3D-visualisation tools like MathGL3D from J.-P. Kuska or Geomview from the Geometry Center, to rotate surfaces in real-time. This property is as good as a solid model of a surface in our hands: we can turn it around and look at it from every direction. The following sequence of pictures was generated with MathGL3D:

```
Get["MathGL3d`OpenGLViewer`"]

trifolium = (x^2 + y^2) * (y^2 + x * (x + 2)) - 8 * x * y^2 ;
MVShow3D[mapOntoSphere[trifolium],
 MVLineTubeSize → 0.008, MVTubeSegments → 5,
 MVPointSphereSize → 0.05]
```

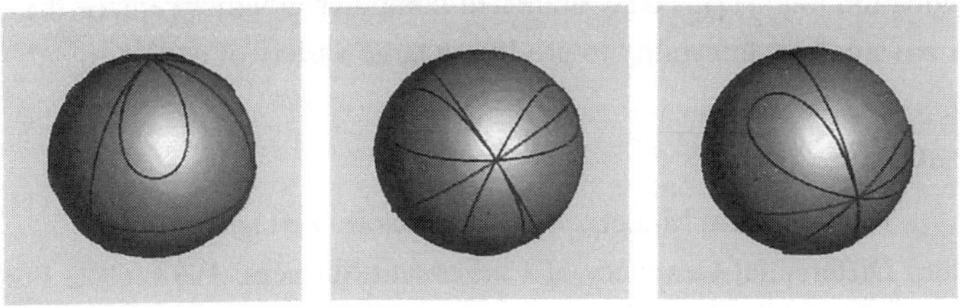

8 Steiner's Roman Surface as a model for the real projective plane

In 1844 Jacob Steiner discovered a surface, which now bears his name, which is another model of $\mathbb{PR}^2$. If (x, y, z) is a point of S^2, then we can map the sphere by $r:(x, y, z) \to (x \cdot y, y \cdot z, z \cdot x)$ onto Steiner's Roman surface. Again S^2 is a two-sheeted covering of the Roman surface. We need not identify points on the Roman surface, like in the hemisphere model, or the unit-disk model, at the price of self-intersection. For details see [2]. We will map the *devil's curve* onto the Roman surface using the external program Geomview:

```
Needs["Graphics`Geomview`"];
Show[mapOntoRoman[devil], Boxed -> False]
```

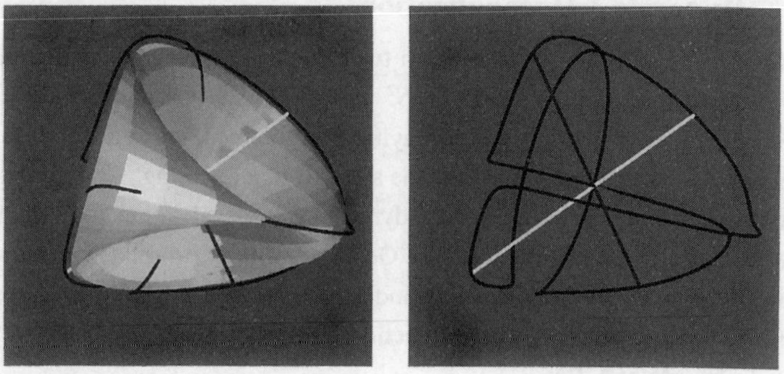

The surface itself is omitted in the right-hand figure above, so we can see the *devil's curve* on the Roman surface as a space curve.

9 Conclusions

Visualization of mathematical objects is an important pedagogical tool. Mathematical cabinets with clay, wood and wire models served in the 19th and early 20th centuries to strengthen the imagination (see [1]). Today animated computer graphics replace the old technique. A major advantage is the ability to produce a large variety of examples.

10 References

[1] Fischer, G.: Mathematical Models, 1986, Akademie-Verlag Berlin

[2] Gray, A.: Modern Differential Geometry of Curves and Surfaces, 1993, CRC Press, Inc.

[3] Kreminski, R.: Visualising the Hopf Fibration, *Mathematica* in Education and Research, Vol. 6, No. 1, 1997, Telos

[4] Welke, S.: Stereographic projection of plane algebraic curves onto the sphere, Innovation in Mathematics, Proceedings of the Second IMS, (eds. V. Keränen and P.Mitic), July 1997, Computational Mechanics Publications

Simpler Games: Using Cellular Automata to Model Social Interaction

Seth J. Chandler
University of Houston Law Center, Houston, TX
SChandler@uh.edu

1 Abstract

In 1997, Professor Randal C. Picker published a pathbreaking article titled "Simple Games for a Complex World" in the University of Chicago Law Review. Drawing on earlier "Spatial Games" work done in theoretical biology, his article explained how legal interventions can affect the evolution of behavioral norms among a networked group of individuals who have modest computational abilities but do have an ability to learn from each other. This article shows that the Spatial Games model is mathematically equivalent to a cellular automaton. The reconceptualization of this area permits leveraging of the computational algorithms and insights regarding cellular automata. The article shows how the "coupling topology" of society and the learning algorithms of its members appear to affect the distribution of Wolfram Class I (homogeneous), Class II (regular), Class III (chaotic) and Class IV (complex) behavior in social systems for which the Spatial Games model serves as a useful metaphor. The *Mathematica* code used to reach these conclusions is available on the conference proceedings CD.

2 Picker's Model

Picker's model of the evolution of norms alloys spatial modeling to non-cooperative game theory [5]. Agents evolve based on mimicry of what appears to be particularly advantageous behavior adopted by a subset of their neighbors. In particular, it features a toroidal topology (a uniform two dimensional lattice of sites with periodic boundary conditions) of sites populated by identical agents each of whom assumes a "color" or "strategy" based on its choice from a finite set of behavioral options. By way of example, the strategies ("colors") might be "Use Internet Explorer" or "Use Netscape" or they might be "drive on the left side of the road" and "drive on the right side of the road." Each agent synchronously plays a non-cooperative 2×2 coordination game with an absolutely unique but relatively identical subset of the agent's neighbors referred to as a "payoff neighborhood." An example might be to play a coordination game with one's eight neighbors on the local grid that surrounds each element of the torus (a "Moore neighborhood"). The resulting set of interactions between each agent and the agents in its payoff neighborhood results in a total payoff to the agent. In Picker's model and those of the theoretical biologists before him [4], the total payoff for each agent on each turn is the sum of payoffs from all the games played by that agent on that turn. After these payoffs are computed, each agent then examines a possibly different subset of the population known as an "information neighborhood." Like the payoff neighborhood, the information neighborhood is absolutely unique for each agent but relatively identical. Frequently, the information neighborhood is the union of each agent's payoff neighborhood and the agent itself. The agent makes a new behavioral choice based on the relationship between the behavioral choices of those agents and the payoffs they received. Although different learning algorithms can be used, [6], Picker generally uses the "Imitate Best Neighbor" approach. Under this algorithm, each agent's behavior on its next turn is the behavior of the agent in the information neighborhood that received the highest payoff. This deterministic process iterates indefinitely.

3 Spatial Games on Uniform Lattices are Lightly Disguised Cellular Automata

In the spatial games models, the only sites that affect how a site updates itself ("the affect neighborhood") are the union of sites in the payoff neighborhoods of all the sites in its information neighborhood. The sites in the information neighborhood are what directly affect each site's decision on how to update itself, but the total payoffs at these sites in the information neighborhood are affected by the sites in their payoff neighborhoods. No other site affects the decision of a site on how to update itself. Because both payoff and information neighborhoods are relatively identical for each site, the affect neighborhood for each site must be relatively identical. Also, because the update rule for a site depends only on the values of sites in the affect neighborhood, spatial games of the sort used by Picker may be perfectly and directly described by a cellular automaton. [9]

In Picker's implementation, for example, the affect neighborhood consists of 25 sites, the union of sites in the 3×3 payoff neighborhoods of a 3×3 information neighborhood. As Picker permits two colorings of each site, there are thus 2^{25} ($\approx$33 million) possible configurations that would need to be analyzed to determine the corresponding cellular automaton rule. Even after taking symmetries into account, and as shown in conference proceedings CD, there are still more than 4 million distinct configurations that would require a 4 million binary digit cellular automaton to analyze. Picker's implementation is thus too complex to reveal many general principles of spatial models.

The phenomena Picker observes after his exhaustive study of his original model can be replicated, however, using less elaborate coupling topologies. One can, for example, use one dimensional cellular automata with identical r=1 payoffs and information neighborhoods. As such, one can study a simple Picker model with r=2, k=2 for which there are only 2^5 configurations to consider. (Indeed, if one restricts oneself to cellular automata generated by symmetric games, such as a coordination game, there are only 20 configurations to consider.) Alternatively, one can study a two dimensional cellular automata in which the payoff and information neighborhoods are the four north, east, south and west neighbors – a so-called von Neumann neighborhood. (Figure 1) The affect neighborhood of each site under this coupling topology has 13 sites and there are thus 2^{13} (8192) configurations that need to be analyzed. The *Mathematica* algorithms contained in this article permit one to find the cellular automata specification of a 2×2 two-player game under this "double von Neumann" topology in about a minute. Once these cellular automaton equivalents are discovered, computer simulation of the system can be vastly accelerated.

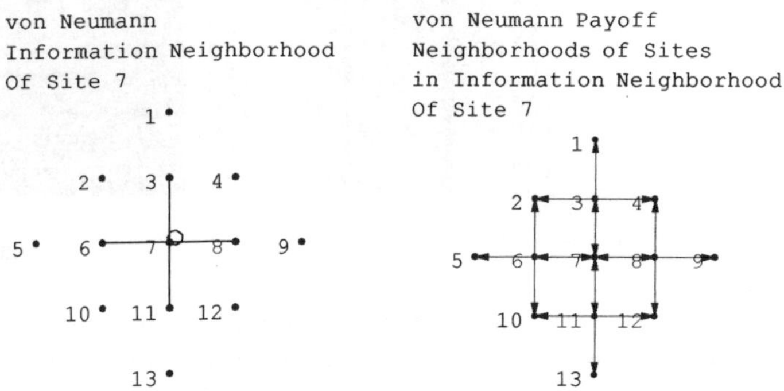

Figure 1

The *Mathematica* code contained on the CD shows the process of how one moves from spatial games played on regular lattices (i.e. those in which each site has a similar affect neighborhood) to a cellular automaton specification. A virtue of the code is that it also permits simulation of spatial games in which more than two strategies are permitted and in which (a) the payoff neighborhood is any undirected graph and (b) the information neighborhood is any directed graph. The restriction of payoff neighborhoods to undirected graphs means that if site B is in the payoff neighborhood of site A then site A must be in the payoff neighborhood of site B. (The information neighborhoods need not be reciprocal; B can be in the information neighborhood of A and A need not be in the information neighborhood of B).

4 Different Coupling Topologies and Different Learning Algorithms

4.1 Regular One Dimensional Lattices

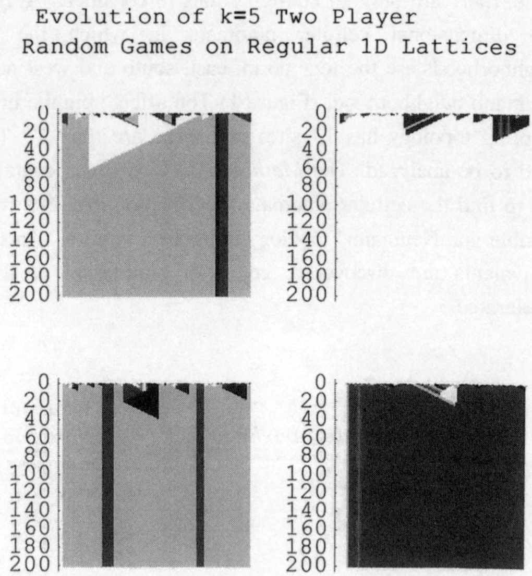

Figure 2

When one runs experiments using such arbitrary non-cooperative games involving up to at least five possible strategies on regular lattices and one uses the typical "Imitate Best Neighbor" learning algorithm, a curious result occurs. Although each experiment generates a cellular automaton, almost all of the automata generated appear to be Wolfram Class I or Wolfram Class II. Wolfram Class I phenomena are characterized by a homogeneous end state in which all the sites have the same "color" (take on the same strategy). Wolfram Class II phenomena are characterized by "separating membranes" in which segregated areas of the population end up either in homogeneous states or in relatively short cycles through a small set of states. Notwithstanding what Stephen Wolfram has shown to be the ubiquity of Class III (chaos) and Class IV (complexity) automata in an r=2 k≥2 one-dimensional setting [9], spatial games clearly have great difficulty in generating them. Figure 2 shows some experiments on r=2, k=5 models.

4.2 *A Best 'on Average' Learning Strategy*

The rarity of Wolfram Class III and Wolfram Class IV behavior is somewhat troubling to those who would see spatial models as a useful metaphor for exploring social interaction. Surely, chaotic and complex behavior is more than occasionally observed in many real social systems. It turns out, however, that all that is needed to generate Wolfram Class III and Wolfram Class IV behavior, is for the sites to adopt a slightly more sophisticated learning algorithm. In the "Imitate Best Average" approach, each site adopts the behavior of the strategy that on average proved best amongst its information neighbors. Thus, if strategy 0 yielded payoffs of 1 and 10 and strategy 1 yielded a payoff of 7, the site would adopt strategy 1 on its next turn because 7 is higher than 5.5. Figure 3 shows Class III behavior emerging in a one dimensional lattice in which the payoff and information neighborhoods have a radius of one and there are only two strategies.

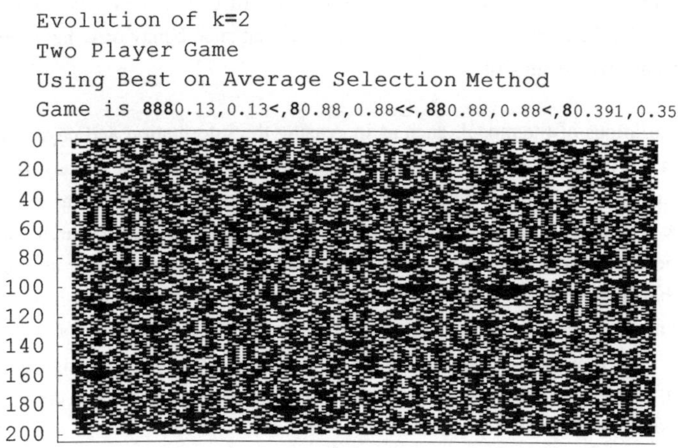

Figure 3

This modest extension preserves the parsimonious character of the model while permitting it to be more of a "field theory" for social interaction. An implication of this finding, as in all settings in which Class III or Class IV behavior emerges, is that the consequences of government interventions in society, such as deliberately flipping some group of behaviors via a pilot project, may be difficult to predict.

4.3 Regular Two Dimensional Lattices

Not only can the full range of system behaviors be generated by adopting a slightly more complex learning algorithm, they can also be generated by a slightly more complex coupling topology. Graphics in the appendix to the CD article show what appears to be Class IV behavior emerging as the result of a 2×2 two-person game played with von Neumann payoff and information neighborhoods.

4.4 Random Networks

Even when the lattice structure is itself random, the behavior produced by such Picker Models appears almost invariably to be Class I or Class II. As shown and discussed further on the conference CD, the same also appears to be true for "small world" graphs of the sort studied by Duncan Watts [7][8] and others [3]. If one looks at the sites in the random lattice, after going through brief transients, they all have either uniform behavior or periodic behavior in which the period involved is quite short. While it cannot be proven that Class III or Class IV never appear in such models, the many experiments run suggest that such behavior is far rarer than mere chance would suggest.

Many variations of the model, such as increasing the information and payoff radius, or using the product (rather than the sum) of the various games as the payoff do not appear to alter this fundamental result. Determining the precise reason for the absence of Class III or Class IV automata still requires further research. (It may occur because a learning algorithm in which all parties copy behavior similarly is somewhat analogous to "locally greedy" optimization algorithms, in which a local extremum is sometimes reached too swiftly and from which the process cannot escape. [2]) Playing random games using the "best on average" algorithm on random networks proves capable of generating complexity just as was the case for regular one dimensional networks.

4.5 *Deterministic Misperception*

Another method for creating chaos or complexity is systematic misperception. Suppose that for at least one configuration of neighbors, each site does not invariably copy the behavior of their most profitable neighbor but does something else instead, perhaps the opposite of that behavior. Figure 4 shows the result for a random two-strategy game {{{0.73, 0.78}, {0.29, 0.21}}, {{0.79, 0.13}, {0.43, 0.14}}} in which sites usually copy their most profitable neighbor, but when all sites in the affect neighborhood are "zero", the site experiments by trying "one" on its next turn. (Such a system bears, perhaps, a slight conceptual resemblance to continuous cellular automata described by Wolfram [9] in which a site takes on the fractional part of some function of the neighbors site values.) The result, depending on initial conditions, can be Class III or possibly Class IV. The impact of government interventions suggested such as pilot projects will thus be difficult to foretell.

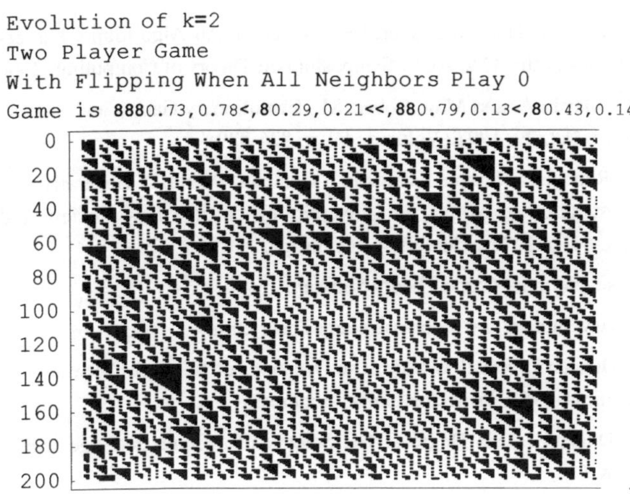

Figure 4

5 Conclusion

Classic models based on constrained optimization have been a powerful tool in economics and in the application of that discipline to the study of legal rules. This methodology has difficulty, however, in addressing scenarios in which numerous agents interact in a complex fashion and is probably unrealistic when complex, multi-dimensional global optimization is required. The "spatial games" metaphor for the world in which people are connected in a complex fashion and engage in copying behavior rather than elaborate optimization computation is an important and promising entry amongst a variety of newer techniques that promise a heightened capability to analyze and modify complex societies. This article demonstrates that these models are, at their core, simply cellular automata. As a result, many variants of spatial games will be capable of undertaking the universe of computations in which cellular automata can engage. Much additional study is needed to address more fully all the possibilities raised by this form of modeling. It would also be useful to develop further extensions of spatial games to situations in which the network of connections between members in society was endogenous.[1][10] This article confirms, however, the promise of the methodology to depict the range of human interactions more realistically and more directly.

6 References

[1] Gaylord, Richard J. & D'Andria, Louis, Simulating Society: A *Mathematica* Toolkit for Modeling Socioeconomic Behavior, 1998, Springer-Verlag

[2] Kennedy, James and Eberhart, Russell C. Swarm Intelligence (Chapter 5), 2001, Morgan Kaufmann.

[3] Kleinberg, Jon, The Small-World Phenomenon: An Algorithmic Perspective in Proceedings of the 32nd ACM Symposium on Theory of Computing, 2000.

[4] Nowak, Martin A. and May, Robert M., Evolutionary Games and Spatial Chaos, in Nature, Oct. 29, 1992 (Volume 359, p. 826-829)

[5] Picker, Randal C., Simple Games in a Complex World: A Generative Approach to the Adoption of Norms, in University of Chicago Law Review, Fall 1997 (Volume 64, p. 1225), University of Chicago.

[6] Siebrasse, Norman, Game Theory Software (http://www.unb.ca/law/Siebrasse/GameTheory.htm)

[7] Watts, Duncan J., Small Worlds: The Dynamics of Networks between Order and Randomness, 1999, Princeton University Press.

[8] Watts, Duncan J., Six Degrees: The Science of a Connected Age, 2003, W.W. Norton & Co.

[9] Wolfram, Stephen. A New Kind of Science, 2002 (particularly Chapters 4, 6, and 11), Wolfram Media.

[10] Zimmermann, Martín, Eguíluz, Victor M., and San Miguel, Maxi, Cooperation, adaptation and the emergence of leadership, in Economics with Heterogeneous Interacting Agents, 2001, Springer-Verlag.

A *Mathematica* Games Toolkit

Robert Cowen
Queens College, CUNY, USA
cowen@forbin.qc.edu
Robert Dickau
InstallShield Corp., USA
pops@prairienet.org

1 Abstract

We provide a toolkit for programming impartial combinatorial games in *Mathematica*. The toolkit consists of a notebook, GamesEngine.nb, which implements the Sprague-Grundy algorithm, along with some generic graphic elements for simple games. Particular games can then be easily programmed, as we shall demonstrate. A website has been established which contains the GamesEngine notebook, several examples of combinatorial games and our earlier paper, Playing Games with *Mathematica*, which can be downloaded. It is intended that the website will act as a repository of interesting games and techniques for exploring them with *Mathematica*. Finally, the role of games in mathematics education is discussed.

2 Impartial Combinatorial Games

Combinatorial games are finite, two-person games in which players alternate in choosing their moves, and the last player to make a move is declared the winner. Moreover, there are no 'draws' and no chance moves (dice, etc.). A combinatorial game is called *impartial* if both players have the same set of moves available in a given position. (If this is not the case, the combinatorial game is termed *partizan.)* Thus, chess is not an impartial combinatorial game on several counts; games can be drawn, the two players have different sets of moves in a given position, etc. Tic-tac-toe fails because draws are allowed. Monopoly has dice, etc. A good example of an impartial combinatorial game is the game of *Nim*. In Nim, a finite collection of heaps of counters is presented. Players take turns removing some of the counters from any one heap of their choice; it is not permitted to remove zero counters. The player who removes the last counter wins. There is a well-known algorithm for choosing a 'good' move, if one is available, in an impartial combinatorial game, called the *Sprague-Grundy* algorithm. It is highly recursive in nature, and easily programmed in *Mathematica,* as we have shown in our paper, [2], which also provides an introduction to these games. Much has been written about these games; for example see [1], [6] and A. S. Fraenkel's extensive bibliography [5].

3 A *Mathematica* Toolkit

In our paper, [2], we showed how to use *Mathematica*'s recursive capabilities to easily program the Sprague-Grundy algorithm for impartial combinatorial games and we used the program to investigate the game, CRAM, in which players place dominoes on a rectangular board ruled into squares. Dominoes cover two squares and can be placed horizontally or vertically. The positions in this particular game had to be represented in some way—we chose matrices of zeros and ones, and we also provided a pleasing output of a board with red and blue dominoes.

Here we provide a separate package, GamesEngine.nb, which contains the programmed Sprague-Grundy algorithm, along with some variations which randomize the choices to be made to produce different games, as well as a quick version of the Sprague-Grundy algorithm which often produces faster results. In addition, the GamesEngine provides some rudimentary graphics for games played on a strip ruled into squares, as well as for games on rectangular boards ruled into squares. The GamesEngine notebook should be loaded and evaluated before running individual games.

Individual games can then be rendered by choosing a representation of a game position and supplying a PossibleMoves function which designates the other positions one can move to from a given position. If the game is played on a strip or rectangular board, the graphics in the GamesEngine can usually be utilized. Of course, more pleasing graphics can also be supplied if desired. We shall illustrate this process with two games, 'Strip Dominoes' and 'Queens'.

4 Strip Dominoes

Strip Dominoes is played on a strip which is ruled into a number of squares. Players have dominoes which can cover two unoccupied squares. They alternate moves, and the last player to place a domino wins. We represent positions as lists of zeros and ones; the initial position is just a list of zeros. A move then consists of replacing two adjacent zeros by ones.

```
In[24]:= PossibleMoves[boardstate_] := ReplaceList[
           boardstate, {X___, 0, 0, Y___} -> {X, 1, 1, Y}]

In[25]:= initial[n_] := Table[0, {i, n}]

In[26]:= PlayTheGame[initial[13]]

Out[26]= {{0, 0, 0, 0, 0, 0, 0, 0, 0, 0, 0, 0, 0},
          {0, 0, 0, 1, 1, 0, 0, 0, 0, 0, 0, 0, 0},
          {1, 1, 0, 1, 1, 0, 0, 0, 0, 0, 0, 0, 0},
          {1, 1, 0, 1, 1, 0, 1, 1, 0, 0, 0, 0, 0},
          {1, 1, 0, 1, 1, 0, 1, 1, 1, 1, 0, 0, 0},
          {1, 1, 0, 1, 1, 0, 1, 1, 1, 1, 1, 1, 0}}
```

4.1 *Strip Dominoes (with graphics)*

We introduce some graphics elements to provide more attractive gameboards.

```
In[27]:= color[0] = GrayLevel[0.8];
         color[1] = GrayLevel[0.3];
         $Player = 0;
         board[n_Integer?Positive] :=
           Table[Line[{{i, 0}, {i + 1, 0},
              {i + 1, 1}, {i, 1}, {i, 0}}], {i, 1, n}];
         domino[M_List] := Module[{L}, L = Sort[M];
            $Player = 1 - $Player; {color[$Player], Rectangle[
              {L[[1]] + 1 / 4, 1 / 4}, {L[[2]] + 3 / 4, 3 / 4}]}];
```

4.2 *Some Games*

```
In[32]:= thegame = RandomPlayTheGame[initial[13]];
         $Player = 0;
```

```
In[33]:= dominoes = Map[domino[Flatten[#]] &,
           Map[Position[#, 1] &, Map[#[[2]] - #[[1]] &,
             Partition[thegame, 2, 1]]]];
```

```
In[34]:= Map[Show[Graphics[{board[13], #}],
           AspectRatio → Automatic] &,
         Table[Take[dominoes, turns],
           {turns, 0, Length[dominoes]}]];
```

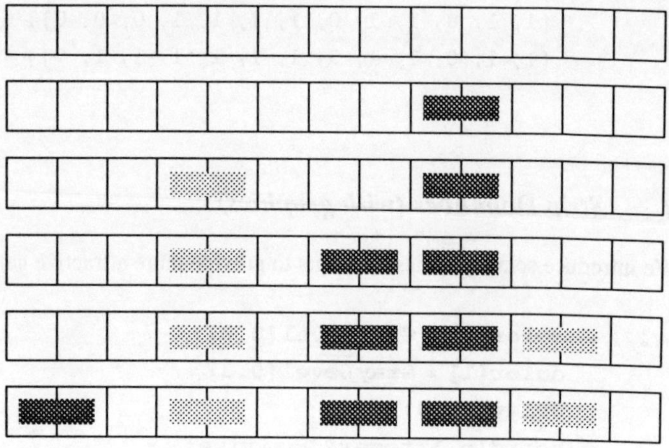

5 Queens

In this game, a rectangular board is given on which players can place non-attacking queens. The last player who can place such a queen wins. initial[n] gives the starting position—a chess board of size n×n with no queens on the board. For any board with queens, we first introduce the Shadow function, which puts a 1 on all squares under attack by any of the queens, while leaving a 0 on "safe" squares. We then define the PossibleMoves function, which adds a queen to any safe square.

```
In[35]:= initial[n_] := Table[0, {i, n}, {j, n}]
```

```
36]:= Shadow[m_] := Module[{w = m, len = Length[m]},
            Do[
                If[m[[i, j]] == 1,
                Do[
        w[[i, k]] = 1; w[[k, j]] = 1, {k, len}];
                        Do[w[[i - k, j + k]] = 1,
        {k, -Min[len - i, j - 1], Min[i - 1, len - j]}];
            Do[w[[i + k, j + k]] = 1,
        {k, 1 - Min[i, j], len - Max[i, j]}];

                ], {i, len}, {j, len}
            ]; w
            ]
```

```
In[37]:= PossibleMoves[m_] :=
       Module[{z = Shadow[m], w = m, moves = {}},
            Do[If[z[[i, j]] == 0, w[[i, j]] = 1;
         moves = Append[moves, w]; w = m],
        {i, Length[m]}, {j, Length[m]}
            ]; moves
            ]
```

This time we make use of the graphics provided in the GamesEngine, rather than providing additional graphics capability. We first use PlayTheGame and then RandomPlayTheGame on the same initial position. The question of how many Queens are necessary to complete a particular game on an n×n board is, in general, not known.

```
In[38]:= Map[showsboard, PlayTheGame[initial[5]]]
```

Out[38]= {

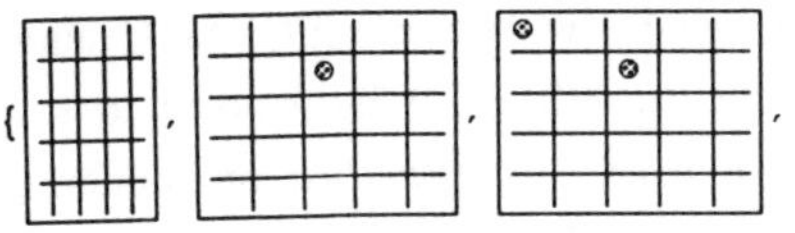

Out[38]= 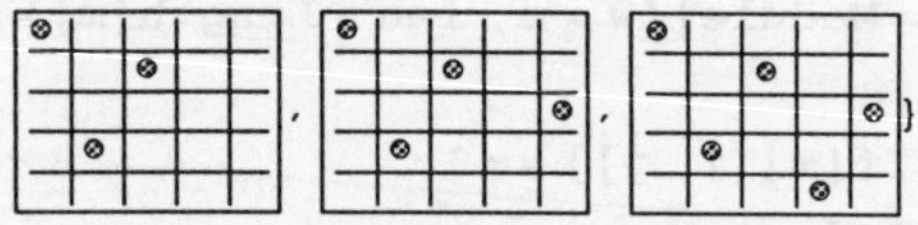

In[42]:= **Map[showsboard, RandomPlayTheGame[initial[5]]]**

Out[42]= 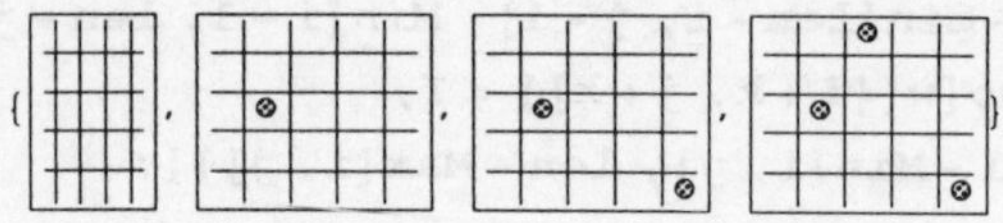

6 The Website

We have started a website to provide a resource for the games community. It currently contains the GamesEngine and our paper, [2], along with several games. We hope that others will contribute more games, better programming, better algorithms, improved graphics, etc. We want the site to be of use to all who are interested in combinatorial games. For example, it would be nice to allow interactive participation in game playing by simply clicking on graphic elements to indicate a move. Also, there are many unsolved problems in combinatorial game theory (see [1]); large computations can often be of some help and *Mathematica* could be of value here. In [2] we used *Mathematica* to work out which player wins in CRAM, with proper play, on the 5 by 5 board; the winner on the 7 by 7 game has yet to be determined.
The address of the website is http://www.qc.edu/~cowen/.

7 Games in Mathematics Education

Games are an ideal vehicle for interesting students in mathematics, as Martin Gardner has so ably demonstrated through his column in Scientific American and his many articles and books. Not only are they intrinsically appealing, but they raise many challeging problems which are often solvable without much formal training. Combining them with *Mathematica* increases the capabilities of the student to solve interesting problems and provides motivation to learn *Mathematica* programming. At Queens College, CUNY, John Kennedy and I, (Robert Cowen), offer a course called "Discovering Mathematics with *Mathematica*". The course uses mathematical problems to motivate the teaching of *Mathematica* programming techniques. Students learn programming in a strong contextual framework, which leads to better understanding of both the mathematics involved and the programming techniques. Student projects are emphasized, which motivates students to learn new techniques and reinforces old ones as they write the programs they need. Games are an important part of this course and we have our students investigate various games with the computer. In some cases the students invent the games they investigate. Students are required to do an individual project, which they hand in and also present to the class; often it will involve a game and new discoveries are almost always made. The Lab Manual, [4], for our course has a chapter devoted to games and another devoted to the Lights Out Puzzle. (John and I also published an article on it—see [3].) In fact, the original motivation for my interest in games and *Mathematica* was the desire to provide material for this class in *Mathematica* programming.

8 References

[1] Berlekamp, E.R., Conway, J.H. and Guy, R.K., *Winning Ways for Your Mathematical Plays*, Academic Press, London, 1982

[2] Cowen, R. and Dickau, R., *Mathematica* Plays Games, in *Mathematica in Education and Research*, vol. 7 (1998), no. 3, 5–10

[3] Cowen, R. and Kennedy, J., The lights out puzzle, in *Mathematica in Education and Research*, vol. 9 (2000), no. 1, 28–32

[4] Cowen, R. and Kennedy, J., *Discovering Mathematics with Mathematica*, Erudition Books, North Chelmsford, MA., 2001

[5] Fraenkel, A.S., Combinatorial Games: Selected Bibliography with a succinct gourmet introduction, in *Games of No Chance*, MSRI Publ. Vol.29 (ed. Nowakowski, R.J.), 493–537, Cambridge Univ. Press, Cambridge, 1998

[6] Guy, R.K., Impartial Games, in *Games of No Chance*, MSRI Publ. Vol.29 {ed. Nowakowski, R.J.), 61–78, Cambridge Univ. Press, Cambridge, 1998

Fractal Dimension and Musical Complexity. From Bach to the Blues

David Fowler
University of Nebraska-Lincoln, USA
dfowler@unl.edu

1 Abstract

A "musical coastline" can be derived from an instrumental solo by plotting pitch against time. The result is a discrete approximation to a (possibly) fractal curve. The fractal dimension of the curve yields some measure of the complexity of the musical phrase. Examples are taken from a violin sonata by J.S. Bach and from solos by three blues guitar players. To establish the reliability of the programming code, mathematical curves of known dimension are included in the analysis.

2 Introduction

Although fractal dimension is a widely-used mathematical tool, relatively little use has been made of it in analysing music from a complexity-theoretic viewpoint. Typical studies include [1], [2], [3]. This report parallels the method described in Mandelbrot's classic paper "How Long is the Coastline of Britain?" [4]. His familiar image is shown in Figure 1. Mandelbrot used the analogy of walking a pair of dividers along a map. His definition of fractal dimension is commonly called the "divider dimension."

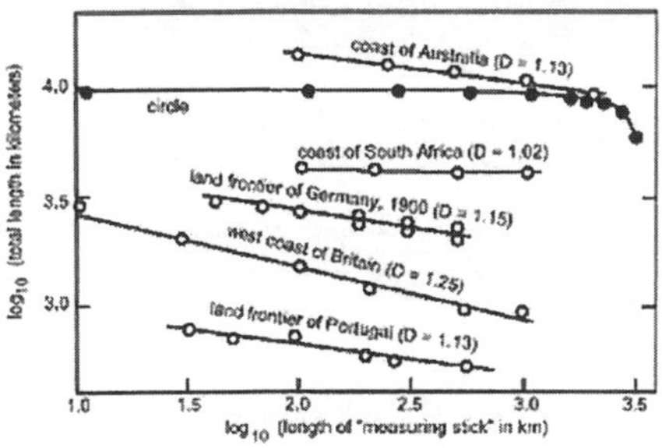

Figure 1. Mandelbrot' s analysis of natural coastlines and boundaries.

3 Defining a "Musical Coastline"

Musical notation can be thought of as a 300-year old object-oriented programming language that is executed at runtime by a musician, (or, in our time, possibly by a computer). In this notation system, horizontal lines correspond to audio frequencies, and symbols ("notes") representing duration are plotted against a temporal axis running from left to right horizontally. Replacing the notes by a connected series of points produces the graph shown in Figure 2.

Figure 2. Construction of a "musical coastline" from traditional notation.

This procedure is applied to the third movement of J. S. Bach's G-minor sonata for unaccompanied violin. Figure 3 shows a portion of the work in Bach's autograph, and Figure 4 shows a portion of the resulting representation. The selection begins approximately at the middle of the top line.

Figure 3. A portion of Bach's score for unaccompanied violin.

Traditional music notation can also be represented using the Musical Instrument Digital Interface (MIDI) standard, in which frequency information is converted from a standard pitch to an integral value. The values for the set of notes corresponding to the movement of the Bach sonata begins with the sequence {79, 79, 82, 82, 79, 79, 74, 74, 79, 79, 74, 74, 70, 70, 74, 74, ... }. The complete movement of the sonata contains 998 notes. A picture of the first 48 notes is readily obtained using *Mathematica*'s `ListPlot` function.

```
ListPlot[Take[bachLine, 48], PlotJoined → True];
```

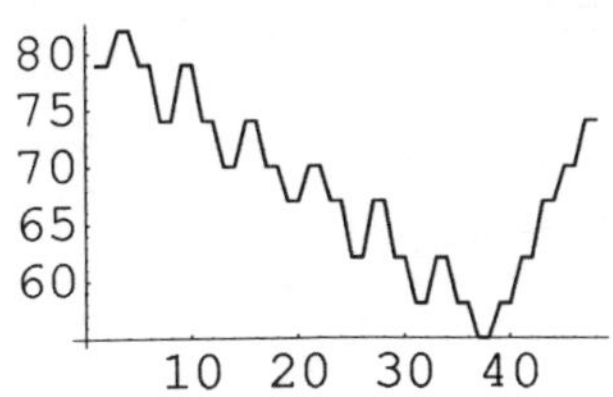

Figure 4. The first 48 notes of the Bach score.

4 Calculating the Divider Dimension

Mandelbrot observed that the length of a coastline, $L(\epsilon)$, is dependent on the length ϵ of the measuring chain, or the distance set on a divider marking off points across a map. He studied the behavior of the ratio $\log L(\epsilon) / \log \epsilon$ as $\epsilon \to 0$. In the following sections, a similar analysis is applied to the "coastlines" generated by the Bach solo and solos by three blues guitar players: B.B. King, Buddy Guy and Gary Moore. The *Mathematica* code used in these analyses is also applied to two mathematical curves of known fractal dimension - a sine curve and the familiar von Koch fractal curve. The von Koch curve can be generated by a number of different algorithms. For our analysis it is essential that the points in a more finely-grained iteration of the curve are regenerated in consecutive order, rather than added to existing points. For example, the third iteration of `kochS` contains 128 points, and point number 64 is at the central apex of the curve. The fourth iteration contains 512 points, and the 64th point is now at the apex of a left branch of the curve, as shown in Figure 5.

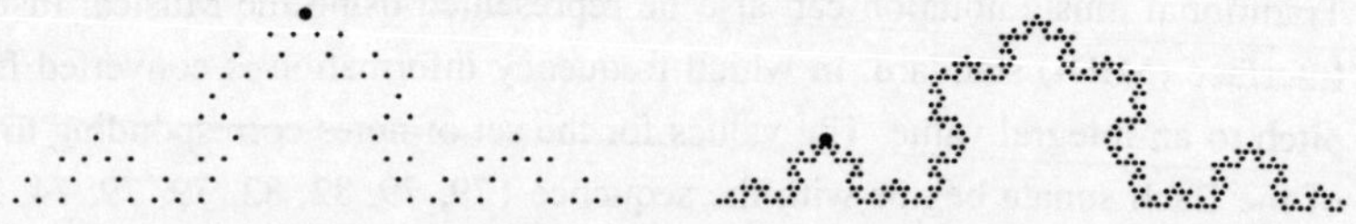

Figure 5. The 64th point in the third and fourth iterations of the von Koch curve.

The function **`testDist`** corresponds to the divider action across a map and can be shown graphically for different divider settings. Figure 6 shows divider settings of 0.2 and 0.053, both applied to the 4th iteration of **`kochS`**.

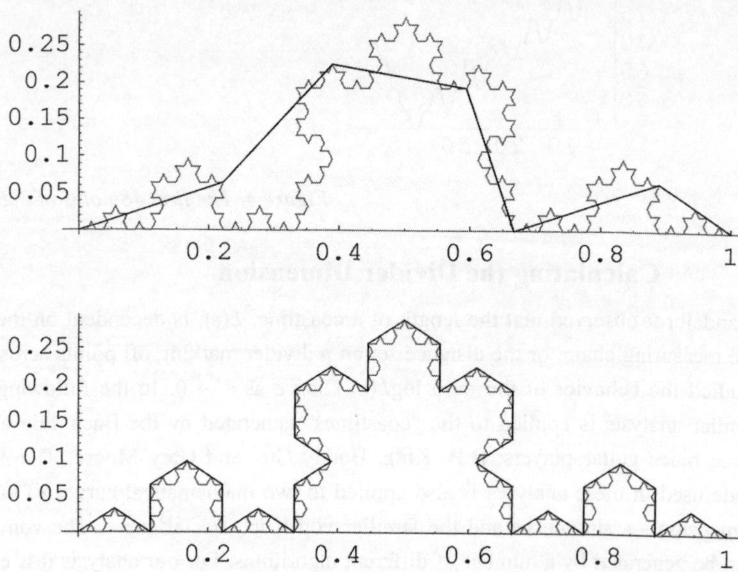

Figure 6. Graphic representation for two settings of the divider function.

The **`testDist`** function can be applied to any normalized data set. Here is the complete data set for the Bach solo, measured with a divider setting of 0.4

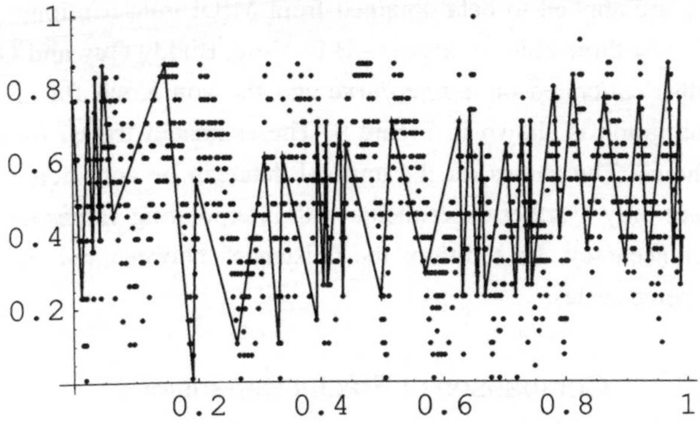

Figure 7. A divider path through the data set for the Bach violin sonata.

The function **returnValues[curve_,divStep_]** returns the length of a curve as estimated by a given divider-step value. We can plot the logarithms of these values as the divider step decreases and then fit a regression line to this plot. *Mathematica* provides useful built-in functions for such a purpose. The slope of the regression line gives an estimation of the fractal dimension. Figure 8 shows a plot of values for a sixth iteration von Koch curve with divider settings from 0.24 to 0.01 in steps of -0.01.

```
logDataPlot = ListPlot[logData];
```

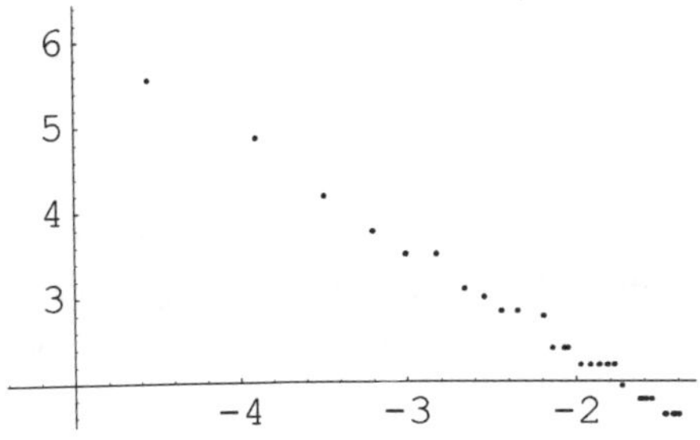

Figure 8. Log-log plot for von Koch curve.

```
In[23]:= Fit[logData, {1, x}, x]

Out[23]= -0.198561 - 1.26315 x
```

The above functions are applied to data obtained from MIDI transcriptions of the Bach violin score and solos by three blues guitarists - B.B. King, Buddy Guy and Gary Moore. The functions are also calibrated on a sine curve and the von Koch fractal curve. The resulting comparison graph is shown in Figure 9. The estimated fractal dimensions for each data set are shown. The values for the musical data can be considered accurate to the first decimal place only. Further experiments with interpolating values between notes may lead to greater accuracy. In addition, the "distance" between two notes may be recalculated for tempered scales.

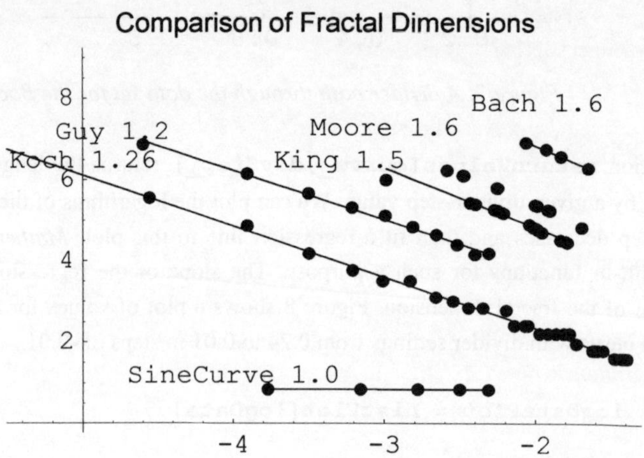

Figure 9. Musical coastline comparisons.

5 Conclusion

This preliminary report looks for evidence of fractal dimension in both classical and modern vernacular music. Evidence of fractal structure in music may add a useful tool to recent mathematical approaches to music [5] and studies in musical cognition [6]. Fractal analysis of music, in particular the solos of blues guitar players, may be an interesting complement to Eglash's study of fractals in African art and culture [7].

Fractal analysis is not without its critics. In a mathematical curve such as the von Koch curve one can study fractal structure at arbitrary depths of recursion. Natural and humanly-constructed patterns may not always have the necessary depth to adequately determine whether the structure is truly "fractal." See the debate on this subject between Avnir and Mandelbrot in [8].

6 References

[1] Hsu, K.J. and A. Hsu, A., "Fractal Geometry of Music," *Proceedings of the National Academy of Science*, USA. **87** (1990), 938-941.

[2] Hsu, K.J. and A. Hsu, A., " Self-similarity of the '1/f noise' called music," *Proceedings of the National Academy of Science* USA **88** (1991), 3507-3509.

[3] Bigerelle, M. and Iost, A. "Fractal dimension and classification of music." *Chaos, Solitons & Fractals,* **11**, p 2179. Elsevier.

[4] Mandelbrot, B. "How long is the coastline of Britain?" *Science* **155** (1967) 636-638.

[5] Assayag, G., Feichtinger, H. G. and Rodrigues, J. F. (Eds). *Mathematics and Music.* (2002). Springer-Verlag.

[6] Wallin, N.L., Merker, B. and Brown, S. (Eds). *The Origins of Music.* (2001) MIT Press.

[7] Eglash, R. *African Fractals: Modern Computing and Indigenous Design.* (1999). Rutgers University Press.

[8] Biham, O., Malcai, O., Lidar, D. and Avnir, D. "Is Nature Fractal? - response to M. Mandelbrot and P. Pfeifer", *Science*, **279** , 785-786 (1998).

Editor's note

Isn't modern music notation rather older than that? I'd guess more like 800 years old!

Functional Logic Origami Programming with Open CFLP

Tetsuo Ida
Institute of Information Sciences and Electronics
University of Tsukuba, Japan
Tetsuo.Ida@acm.org
Mircea Marin
Radon Institute for Computational Applied Mathematics
Austrian Academy of Sciences
mmarin@risc.uni-linz.ac.at

1 Abstract

Origami is a Japanese traditional art of paper folding. Recently, Origami became a topic of active research due to its relation to art, geometry, theorem proving, and declarative programming. It has been recognized that several interesting mathematical problems can be described in an elegant way by paper folds. In this paper we describe an origami programming methodology based on functional logic programming. We use two software components. The first one provides primitives to construct, manipulate and visualize paper folds. The second one, called Open CFLP, solves systems of equations whose operations are defined by conditional rewrite rules. We show that paper folding constructs can be expressed as systems of equations which are then solved by Open CFLP. Both components of our programming environment are implemented in *Mathematica*. We illustrate functional logic origami programming with some interesting examples.

2 Introduction

In this paper we describe Origami programming based on a constraint functional logic programming system called Open CFLP. There are two stages in functional logic Origami programming. The first one is to define operations to construct, manipulate and visualize paper folds, and the second is to program origami constructions with Open CFLP in a functional logic style.

The *Mathematica* user can program an origami construction by invoking certain folding functions with parameters that determine the paper fold. Moreover, the final result, as well as the results of the intermediate steps, of these operations can be visualized and manipulated. This part is described in [5] and the nucleus of its implementation consists of six basic fold operations. Each operation yields a result which satisfies a set of constraints which can be encoded as geometric relations in the Euclidean space $\mathbf{R}^2$. Therefore, we can exploit a constraint functional logic programming style, where the operations are defined by conditional rewrite rules. The advantages of such a programming style are:

1. (Constraint) functional logic programs can be regarded as logical specifications of the program behavior. Therefore, the origami construction steps are easy to understand and verify.

2. We can solve systems of equations which are defined by operators defined by functional logic programs. In the context of origami programming, this amounts to querying whether certain geometric properties of origamis, e.g. analytic characterizations of their defining points, segments or planes, hold.

Solving such systems of equations can be achieved by combining constraint solvers. The combined solvers interpret the functional logic operators and can solve the constraints which describe the geometric properties of origamis. To this end, we employ a collaborative constraint solving system called Open CFLP [3,4]. To Origami programmers, Open CFLP provides support for:

1. constraint functional logic programming, i.e. defining functional logic operators by means of conditional rewrite rules;

2. collaborative constraint solving, i.e. defining complex constraint solving procedures by combinations of elementary constraint solvers.

The range of problems that can be modeled in this way is surprisingly diverse. For instance, origamis can be regarded as geometric tools to solve quadratic or cubic equations, or to solve the problem of trisecting an angle, (which requires solving a cubic equation). It is well known that solving cubic equations cannot be done by ruler and compass construction.

Later we will illustrate how the origami operations can be programmed with Open CFLP, how origamis can model interesting problems, and how we can query Open CFLP to solve them.

3 Origami Axioms

We have adopted the six origami axioms due to Huzita [2]. The types of operations involved in each axiom are indicated within square brackets. "Th" denotes a "through" operation, and "Br" denotes a "bring" operation. They are used as part of the names of functions to be defined.

(O1) [Th] Given two points *P* and *Q*, make a fold along a line connecting them.

(O2) [Br] Given two points *P* and *Q*, make a fold to bring one of the points onto the other.

(O3) [Br] Given two lines *m* and *n*, make a fold to bring one of the lines onto the other.

(O4) [Th] Given a point *P* and a line *m*, make a fold along the line perpendicular to *m* that passes through *P*.

(O5) [BrTh] Given two points *P* and *Q* and a line *m*, make a fold that brings *P* onto *m* and passes through *Q*.

(O6) [BrBr] Given two points *P* and *Q* and two lines *m* and *n*, make a fold that brings *P* and *Q* onto *m* and *n*, respectively.

The specification of these operations can be defined by a functional logic program. We illustrate how to do this with Open CFLP.

```
In[1]:= << cflp`FrontendCFLP`;

In[2]:= (* Declare type constructors *)
        Instance[TLine : Eq]; Instance[TPt : Eq];

In[3]:= Instance[TSegm : Eq]; Instance[TPolyPair : Eq];
        Instance[TSPPair : Eq];

In[4]:= DataConstructors[{
            Line : ℝ ⟶ ℝ ⟶ ℝ ⟶ TLine,
            Pt : ℝ ⟶ ℝ ⟶ TPt,
            SPPair : TLine ⟶ TPt ⟶ TSPPair,
            PolyPair : list[TPt] ⟶ list[TPt] ⟶ TPolyPair,
            Segm : TPt ⟶ TPt ⟶ TSegm}];
```

```
In[5]:= DefinedSymbols[{
         DoubleVerticalBar : TLine⟶TLine⟶B,
         UpTee : TLine⟶TLine⟶B,
         DownRightVector : TPt⟶TLine⟶B,
         SegmToLine : TSegm⟶TLine,
         PerBisector : TPt⟶TPt⟶TLine,
         MidPt : TPt⟶TPt⟶TPt,
         SymPt : TPt⟶TLine⟶TPt,
         BrTh : TPt⟶TSegm⟶TPt⟶TSPPair,
         BrBr : TPt⟶TSegm⟶TPt⟶TSegm⟶TLine}]
```

Open CFLP is enabled by loading the package `FrontendCFLP.m`. The user can declare his own type constructors (`TLine` for lines, `TPt` for points, etc.) via `Instance[]` calls, and the associated data constructors (`Line` for `TLine` type, `Pt` for `TPt` type, etc.) via `DataConstructors[]` calls. Eq is the class of types with equality, **R** is the built-in type of real numbers, and **B** is the built-in type of booleans. A line described by the equation $a\ x + b\ y + c = 0$ corresponds to the construct `Line[a,b,c]`. Similarly, user defined operators are declared via `DefinedSymbols[]` calls.

The meaning of user defined symbols is specified via a (constraint) functional logic program. Such a program is a set of conditional rewrite rules of the form *f[args]→rhs* or *f[args]→rhs⇐cond,* where *cond* is a list of boolean expressions which denotes their logical conjunction. Extra variables are allowed in the conditional part, where they are supposed to be existentially quantified. An expression *f[args]→rhs⇐cond* partially defines *f,* to reduce any instance *f[args]/.θ* to *rhs/.θ* if *cond/.θ* holds.

Please note the semantic differences between conditional rewrite rules and *Mathematica* transformation rules, and that conditional rewrite rules must be well typed.

The program `Prog` shown below illustrates how origami operations can be defined in a functional logic style.

```
In[6]:= Prog = FLPProgram[{
          Line[a, b, c] ∥ Line[a1, b1, c1] ⇐ a b1 - a1 b ≈ 0,
          Line[a, b, c] ⊥ Line[a1, b1, c1] ⇐
           a a1 + b b1 ≈ 0,
          (Pt[x, y] ⇁ Line[a, b, c]) ⇐ a x + b y + c ≈ 0,
          SegmToLine[Segm[Pt[x1, y1], Pt[x2, y2]]] →
           Line[y2 - y1, x1 - x2, x2 y1 - x1 y2],
          MidPt[Pt[px, py], Pt[qx, qy]] →
           Pt[(px + qx) / 2, (py + qy) / 2],
          PerBisector[Pt[a1, b1], Pt[a2, b2]] → Line[
            2 (a2 - a1), 2 (b2 - b1), a1^2 - a2^2 + b1^2 - b2^2],
          SymPt[Pt[px, py], Line[a, b, c]] →
           Pt[((b^2 - a^2) px - 2 a b py - 2 a c) / (a^2 + b^2),
            ((a^2 - b^2) py - 2 a b px - 2 b c) / (a^2 + b^2)],
          BrTh[P, seg, Q] → SPPair[n, R] ⇐
           {PerBisector[P, R] ≈ n,
            Q ⇁ n, R ⇁ SegmToLine[seg]},
          BrBr[P, s1, Q, s2] → l ⇐
           {l ≈ Line[a, 1, c], P2 ≈ SymPt[P, l],
            Q2 ≈ SymPt[Q, l], SegmToLine[s1] ▶ l1,
            P2 ⇁ l1, SegmToLine[s2] ▶ l2, Q2 ⇁ l2},
          BrBr[P, s1, Q, s2] → l ⇐
           {l ≈ Line[1, 0, c], P2 ≈ SymPt[P, l],
            Q2 ≈ SymPt[Q, l], SegmToLine[s1] ▶ l1,
            P2 ⇁ l1, SegmToLine[s2] ▶ l2, Q2 ⇁ l2}}];
```

`DoubleVerticalBar` and `UpTee` test whether the lines given as arguments are parallel or perpendicular respectively. `P⇁l` yields True iff `P` belongs to the line `l`. `MidPt[P,Q]` yields the midpoint between `P` and `Q`. `SegmToLine[seg]` yields the support line of segment `seg`. `BrBr[P,s,Q,t]` is, in essence, the implementation of axiom [O6]. The operators ⊳ and ≈ are predefined in Open CFLP, and their meaning is given by: $term_1 \approx term_2$ if $term_1$ and $term_2$ evaluate to some common value, and *term*⊳*val* if *term* evaluates to *val*. The meaning of the other user defined operations is easy to read off from their implementation.

4 Applications

Problem 1. Solve the quadratic equation $x^2 + p\,x + q = 0$ by Origami.

The following method is due to [1]. We consider a sheet of paper whose corners are:

```
In[7]:= •O = Pt[0, 0]; •P = Pt[-p, 0];
        •X = Pt[-p, q]; •Q = Pt[0, q];
```

To make our description concrete, we assume p=4 and q=3. If we consider the points •U=Pt[0,1], •M=MidPt[•X, •U] and the segment seg determined by •P and •O,

```
In[8]:= p = 4; q = 3; •U = Pt[0, 1];
        •M = MidPt[•X, •U]; seg = Segm[•P, •O];
```

then it can be shown that the equation SPPair[n,Pt[X,0]]≈BrTh[•U,seg,•M] in variables n and X holds iff X is a solution of $x^2 + p\,x + q = 0$, and n is the line through •M which brings point •U to Pt[X,0] on seg. With Open CFLP, we can specify the problem of computing X as follows:

```
In[9]:= Prob1 = exists[{X},
            SPPair[n, Pt[X, 0]] ≈ BrTh[•U, seg, •M]];
```

Finally, we inform the system what solvers to employ and how to combine their solving capabilities in order to get the desired result. For Prob1 we need only 2 solvers: a solver which can interpret the user defined operators, and a solver which can solve polynomial equations over **R**. The basic distribution of Open CFLP provides default implementations for both solvers: LNSolver.m for equational reasoning with user defined solvers, and PolynSolver.m for systems of polynomial equations.

```
In[10]:= polyn = MkLocalSolver["PolynSolver`"];
         flp = MkLocalSolver["LNSolver`"];
```

Some solvers must be configured in order to work properly. To solve Prob1, we configure flp to reason over the program Prog and compute with the calculus LCNCd, (deterministic lazy conditional narrowing calculus). We apply the functional logic solver flp to the query Prob1. The solver flp reduces Prob1 to an equivalent problem without user defined symbols, and then the solver polyn computes the bindings for n and X by solving the intermediate problem. The sequential application of these solvers is indicated by the combinator seq.

```
In[11]:= ConfigSolver[flp,
            {Program → Prog, Calculus → "LCNCd"}];
         ApplyCollaborative[seq[{flp, polyn}], {Prob1}]

Out[12]= {{X → -3}, {X → -1}}
```

Problem 2. Consider a rectangular paper whose corners are given by

```
In[13]:= ▪A = Pt[a, -b]; ▪B = Pt[a, b];
         ▪C = Pt[-a, b]; ▪D = Pt[-a, -b];
```

The points F1, F2 and the segments seg1, seg2 are defined by

```
In[14]:= F1 = Pt[a, 0]; F2 = Pt[0, b];
         s1 = Segm[▪D, ▪C]; s2 = Segm[▪D, ▪A];
```

Then BrBr[F1,s1,F2,s2] computes a line of the form Line[$\sqrt[3]{a/b}$,1,w]. We illustrate how to use this property to compute $\sqrt[3]{7/11}$ with origami.

```
In[18]:= a = 7; b = 11;
         Prob2 = exists[{l}, l ≈ BrBr[F1, s1, F2, s2]];
```

We solve this problem by a sequential collaboration of two constraint solvers: flp and a solver based on variable elimination with back substitution, (based on the *Mathematica* implementation ElimSolver.m provided with the distribution of Open CFLP). The call:

```
In[19]:= elim = MkLocalSolver["ElimSolver`"];
         ApplyCollaborative[seq[{flp, elim}], {Prob2}]
```

Out[20]= $\{\{1 \to \mathrm{Line}[(\frac{7}{11})^{1/3}, 1, 7^{2/3}\, 11^{1/3}]\}\}$

yields the solution $\{\{\texttt{l} \to \texttt{Line}[(\frac{7}{11})^{1/3}, 1, 7^{2/3}\,11^{1/3}]\}\}$.
The corresponding paper folding step is illustrated below.

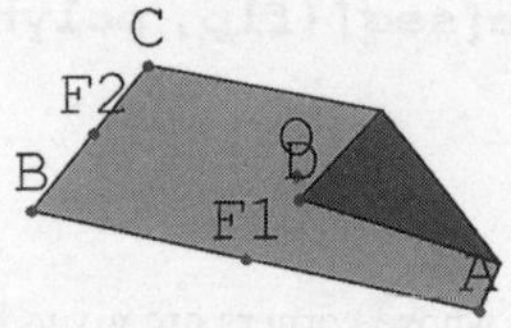

The construction of a segment of length $\sqrt[3]{7/11}$ is straightforward. We can construct (via two successive applications of `[Th]`) a fold l_1 through the origin •O which is parallel with `l`. Similarly, we can construct a fold l_2 through `Q=Pt[1,0]` which is parallel with `F1•O`. If `R` is the intersection of l_1 with l_2, then the length of the segment `QR` is $\sqrt[3]{7/11}$.

5 Conclusion

Functional logic origami programming gives a succinct way to construct paper folds. Moreover, we can analyze the geometric properties of origamis with Open CFLP.

References

[1] Geretschläger, R., Geometric Constructions in Origami (in Japanese, translation by Hidetoshi Fukagawa), Morikita Publising Co., 2002.

[2] Huzita, H., Axiomatic Development of Origami Geometry, Proceedings of the First International Meeting of Origami Science and Technology, pp. 143--158, 1989.

[3] Kobayashi, N., Marin, M., and Ida, T., Collaborative Constraint Functional Logic Programming System in an Open Environment. *IEICE Transactions on Information and Systems*, E86-D(1):63--70, January 2003.

[4] Marin, M., Ida, T., and Schreiner, W., CFLP: A *Mathematica* Implementation of a Distributed Constraint Solving System. *The Mathematica Journal*, 8(2):287-300, 2001.

[5] Takahashi, H., and Ida, T., Origami Programming Environment, in International *Mathematica* Symposium, July 2003. To appear.

A webMathematica Application for a Sports Records Database

Chikara MIYAJI
Japan Institute of Sport Science
Chikara.MIYAJI@jiss.ntgk.go.jp

1 Abstract

This paper describes an application of webMathematica for constructing a web site for a sports records database. The web site has: a visual interface for the record data; a flexible searching mechanism; an interaction mechanism for user events. JDBC is used to access the database, and the data is directly imported into *Mathematica*. For the visual interface, Scalable Vector Graphics-SVG is used and the SVG is processed by *Mathematica* as expressions. *Mathematica*'s list manipulation functions enable flexible handling of XML tags. From the experience of the web site construction, the author concludes that webMathematica is suitable as a main processing center of web resources, such as HTML pages, graphics, XML, and databases.

2 Introduction

Our sports records database is a collection of various sports records such as for the Olympic Games, World Cups, and world class competitions. From the data, scientists involved in training can see the trends of the records, athletes and coaches can compare the records with others, and the media can check world records. The records are not only for sports science, but are also an important source for various users. The web site for sports records data should be user friendly and not pedantic. Although there are many sites for sports records on the net, these sites only show a list of numbers and lack a visual interface. User-friendly sports records data sites should have these capabilities: visualize the sports record data; versatile searching method; interactive user input. The author's web site for sports records database aims to have such capabilities.

3 Construction

The system comprises three layers; database, webMathematica, and browser. MySQL is used for the database engine, and JDBC is used for communication between the database and *Mathematica*. webMathematica visualizes the data. For the graphics on the web, SVG is used. The figure below shows the system construction.

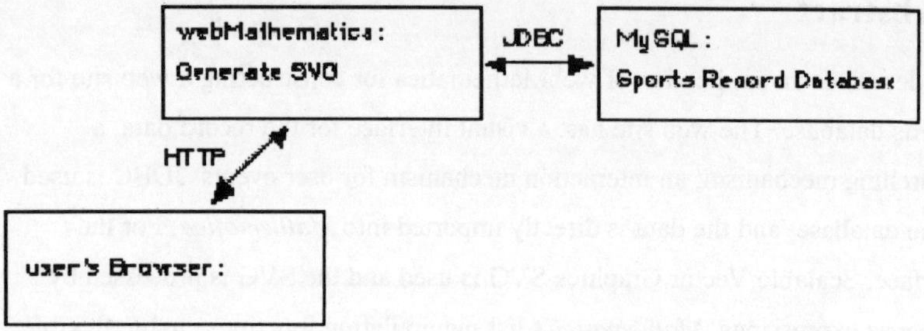

Figure 1. Three layers of the system. Database, Mathematica, and Browser.

3.1 Interface

There are several ways to communicate with the database on webMathematica. Wakamatsu [1] discusses methods involving PHP, Database Access Kit, and JDBC. For seamless connection with *Mathematica*, the author chooses JDBC as the connection method, and uses a JDBC driver for MySQL, Connector/J [2]. A small amount of Java codes was created for the interface between *Mathematica* and the database. A method, **evalsql(String sql)**, sends SQL commands and returns the result as a matrix. Inside the method, a loopback link is used to construct the result.

Program.1 Java interface to JDBC.

```
public int evalsql(String sql)
{...
// send SQL statement to database
{
      = conn.prepareStatement(sql);
      = stmt.executeQuery();
      rsmd = rs.getMetaData();
      = rsmd.getColumnCount();
      } catch(java.sql.SQLException e){
      ...
// get the result and write them on to loopback link
{
      (rs.next()){
            .putFunction("List", colcount);
            (int i=1; i <= colcount; ++i)
                  .put(rs.getString(i));
                  ++len;
            ( link.wasInterrupted()){
                  .close();
                  (0);
                  }
            }
      ...    //the result to Mathematica
{
.beginManual();
.putFunction("List", len);
.transferToEndOfLoopbackLink(loop);
      } catch(MathLinkException ex) {
      .close();
      (-6);
      }
.close();(0);
}
```

For example, these three lines of Mathematica code get the database result as a matrix.

```
zz = JavaNew["db"]
zz@connect["jdbc:mysql://127.0.0.1/result_db", "miyaji","
***"];
zz@
evalsql["select def_details from event_info where
eiid='49'"]
```

3.2 A Hyperlink on the SVG Graphics Object

SVG is a W3 standard 2-D graphics protocol that uses XML. The user can enlarge or shrink the graphics without any distortion. webMathematica has a built-in function called **`SVGShow[<graphics>]`** to make an SVG file and link it from a *Mathematica* graphics object.

To make the Sports records database interactive, a hyperlink to the SVG object is added. More precisely, there are plotted points of the records in graphical form, and each point has a hyperlink to the detailed description of the record such as name or date of the record.

A circle object is used to represent a point. Here is the difference between original circle object and hyperlinked object.

This is the original XML object of the plotting point in SVG output.

```
<circle stroke='none'
='1.0
='1.23607'
='0.0141874'/>
```

To add a hyperlink to this object, an xlink tag is wrapped around the object.

```
<a xlink:href="http://127.0.0.1:8080/
  webMathematica/MSP/Examples/detail?did=214"
target="_blank">
<circle stroke='none'
='1.0
='1.23607'
='0.0141874'/>
</a>
```

When this circle is clicked, a new webMathematica page "detail.msp" is opened with a 'did=214' argument. From this *did* information, the new page can display the detailed information of the record.

For all circle objects, an xlink tag and 'did' information must be added. To perform this task, list manipulation functions are used. These are the detailed procedures to perform this processing.

(1) Create an SVG graphic as a String.

```
xs = ExportString[g,"SVG"]
```

(2) Read it as a XML string.

```
xm = ImportString[xs,"XML",
  ConversopnOptions->{"ReadDTD"->False}]
```

Now, a circle tag is represented as an XMLElement expression.

```
XMLElement["circle", {"stroke"->"none", "cx"->"1.0",
   "cy"->"1.23607", "r"->"0.0141874"}, {}]
```

(3) Use **ReplaceAll** to add an xlink expression around circle expression.

```
XMLElement["a", "{xlink:href" ->
   "http://127.0.0.1:8080/
    webMathematica/MSP/Examples/detail?did=214",
   "target"->"-blank",
      XMLElement["circle", {"stroke"->"none", "cx"->"1.0",
       "cy"->"1.23607", "r"->"0.0141874"}, {}]} ]
```

(4) Export these modified XML expression to a file.

3.3 Files

This is a fragment of the final .msp file for plotting the data. The above processing is integrated into a Function **func[]**, and **xSVGShow[]** uses it.

Program 2: A fragment of .msp file for plotting the data.

```
...
<%Mathlet Get["Graphics`Graphics`"]; %>
<%Mathlet Get["recordDB`recdb`"]; %>
<%Mathlet LoadJavaClass["db"]; %>
<%Mathlet
```

```
preparePlotSQL[sql_String, x_String, y_String,
rrid_String] :=
      "SELECT " <> x <> "," <> y <> "," <> rrid <> "
from" <>
[sql, StringPosition[sql,
      "from"][[1, 2]]];
func[g_Graphics, rrids_] :=
[{zz,i=0,add,makelink},
[s_String,n_]:=ToString[ToExpression[s]+n];
[n_]:=
"http://xxx.jiss.ntgk.go.jp/webMathematica
/MSP/detail?dattype=rrid&did="<>
[rrids[[n]]];
zz=ImportString[ExportString[g,"SVG"],"XML",
ConversionOptions->{"ReadDTD"->False}]
/.XMLElement["circle",{"stroke"->"none",
"cx"->xx_,"cy"->yy_,"r"->rr_},{}]
                  ->addlink[xx,yy,rr];
              zz/.addlink[xx_,yy_,rr_]:>
        XMLElement["a",{"xlink:href"->
    makelink[++i],"target"->"_blank"},
      {XMLElement["circle",
      {"stroke"->"none","cx"->xx,"cy"->yy,
      "r"->add[rr,0.005]},{}]}]] %>
 <%Mathlet zz = JavaNew["db"]; %>
 <%Mathlet zz@connect["jdbc:mysql://127.0.0.1/result_db",
"miyaji", "XXX"]; %>
 <%Mathlet dd = zz@evalsql[preparePlotSQL[$$qq, $$x, $$y,
"rrid"]]; %>
 <%Mathlet zz@close[]; %>
<h1>plot the data</h1>
<%Mathlet ff=Fit[ret, If[Length[ret]>2,{1,x,x^2},{1,x}],
x] ;%>
<%Mathlet g1=Plot[ff,{x, min, max},
PlotStyle->RGBColor[1.0,0.0,0.0]]; %>
<%Mathlet g2=ListPlot[ret,
Frame->True,FrameLabel->{$$x,$$y},
ImageSize->{580,450}]; %>
<%Mathlet MSPBlock[{},xSVGShow[Show[g1,g2],func[#,rrids]
&]] %>
```

4 Results and Discussion

When the record data is displayed as graphics, it becomes easier to find small differences, and also enables us to show a large amount of data at once. Graphics are suitable for seeing the trend of particular athletes performances. But one of the problems of our site is its speed. Plotting SVG graphics sometimes takes time, for example, more than 10 sec, for a large amount of data. The bottleneck might be *Mathematica*'s slow in/out of strings, which is used three times in the **xSVGShow** function.

5 References

[1] Wakamatsu, , B. A Comparison of database access method for webMathematica, in Symbolic Manipulation (ed. Noro,E.N.), May 2001.

[2] MySQL JDBC driver,. <http://www.mysql.com/products/connector-j/index.html>

An Origami Programming Environment

Hidekazu Takahashi
Kawase High School, Japan
hidekazu@pop.biwako.ne.jp
Tetsuo Ida
Institute of Information Sciences and Electronics
University of Tsukuba, Japan
Tetsuo.Ida@acm.org

1 Abstract

This paper describes the implementation of an origami programming environment. The origami programming environment enables us to construct sophisticated origamis by calling origami folding functions with parameters necessary to define a fold. The *Mathematica* notebook with our origami environment package can become the place for constructing origamis step-by-step. The origami programming environment goes further than just simulating origami found in the literature, as it performs intriguing combinations of symbolic, numeric and graphic computations. We show examples taken from well-known pieces of origami art and from a mathematical origami construction.

2 Introduction

Origami is a Japanese traditional art created by means of paper folding. "Ori" in origami means folding and "gami" means paper in Japanese. The combined word "origami" means a folded paper, an act of folding an origami, or the art of folding an origami, depending on the context. Origami attracts not only lovers of paper art but mathematicians and computer scientists since, in the last decade, it was rediscovered as a highly sophisticated and powerful technique of constructing geometrical objects.

This paper describes the implementation of an origami programming environment in *Mathematica*. In our programming environment, instead of actually folding paper, we can program origami constructions. We invoke folding functions with parameters that determine the paper fold. We then view the result of the paper folding. We perform these operations repeatedly until we complete the origami construction.

The organization of the paper is as follows: We first give representations of basic geometrical objects used for representing origami. Then we illustrate origami programming with the construction of the trisection of an angle. Furthermore, we show two art pieces constructed in our origami environment.

3 Basics

We assume the notions of point, line, plane, (line) segment and polygon, and give the representations of those geometric objects in the following:

Origami

An origami is viewed as a collection of planes. An origami at step t is represented as a list $\{t$, *planes*, *order*$\}$, where *planes* is a list of planes (see below) and *order* is the overlap relationship between the planes that constitute the origami. Origamis are stored in the array Origami. We thus have the following identity: Origami[t]=$\{t$, *planes*, *order*$\}$.

Plane

A plane is represented as a list $\{$*id*, *poly*, *neighbors*, *mpoints*$\}$, where *id* is an integer identifying the plane, *poly*= $\{P_1, ..., P_k\}$ represents a polygon consisting of vertices P_1, ..., P_k, *neighbors* is a list of neighboring planes $\{o_1, ..., o_k\}$, where o_j is the *id* of the plane that is adjacent to the edge $\overline{P_j P_{j+1}}$, (we let $P_{k+1} = P_1$), and *mpoints* is a list of marked points (see below for the definition). As in the origami structure, we have the following identity: Plane[*id*] = $\{$*id*, *poly*, *neighbors*, *mpoints*$\}$.

Polygon

A polygon is represented in two ways in the system: vertex representation and edge representation. Given a sequence of vertices P_1, ..., P_k, ordered counter-clockwise, the vertex representation of a polygon is a list $\{P_1, ..., P_k\}$, and the edge representation of a polygon is a list $\{\{P_1, P_2\}, ..., \{P_k, P_1\}\}$.

Point and Marked point

A point whose x-, y- and z- co-ordinates are u, v and w respectively is represented as a list $\{u, v, w\}$. The z- co-ordinate is omitted when we are working only in 2D in the z=0 plane. A marked point is a structure that is used to refer to a designated point at some steps of the origami construction. It consists of a pair; a point name and co-ordinates.

Line, Segment and Ray

A segment between points *P* and *Q* (inclusive) is represented as a list {*P*, *Q*}. Segments are regarded as undirected. A directed segment is called a ray. A line $ax+by+c = 0$ on the z=0 plane is represented as Line[*a*, *b*, *c*]. A line is said to extend segment or ray {*P*, *Q*} if it passes through the points *P* and *Q*.

4 Origami Programming

An origami is folded along a line called a crease. The crease can be determined by the points it passes through or by the points (and/or lines) it brings together. The former type is called through action (abbreviated to Th), and the latter type, bring action (abbr. Br). We can also combine the two actions to get the following six basic fold operations, formally stated as Origami Axioms [1, 2] of Huzita. It is shown that with the six basic operations we can construct 2D geometrical objects that can be constructed by a ruler and compass. Actually, origami is more powerful than the ruler and compass method, since origami can construct objects that are not possible with ruler and compass. One of them is the trisection of a given angle, which we show later. Our first task in implementing the origami environment is to define the six basic operations.

4.1 Origami Axioms

Origami axioms are as follows: (The type of action involved in each axiom is indicated by the square brackets.)

(O1) [Th] Given two points *P* and *Q*, we can make a fold along a line connecting them.

(O2) [Br] Given two points *P* and *Q*, we can make a fold to bring one of the points onto the other.

(O3) [Br] Given two lines *m* and *n*, we can make a fold to bring one of the lines onto the other.

(O4) [Th] Given a point *P* and a line *m*, we can make a fold along the line perpendicular to *m* that passes through *P*.

(O5) [BrTh] Given two points *P* and *Q* and a line *m*, we can make a fold that brings *P* onto *m* and passes through *Q*.

(O6) [BrBr] Given two points *P* and *Q* and two lines *m* and *n*, we can make a fold that brings *P* and *Q* onto *m* and *n*, respectively.

From an algorithmic point of view, these axioms each imply two operations; finding a line and folding the origami along the line.

As an example, we discuss a function BrBr[*P_*, *m_*, *Q_*, *n_*] that computes a line such that the fold along it brings *P* onto *m* and *Q* onto *n*. For given parameters, there are at most three such lines, and BrBr returns a list of them. Once BrBr is defined, it is easy to define a function FoldBrBr[*id_*, *P_*, *m_*, *Q_*, *n_*] that makes a fold on plane *id* of the current origami along the line that BrBr[*P, m, Q, n*] computes. FoldBrBr[*id, P, m, Q, n*] realizes Axiom (O6). The fact that BrBr[*P, m, Q, n*] gives at most three lines suggests that the method for finding the lines involves solving a cubic equation. Indeed, it is known that trisecting an angle, which requires solving a cubic equation, is possible with origami, whereas it is not possible with ruler and compass [3]. We will next show how to program the trisection of an angle in our origami environment.

4.2 Trisecting an angle

The method that we show is due to H. Abe, as described in [4, 5]. First, we define an origami, which is a rectangle whose corners are designated by the points at a={0,0}, b={4,0}, c={4,4}, and d={0,4}.

```
BeginOrigami[{a = {0, 0}, b = {4, 0}, c = {4, 4}, d = {0, 4}}];
```

To facilitate identifying the points during the origami construction, we mark points a, b, c and d by names A, B, C, and D, respectively. We further introduce an arbitrary point e={3, 4}, marked by E on the origami. Our problem is to trisect the angle ∠EAB.

```
MarkPoint[{{"A", a}, {"B", b},
   {"C", c}, {"D", d}, {"E", e = {3, 4}}}, 1];
```

To see what the origami looks like, we invoke function ShowOrigami. ShowOrigami[ ] displays the origami at the current step.
Let f , g, h and i be the co-ordinates of the midpoints of line segments AD, BC, AF, and BG, respectively. We mark those points by F, G, H, and I, respectively.

```
MarkPoint[
  {{"F", f = Midpoint[a, d]}, {"G", g = Midpoint[b, c]},
   {"H", h = Midpoint[a, f]}, {"I", i = Midpoint[b, g]}}];
g = ShowOrigami[ShowId → False];
```

We can also show the origami in 3D to enhance the quality of the image. The function InjectToZ0, below, places the 2D point on the *z*=0 plane.

```
Show[{g, Graphics3D[{Hue[0], Line[InjectToZ0[{e, a}]],
     Line[InjectToZ0[{h, i}]]}]}] ;  (* fig.1 *)
```

We then make a fold along the line computed by BrBr[f, {a, e}, a, {h, i}], such that point F is brought onto line $\overline{AE}$ and point A is brought onto line $\overline{HI}$.

```
FoldBrBr[1, f, {a, e}, a, {h, i}];
    "Which line(1,2,3)?"
```

In this case there are three lines that can make this fold possible. The system responds by asking us along which line we want to make a fold. Let us answer this question by setting the 6th parameter to 1, i.e. choosing the first line.

```
FoldBrBr[1, f, {a, e}, a, {h, i}, 1, MarkCrease → True];
ShowOrigami[ShowId → False]; (* fig .2 *)
```

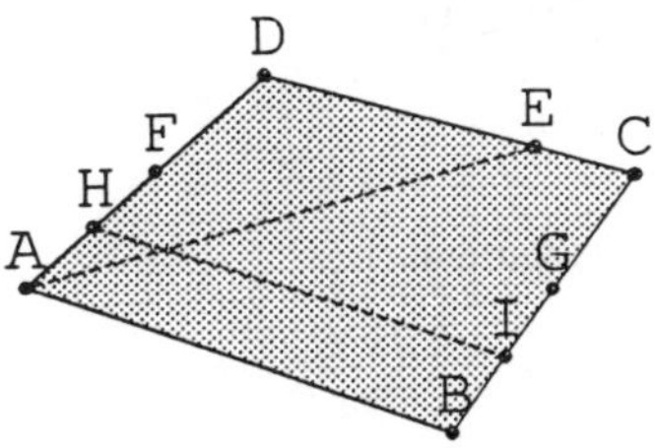

Figure 1

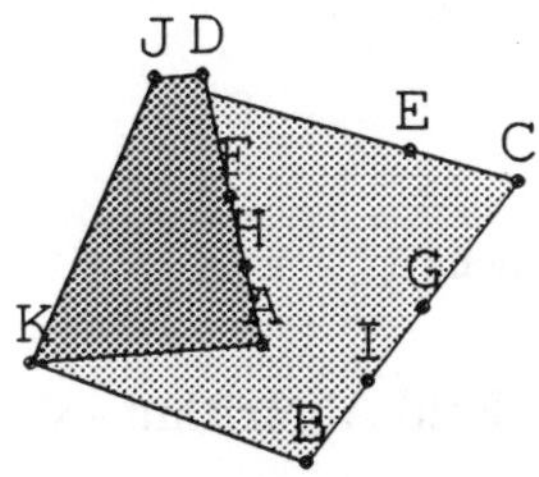

Figure 2

Let a_1 , h_1 and f_1 be the co-ordinates of points A, H and F after the fold. Points j and k are the end points of the crease generated by FoldBrBr[..., MarkCrease→True]. Points a_1 , h_1 , f_1 , j and k are marked by A_1 , H_1 , F_1 , J and K, respectively. We can revert to the origami at step 1 by calling BackToOrigami[1].

```
{a1, h1, f1, j, k} = Drop[First[MarkPointPosition[#]], -1] & /@
    {"A", "H", "F", "J", "K"};
BackToOrigami[1];
MarkPoint[Transpose[{{"A1", "H1", "F1"}, {a1, h1, f1}}]];
g1 = ShowOrigami[ShowId → False]; (* fig .3 *)
```

We now see that lines $\overline{AH_1}$ and $\overline{AA_1}$ trisect ∠EAB.

```
Show[{g1, Graphics3D[{Line[InjectToZ0[{j, k}]],
      {Hue[0], Line[InjectToZ0[{a, e}]],
       Line[InjectToZ0[{h, i}]]}, {Hue[0.6],
       Line[InjectToZ0[{a, f1}]], Line[InjectToZ0[{a, h1}]],
       Line[InjectToZ0[{a, a1}]]}}]}]; (* fig.4 *)
```

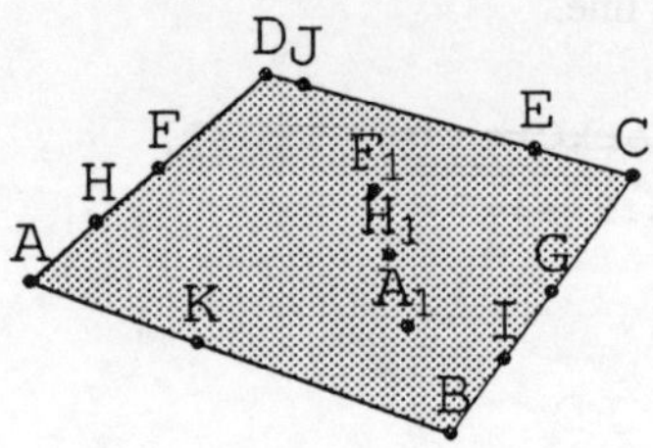

Figure 3

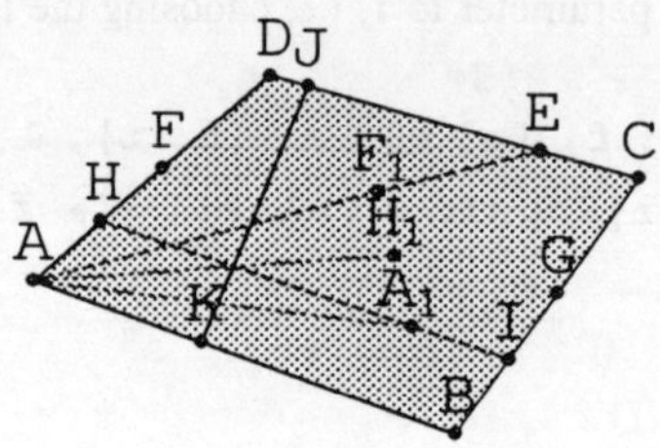

Figure 4

5 Basic Fold Algorithm - sketch

A one-step fold of an origami consists of finding a crease and making a fold along the crease. The fold is performed by the following algorithm: Given Origami[*t*]={*t, planes, order*}, and a *plane* ={*id*, *poly*, *neighbors*, *mpoints*} in *planes*, the algorithm performs the following:

1. Divide the polygon *poly* into two polygons *poly1* and *poly2,* where *poly1* is to be rotated by the fold and *poly2* is not rotated. We take *poly2* = *poly* if *poly* is not divided.

2. Collect all the planes that are to be rotated together with *poly1,* using the neighborhood relation *neighbors*. Let *moved* be the list of the collected planes.

3. Rotate all the planes in *moved* and the marked points in *mpoints* if they are on the moved planes. Let *newplanes* be the list of the planes after the rotation.

4. Compute the overlay relation *neworder* of all the planes in *newplanes*.

5. Set Origami[*t+1*]={*t+1*, *newplanes*, *neworder*}.

When we deal with 2D origamis performed by Origami Axioms, it is sufficient to consider two modes of folds, i.e. mountain fold and valley fold, which rotate the planes by angle π or $-\pi$. However, for art origami we need more sophisticated folds. We have implemented the inside reverse fold, (which is called *naka wari ori* in Japanese, meaning "fold by splitting in the middle"). Due to lack of space we do not describe the algorithm for the inside reverse fold, but we can see the necessity of this fold in the crane example below.

6 Examples

The construction of these examples requires 3D processing, as the origami is folded by an angle other than π or $-\pi$.

6.1 *Airplane*

We begin by declaring an origami:

```
BeginOrigami[{{0, 0}, {4, 0}, {4, 4}, {0, 4}}];
ShowOrigami[ShowId → False];
```

Then we execute the following:
(2) ValleyFold[{{0,2},{2,0}}, π]
(3) ValleyFold[{{2,4},{0,2}}, π]
(4) ValleyFold[{{4,2},{0,2}}, π]
(5) ValleyFold[{{0,2},{4,6-4Sqrt[2]}}, $\pi/2$]
(6) MountainFold[{{0,2},{4,6-4Sqrt[2]}}, $\pi/2$]
(7) ValleyFold[{{0,2},{4,2}}, $\pi/12$]
We have the following sequence of origamis. The last origami is the rotation of the origami in step 7.

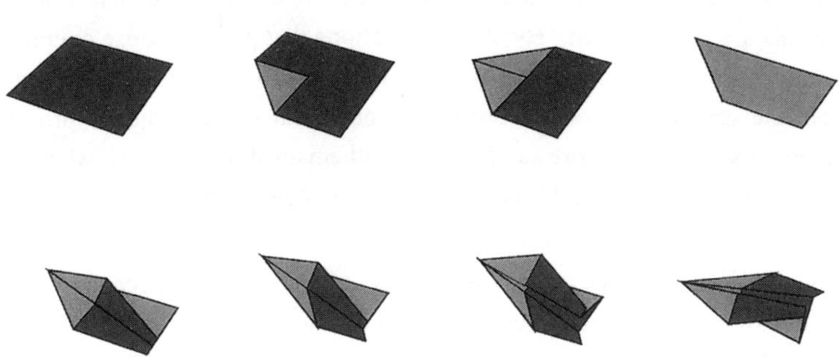

6.2 *Crane*

We show the crane construction below. The construction requires 18 steps of folds. The folds at steps 2-7 and 14-16 are inside reverse folds.

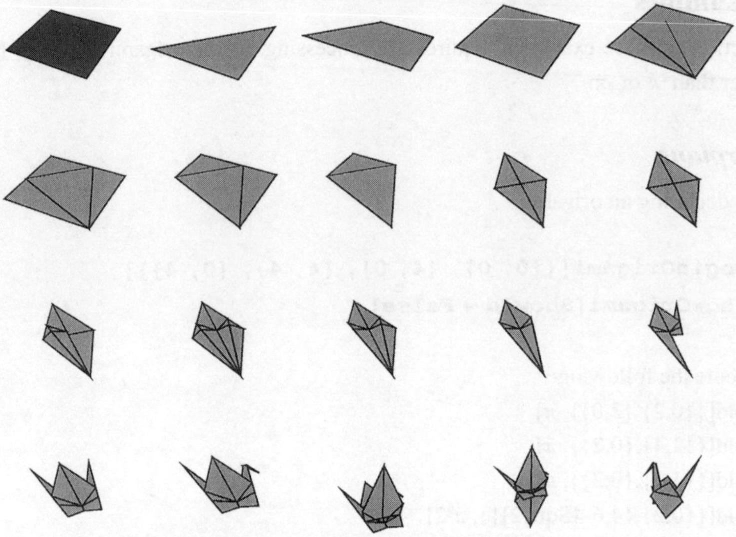

7 Concluding Remarks

We have shown the origami programming environment by illustrating basic and sophisticated examples taken from well-known origami constructions. Origami is constructed purely by means of geometrical computation: solving equations and simplifying symbolic equations. Mountain fold, valley fold, inside reverse fold and display of origamis are implemented. By calling functions that perform those operations we can program origami constructions easily. Pedagogically, origami is also interesting since it reveals the essence of constructions of geometrical objects and of proof of geometrical theorems without resort to difficult mathematical techniques, which are often far away from the minds of most high school and university students.

References

[1] Humiaki Huzita, Axiomatic Development of Origami Geometry, Proceedings of the First International Meeting of Origami Science and Technology, pp. 143--158, 1989.

[2] Thomas Hull, Origami and Geometric Constructions, http://web.merrimack.edu/~thull/geoconst.html, 1997.

[3] Tzer-lin Chen, Proof of the impossibility of trisecting an angle with Euclidean tools, Math. Mag. 39, pp. 239-241, 1966.

[4] Koji Fushimi, Science of Origami, a supplement to Saiensu, Oct. 1980, p. 8 (in Japanese)

[5] Robert Geretschläger, Geometric Constructions in Origami, (in Japanese, translation by Hidetoshi Fukagawa), Morikita Publishing Co., 2002.